Essentials of Metallurgy

Essentials of Metallurgy

Edited by **Ricky Peyret**

NY RESEARCH
P R E S S

New York

Published by NY Research Press,
23 West, 55th Street, Suite 816,
New York, NY 10019, USA
www.nyresearchpress.com

Essentials of Metallurgy
Edited by Ricky Peyret

International Standard Book Number: 978-1-63238-469-0 (Hardback)

Printed in the United States of America.

Contents

Preface   IX

Chapter 1   **The Uptake of Copper(II) Ions by Chelating Schiff Base Derived from 4-Aminoantipyrine and 2-Methoxybenzaldehyde**   1
Chuan-Wei Oo, Hasnah Osman, Sharon Fatinathan and Maizatul Akmar Md. Zin

Chapter 2   **Extractive Metallurgy and National Policy**   10
Fathi Habashi

Chapter 3   **Factors Research on the Influence of Leaching Rate of Nickel and Cobalt from Waste Superalloys with Sulfuric Acid**   14
Xingxiang Fan, Weidong Xing, Haigang Dong, Jiachun Zhao, Yuedong Wu, Bojie Li, Weifeng Tong and Xiaofeng Wu

Chapter 4   **Liquid-Liquid Extraction of V(IV) from Sulphate Medium by Cyanex 301 Dissolved in Kerosene**   19
Ranjit Kumar Biswas and Aneek Krishna Karmakar

Chapter 5   **Comparative Study of Gold Concentration by Elutriation from Different Precious Metal Bearing Ores**   28
Martín A. Encinas-Romero, Guillermo Tiburcio-Munive and Jesús L. Valenzuela-García

Chapter 6   **Axisymmetric Strain Stability in Sheet Metal Deformation**   35
Tatjana V. Brovman

Chapter 7   **Solvent Extraction of Lanthanum Ion from Chloride Medium by Di-(2-ethylhexyl) Phosphoric Acid with a Complexing Method**   41
Shaohua Yin, Wenyuan Wu, Xue Bian, Yao Luo and Fengyun Zhang

Chapter 8   **Developing a Thermodynamical Method for Prediction of Activity Coefficient of TBP Dissolved in Kerosene**   46
Eskandar Keshavarz Alamdari and Sayed Khatiboleslam Sadrnezhaad

Chapter 9   **Argentinean Copper Concentrates: Structural Aspects and Thermal Behaviour**   53
Vanesa Bazan, Elena Brandaleze, Leandro Santini and Pedro Sarquis

Chapter 10   **Kinetics Analyzing of Direction Reduction on Manganese Ore Pellets Containing Carbon**   61
Bo Zhang and Zheng-Liang Xue

Chapter 11   **Alumina/Iron Oxide Nano Composite for Cadmium Ions Removal
from Aqueous Solutions**   66
Mona Mahmoud Abd El-Latif, Amal M. Ibrahim, Marwa S. Showman
and Rania R. Abdel Hamide

Chapter 12   **Transformation Mechanism of Ore Matter in the Weathering**   82
Victor Bragin, Irina Baksheyeva and Margaret Sviridova

Chapter 13   **Influence of Ferric and Ferrous Iron on Chemical and Bacterial Leaching
of Copper Flotation Concentrates**   86
Ali Ahmadi

Chapter 14   **Removal of Chromium(III) from the Waste Solution of an Indian Tannery
by Amberlite IR 120 Resin**   93
Pratima Meshram, Sushanta Kumar Sahu, Banshi Dhar Pandey, Vinay Kumar
and Tilak Raj Mankhand

Chapter 15   **An Exploratory Study of Tridentate Amine Extractants: Solvent Extraction
and Coordination Chemistry of Base Metals with
Bis((1R- benzimidazol-2-yl)methyl)amine**   103
Nomampondo P. Magwa, Eric Hosten, Gareth M. Watkins and Zenixole R. Tshentu

Chapter 16   **Kinetic Analysis of Isothermal Leaching of Zinc from Zinc Plant Residue**   113
Ali Reza Eivazi Hollagh, Eskandar Keshavarz Alamdari, Davooud Moradkhani
and Ali Akbar Salardini

Chapter 17   **Effect of Stacking Fault Energy on the Mechanism of Texture Formation during
Alternating Bending of FCC Metals and Alloys**   124
Natalia Shkatulyak

Chapter 18   **Sono-Photo Fenton Treatment of Liquid Waste Containing
Ethylenediaminetetraacetic Acid (EDTA)**   130
S. Chitra, K. Paramasivan and P. K. Sinha

Chapter 19   **Thermodynamic Assessment of the Pt-Sb System**   136
Jinming Liu, Yinghui Zhang and Cuiping Guo

Chapter 20   **Ball Milling and Annealing of Co-50 at% W Powders**   141
A. S. Bolokang, M. J. Phasha and D. E. Motaung

Chapter 21   **Microbial Recovery of Manganese Using *Staphylococcus epidermidis***   147
Alok Prasad Das, Lala Behari Sukla and Nilotpala Pradhan

Chapter 22   **A Statistical Method for Determining the Best Zinc Pregnant Solution
for the Extraction by D2EHPA**   151
Hossein Kamran Haghighi, Davood Moradkhani and Mohammad Mehdi Salarirad

Chapter 23   **Thermal Studies and Analytical Applications of a Newly Synthesized Composite
Material "Polyaniline Stannic Molybdate"**   159
Sajad Ahmad Ganai, Javid Ahmad Banday, Abid Hussain Shalla and Tabassum Ara

Chapter 24   **Microstructural Evolution in Cu-Mg Alloy Processed by Conform**   163
Lianpeng Song, Yuan Yuan and Zhimin Yin

Chapter 25   **Extraction of Palladium from Acidic Chloride Media into Emulsion Liquid Membranes Using LIX 984N-C®**   169
Satit Praipruke, Korbratna Kriausakul and Supawan Tantayanon

Chapter 26   **Efficacy of Bacterial Adaptation on Copper Biodissolution from a Low Grade Chalcopyrite Ore by *A. ferrooxidans***   179
Abhilash, Kapil Deo Mehta and Bansi Dhar Pandey

Chapter 27   **Extraction Kinetics of Ni(II) in the Ni^{2+}-SO_4^{2-}-Ac^- (Na^+, H^+)-Cyanex 272 (H_2A_2)-Kerosene-3% (v/v) Octan-1-ol System Using Single Drop Technique**   186
Ranjit Kumar Biswas, Aneek Krishna Karmakar and Muhammad Saidur Rahman

Chapter 28   **Equilibrium of the Extraction of V(IV) in the V(IV)-SO_4^{2-}(H^+, Na^+)— Cyanex 302-Kerosene System**   195
Ranjit Kumar Biswas and Aneek Krishna Karmakar

Chapter 29   **Aminododecyldiphosphonic Acid for Solvent Extraction of Bismuth Ions**   204
Baghdad Medjahed, M'Hamed Kaid, Mohamed Amine Didi and Didier Villemin

Permissions

List of Contributors

Preface

Metallurgy is the subfield of materials science that studies the physical and chemical properties of metals and their alloys. This field has been helping different industries by forming new and improving already existing materials and alloys. Bioleaching, effluent treatment, pressure leaching, pretreatment of ores, process metallurgy, process modelling and control, pyrometallurgy for non-ferrous metal extraction, rare earth and precious metal extraction, recycling of nonferrous metals and laden wastes are some of the topics discussed in this book. While understanding the long-term perspectives of the topics, the book makes an effort in highlighting their impact as a modern tool for the growth of the discipline. It will help the readers in keeping pace with the rapid changes in this field. The aim of this book is to present the unexplored aspects of this field and develop an in-depth understanding of this field.

This book is the end result of constructive efforts and intensive research done by experts in this field. The aim of this book is to enlighten the readers with recent information in this area of research. The information provided in this profound book would serve as a valuable reference to students and researchers in this field.

At the end, I would like to thank all the authors for devoting their precious time and providing their valuable contributions to this book. I would also like to express my gratitude to my fellow colleagues who encouraged me throughout the process.

Editor

The Uptake of Copper(II) Ions by Chelating Schiff Base Derived from 4-Aminoantipyrine and 2-Methoxybenzaldehyde

Chuan-Wei Oo*, Hasnah Osman, Sharon Fatinathan, Maizatul Akmar Md. Zin

School of Chemical Sciences, Universiti Sains Malaysia, Penang, Malaysia

Email: *oocw@usm.my

ABSTRACT

The Schiff base, 4-[(2-methoxybenzylidene)amino]-1,5-dimethyl-2-phenyl-1H-pyrizol-3(2H)-one (SB), was used for the first time to adsorb copper(II) ions in aqueous solution. Various parameters such as initial pH, agitation period and different initial concentration of copper(II) ions which influenced the adsorption capacity were investigated. The equilibrium adsorption data for copper(II) ions were fitted to Langmuir, Freundlich and Dubinin-Radushkevish isotherm models. The maximum monolayer adsorption capacity of SB as obtained from Langmuir isotherm was 5.64 mg/g. Kinetic data correlated well with the pseudo second-order kinetic model indicating that chemical adsorption was the rate limiting step.

Keywords: Schiff Base; Adsorption; Copper(II) Ions; Isotherm; Kinetics

1. Introduction

Heavy metal ions are the most harmful of the elemental pollutants and are of particular concern because of their toxicities to humans. The human body cannot metabolize heavy metals and it displays bioaccumulation properties causing various diseases and disorders. Metal ions in the environment can accumulate and are biomagnified along the food chain. Therefore, their toxic effects are more pronounced in human. There are many possible sources of these heavy metal ions. These include wastes from metal plating operations, mining operations, tanneries, fertilizer and textiles industries [1,2]. Copper is one of the most useful metals due to its low toxicity, corrosion resistant, workability and electrical conductivity. Copper has antibacterial properties and has a biological role in sustaining life. However, accumulation of copper in human body can cause adverse effect on human health such as stomach and intestinal cancer, liver and also kidney damage [3]. The presences of copper are mainly found in effluents from electroplating and brass manufacturing industries and also in the Cu-based agrichemicals run off from agricultural lands [4]. Many conventional techniques such as ion exchange, reverse osmosis, membrane filtration, precipitation-neutralization etc. have been used to remove heavy metal ions from aqueous solution [5,6]. However, these methods involved high operation cost,

generated secondary toxic sludge and were incompetent for the removal of heavy metal ions at low concentration [7]. Alternatively, adsorption onto solid substrate materials is considered as the most suitable process for the removal of heavy metal ions from solution of high and low concentrations.

Over the years, many studies have been conducted on the removal of heavy metal ions from aqueous solution by using adsorbents such as activated carbon, chitosan beads, silica, ion exchange resins and also chelating agents. Chelating agents have been widely employed as adsorbent because it has moderate coordination sites like nitrogen which shows great affinity towards metal ions [8]. One of the important chelating agents receiving an increased interest recently due to the presences of defined chelating groups is Schiff bases. Schiff bases are obtained through the condensation of aldehydes with amines and contain multidentate coordination sites such as O- and N-donor atoms which have high bonding affinity towards many heavy metal ions [6,8,9].

In this study, the efficiency of 4-[(2-methoxybenzylidene)amino]-1,5-dimethyl-2-phenyl-1H-pyrizol-3(2H)-one (written as SB) synthesized from 4-aminoantipyrine and 2-methoxybenzaldehyde was evaluated for the first time in the removal of copper(II) ions from aqueous sotion. Crucial parameters that affected the removal of copper(II) ions, such as initial pH, contact time and difrent initial copper(II) concentration were investigated.

---

*Corresponding author.

2. Materials and Methods

2.1. Materials

The starting materials such as 4-aminoantipyrine and 2-methoxybenzaldehyde used in this research were purased from Fluka. All the reagents used were of analyticl-reagent grade and were used without further purificaon. Distilled water was used throughout this research.

2.2. Synthesis of 4-[(2-Mthoxybenzylidene)aino]-1,5-dmethyl-2-penyl-1*H*-prizol-3(2H)-one (SB)

The Schiff base, 4-[(2-methoxybenzylidene)amino]-1,5-dimethyl-2-phenyl-1*H*-pyrizol-3(2*H*)-one (SB) was prepared by adding 4-aminoantipyrine (10 mmol, 2.03 g) into 2-methoxybenzaldehyde (10 mmol, 1.36 g) solution that was dissolved in 20 mL of ethanol in the presence of one drop of glacial acetic acid. The mixture was refluxed for 2 hours. The resulting mixture was cooled to room temperature and the resulting precipitate was filtered, washed repeatedly with distilled water and recrystallized from ethanol. The Schiff base (SB) was isolated as a yellow crystalline solid. The structure of SB was shown in **Figure 1**.

FTIR (KBr pellet): ν 3051, 2933, 2833, 1642, 1609, 1593, 1508 cm^{-1}. Calc. for $C_{19}H_{19}N_3O_2$: C, 71.01; H, 5.96; N, 13.08. Found: C, 71.20; H, 5.77; N, 12.80.

2.3. Characterization of the SB

The types of functional groups existing on SB were confirmed by a PerkinElmer Fourier transform infrared (FTIR) System 2000 Model spectrophotometer. Meanwhile, the surface morphology of SB before and after adsorption of copper(II) ions was observed using a Leica Cambridge S360 scanning electron microscope (SEM) coupled with energy dispersive X-ray spectroscopy (EDX).

2.4. Batch Adsorption Experiments

Stock solution of 1000 mg/L of copper(II) ions was prepared using the analytical-reagent grade copper(II) nitrate salt. The stock solution was diluted accordingly to prepare different concentration of copper(II) ions. The

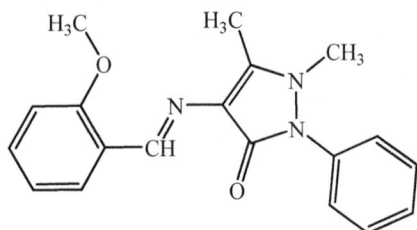

Figure 1. The structure of 4-[(2-methoxybenzylidene)amino]-1,5-dimethyl-2-phenyl-1*H*-pyrizol-3(2*H*)-one (SB).

batch adsorption experiments were conducted in 250 mL Erlenmeyer flasks with 50 mL of standard solution and equilibrated using a shaker. The amount of SB used throughout this research was 0.1 g. Each experiment was carried out in duplicates. All filtrates were analysed using atomic absorption spectrometer (AAnalyst 200 AA, PerkinElmer, USA) at a wavelength of 324.75 nm for copper(II) ions.

The effect of pH on the adsorption of copper(II) ions was studied in a pH range of 1.5 - 7.0. The pH of the 10 mg/L of copper(II) ions was adjusted to different pH using appropriate concentration of HCl and NaOH solutions and equilibrated for 3 h. The effect of agitation time was also studied at different time interval ranging from 5 - 240 min at the optimum pH. The isotherm study was conducted under the optimum condition by varying the initial concentration of copper(II) ions from 5 to 30 mg/L. The amount of adsorption at equilibrium was calculated using the following equation:

$$\text{Adsorption capacity } \left(q_e\right) = \left(\frac{C_o - C_e}{W}\right)V \qquad (1)$$

where C_o is the initial concentration of copper(II) ions (mg/L), C_e is the final concentration of copper(II) ions (mg/L), V is the volume of copper(II) ions solution (L) and W is the weight of SB (g) used.

3. Results and Discussion

3.1. FTIR of the SB

The FTIR spectrum of SB was shown in **Figure 2**. The peaks at 1642 and 1508 cm^{-1} were attributed to the C=O group (conjugated) and the C=N bonds, respectively. The present of imine group, C=N confirmed that a Schiff base was formed from the reaction between 4-aminoantipyrine and 2-methoxybenzaldehyde. The peaks found at 1609 and 1593 cm^{-1} were due to the C=C bonds in the benzene rings of SB. The peaks at 2933 and 2833 cm^{-1} were attributed to the C-H stretching of the methoxy group. Meanwhile the =C-H stretching was evident at 3051 cm^{-1}. The medium broad peak at 3431 cm^{-1} could be due to the present of water molecule on the SB because of its hygroscopic nature.

3.2. Adsorption Studies

3.2.1. Effect of Initial pH

The pH of metal solution is the most important parameter governing metal adsorption. This is due to the fact that pH of the metal solution affects the solubility of metal ions and the surface charge of the adsorbent. In most cases, the metal uptake increased at higher pH when the surface charge of the adsorbent is opposite to that of the metal cation. At this point, the attraction force between the adsorbent and the metal cation will increase. **Figure 3**

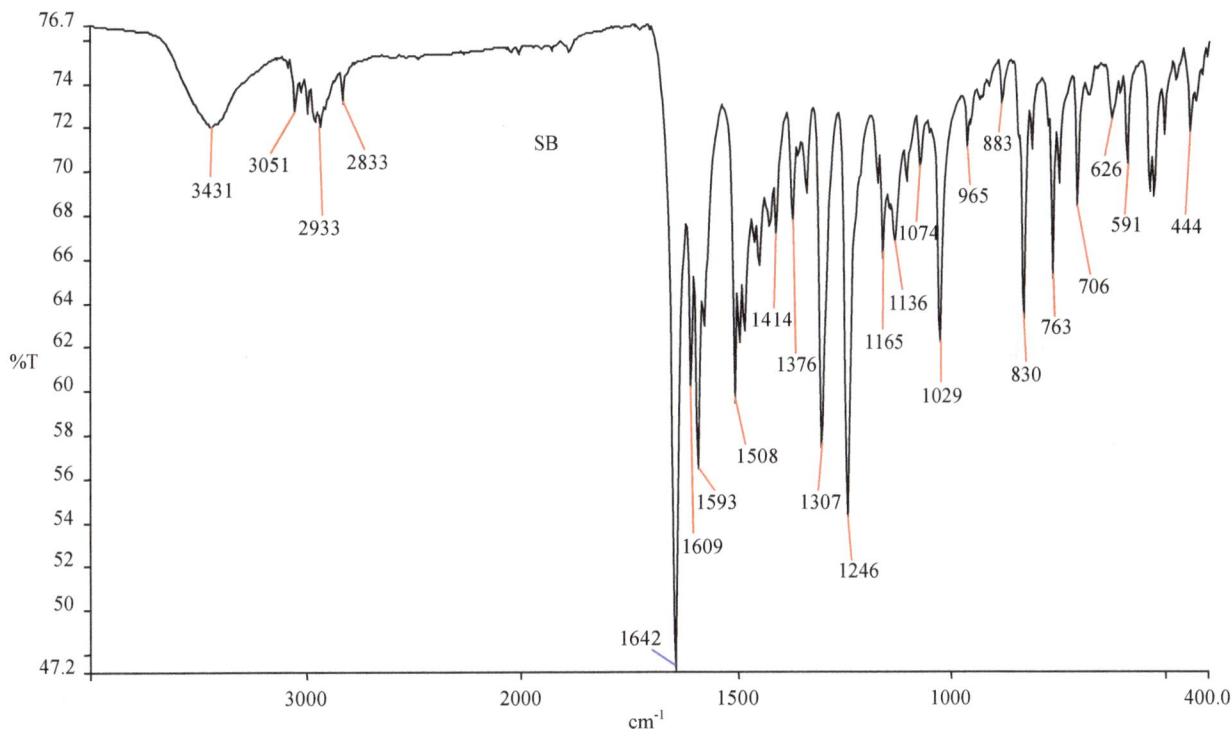

Figure 2. FTIR spectrum of SB.

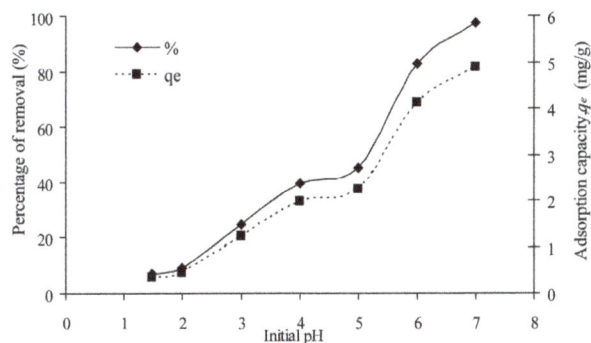

Figure 3. Effect of initial pH on removal of copper(II) ions by SB.

shows the effect of pH on the removal of copper(II) ions by SB. At low pH, the uptake of copper(II) ions was low because the functional groups (O-CH$_3$ and C=N) on the adsorbent surface were protonated giving rise to positively charged adsorption sites. The electrostatic repulsion between the positively charged SB and the copper(II) ions together with the competition from the proton (H$^+$) for the active adsorption sites will contribute to the low adsorption percentage. Furthermore, dissolution of SB was observed at acidic pH of 2 to 4. However, as the pH of the solution increased the adsorption percentage increased as well, whereby at pH 6, 82.66% of copper(II) ions was being adsorbed. At higher pH, the functional groups of the adsorbent were deprotonated creating the negatively charged surface that will attract the copper(II)

ions. Less amount of protons in the solution also reduced the competition between copper(II) ions and the protons for the active adsorption sites. In this study, pH 6 was chosen as the optimum adsorption pH. Although at pH 7, the percentage of removal was much higher, it was not considered as the optimum pH because at higher pH, copper(II) ions will form the insoluble Cu(OH)$_2$ precipitate.

3.2.2. Effect of Contact Time

The uptake rate of copper(II) ions onto SB was investigated as it represented the amount of time required before the adsorption process becomes constant and equilibrium was reached. As shown in **Figure 4**, for the first 90 min, the uptake rate was high and it decreased gradually until it reached equilibrium at 180 min with an adsorption percentage of 83.18%. The high uptake rate observed at initial stage could be due to the adsorption of copper(II) ions onto the external surface of the adsorbent. Meanwhile, the slower uptake rate observed after 90 min is due to the pore diffusion and the quick exhaustion of the available adsorption sites. The adsorption data were then analyzed with pseudo first- and pseudo second-order kinetic models.

3.2.3. Adsorption Kinetics

The adsorption kinetics of metal ions is important for designing the adsorption system and is required for determining the optimum operating conditions for a full-scale batch process [10]. Many adsorption kinetic models

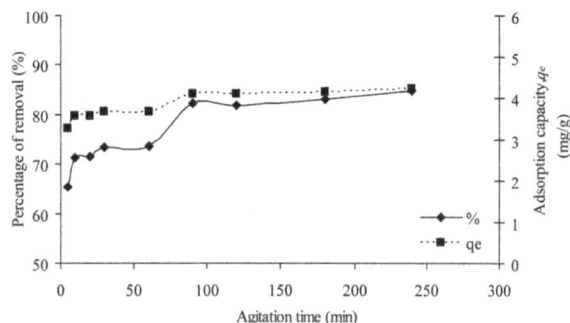

Figure 4. Effect of agitation time on the removal of copper(II) ions by SB.

have been applied to understand the adsorption kinetics and the rate-limiting step. Among all, pseudo first- and pseudo second-order models are the most commonly used to study the adsorption kinetics of heavy metal ions [11]. The pseudo first-order kinetic model is given in its linear form as:

$$\log\left(q_e - q_t\right) = \log q_e - \frac{K_1 t}{2.303} \quad (2)$$

where q_e is the amount of copper(II) ions adsorbed at equilibrium (mg/g), q_t is amount of copper(II) ions adsorbed at time t (mg/g) and K_1 is the pseudo first-order kinetic constant (1/min). Linear plot of $\log(q_e\text{-}q_t)$ against t for the adsorption of copper(II) ions onto SB was shown in **Figure 5**. In many cases, the pseudo first-order kinetic model does not fit well to the whole range of contact time. It would only be applicable over the initial stage of the adsorption process [12]

The pseudo second-order kinetic model is based on the assumption that the rate-limiting step in the adsorption of heavy metal ions is chemisorption involving the valence force through the sharing or exchange of electrons between the adsorbate and the adsorbent [2,13-15]. This kinetic model is represented as:

$$\frac{t}{q_t} = \frac{1}{K_2 q_e^2} + \frac{t}{q_e} \quad (3)$$

where q_e and q_t are the amount of copper(II) ions adsorbed at equilibrium and time t (mg/g), respectively and K_2 is the pseudo second-order kinetic constant (g/mg·min). **Figure 5** showed the linear plots of t/q_t against t.

The pseudo second-order kinetic model gave higher correlation ($R^2 = 0.9991$) than the pseudo first-order kinetic model ($R^2 = 0.8953$). Furthermore, the calculated equilibrium adsorption capacity based on pseudo second-order (4.27 mg/g) was in agreement with the experimental data (4.16 mg/g). Hence, the pseudo second-order is more appropriate to represent the adsorption data of copper(II) ions onto SB. The poor fit observed for pseudo first-order kinetic model could be due to a time lag caused by a boundary layer or the external resistance

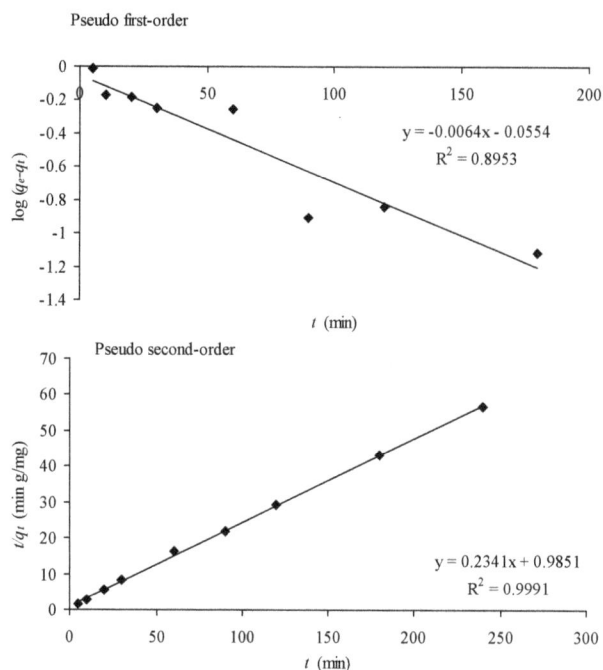

Figure 5. Pseudo first- and pseudo second-order plots for the removal of copper(II) ions with SB.

controlling at the beginning of the adsorption process [16].

3.2.4. Isotherm Studies
The equilibrium adsorption isotherm is essential to be studied as it describes the interactive behavior between the solutes and the adsorbent. Different initial concentration of copper(II) ions ranging from 5 - 30 mg/L was used to study the effectiveness of SB to remove trace amount of copper(II) ions. As shown in **Figure 6**, the adsorption capacity increased with an increase in the initial concentration of copper(II) ions at optimum adsorption pH and agitation time. The adsorption capacity reached equilibrium once the active adsorption sites were saturated at the initial concentration of 25 mg/L of copper(II) ions.

The adsorption equilibrium data were fitted with different isotherm models mainly Langmuir, Freundlich and Dubinin-Radushkevich (D-R) isotherm in order to evaluate the adsorption phenomenon. In this study, non-linear method was used to evaluate the fitness of the isotherm models to the experimental data. The non-linear method is a better way to obtain the isotherm parameters compared to the linear least-square method [17]. The plots based on Langmuir, Freundlich and D-R isotherms were shown in **Figures 7** and **8**, respectively, while the isotherm constants were tabulated in **Table 1**.

3.2.4.1. Langmuir Isotherm
Langmuir isotherm relates the coverage of molecules on a solid surface to the concentration of a medium above

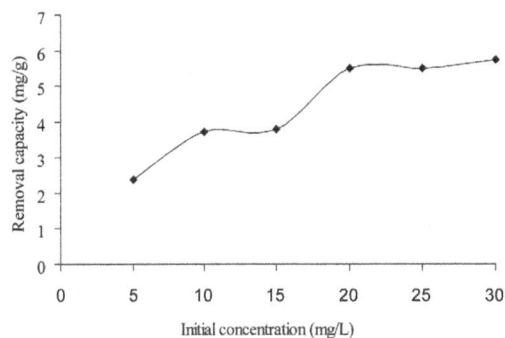

Figure 6. Effect of initial concentration of copper(II) ions on the adsorption capacity of SB.

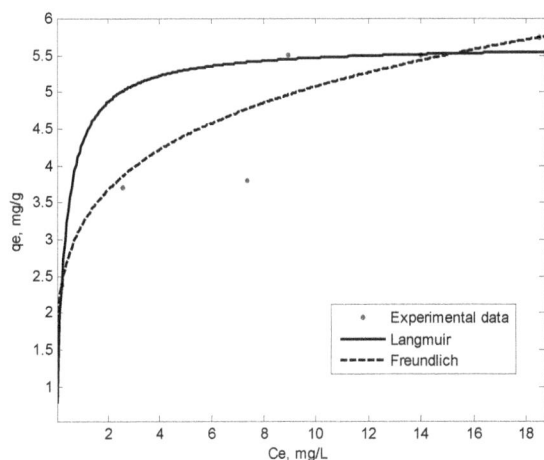

Figure 7. Langmuir and Freundlich isotherms for the adsorption of copper(II) ions onto SB.

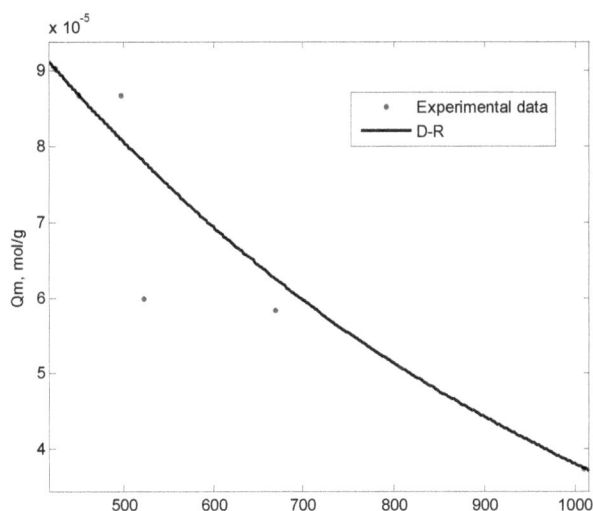

Figure 8. D-R isotherm for the adsorption of copper(II) ions onto SB.

the solid surface at a fixed concentration [11]. It describes monolayer adsorption and is based on the assumption that all the adsorption sites are energetically

Table 1. Isotherm and kinetic constants of different models for the adsorption of copper(II) ions onto SB.

Isotherm models			Kinetic models		
Langmuir	K_L	3.127	Pseudo first-order	K_1	0.0147
	V_m	5.64		q_e (theor.)	0.88
	r	0.9129		r	0.8953
Freundlich	K_F	3.192	Pseudo second-order	K_2	0.0557
	$1/n$	0.2013		q_e (theor.)	4.27
	r	0.9729		r	0.9991
D-R	Q_m	6.9860×10^{-5}		q_e (exp.)	4.16
	K	0.3305			
	E	1.2299			
	r	0.9692			

identical and adsorption takes place on a structurally homogenous adsorbent [2]. This model can be represented as:

$$q_e = \frac{V_m K_L C_e}{1 + K_L C_e} \qquad (4)$$

where C_e is the equilibrium concentration (mg/L), q_e is the amount of copper(II) ions adsorbed at equilibrium (mg/g), V_m is the monolayer adsorption capacity (mg/g) and K_L is the Langmuir equilibrium constant (L/mg). Based on **Figure 7**, the steep initial slope observed for the Langmuir plot reflected high K_L value (3.127 L/ mg) indicating that the adsorbent has high affinity for copper(II) ions [18,19] with a monolayer adsorption capacity of 5.64 mg/g.

3.2.4.2. Freundlich Isotherm
Freundlich isotherm is an empirical equation used for the adsorption on heterogeneous surfaces or the surface adsorption sites of varied affinities [20]. It describes monomolecular layer coverage of adsorbent by the solutes [21]. Freundlich isotherm is given as:

$$q_e = K_F C_e^{1/n} \qquad (5)$$

where q_e is the amount adsorbed at equilibrium (mg/g), C_e is the equilibrium concentration (mg/L), K_F is the adsorption capacity (mg/g) and $1/n$ is the adsorption intensity. Adsorption is said to be favorable when the Freundlich constant $1/n$ is between 0.1 and 1 and represented a heterogenous surface structure of the adsorbent [22,23]. Meanwhile, smaller value of $1/n$ indicates stronger interaction between the adsorbent and the heavy metal ions, while $1/n$ equal to 1 implies linear adsorption leading to identical adsorption energies for all sites [11]. As shown in **Table 1**, adsorption of copper(II) ions onto SB has a $1/n$ value of 0.2013 with K_F equal to 3.192 mg/g. This showed that SB had a strong interaction with copper(II) ions which makes SB a favourable adsorbent for copper(II) ions.

3.2.4.3. Dubinin-Radushkevich (D-R) Isotherm

D-R isotherm, apart from being analogue of Langmuir isotherm, is a more general model than Langmuir isotherm as it does not assume homogenous surface or constant sorption potential [24]. D-R isotherm was derived as the overall adsorption isotherm for the adsorption onto heterogenous solid surface [20]. The D-R isotherm equation is expressed as:

$$Q = Q_m \exp\left(-K\varepsilon^2\right) \qquad (6)$$

where, Q is the amount adsorbed at equilibrium (mol/g), Q_m is the maximum adsorption capacity (mol/g), K is the D-R constant (mol/kJ) which is related to the adsorption energy. The Polanyi potential (ε) is given as:

$$\varepsilon = RT \ln\left(1 + \frac{1}{C_e}\right) \qquad (7)$$

where R is the gas constant in kJ/mol K and T is the temperature in Kelvin. The mean energy of adsorption, E is used to estimate the type of adsorption process and can be calculated by the following equation:

$$E = \frac{1}{\sqrt{2K}} \qquad (8)$$

The adsorption process can be explained by an ion exchange process if the value of E is between 8 and 16 kJ/mol [2,18]. Meanwhile, E values lower than 8 kJ/mol correspond to physical adsorption (physisorption) [25,26]. The E value for the adsorption of copper(II) ions onto SB was calculated as 1.2299 kJ/mol indicating the involvement of physisorption.

3.3. Adsorption Mechanism

The adsorption of solute on any sorbent can happen either through physical bonding, ion exchange, complexation, chelation or a combination of any of these interactions [11]. The adsorbent used in the present study is a Schiff base which has the -OCH$_3$ and C=N functional groups as the active sites to interact with copper(II) ions. At the optimum pH, the oxygen and nitrogen atoms on these active sites behave as the electron donors and can form complex with the copper(II) ions as shown in **Figure 9**. This process can be classified as chemisorption which involves the sharing of electrons between the adsorbate and the adsorbent.

The removal of copper(II) ions by SB, however, was not solely governed by chemisorption. The mean adsorp-

tion energy, E calculated from the D-R isotherm suggested that physical adsorption might have taken place as well. This was proven based on the SEM-EDX analysis carried out on SB before and after the adsorption of copper(II) ions. The SEM-EDX micrographs were shown in **Figure 10**. The SEM image for SB before the adsorption of copper(II) ions (**Figure 10(a)**) shows an even and smooth surface morphology. Many small particles which clustered on the surface of SB were also found. Based on the EDX analysis, the presences of carbon, hydrogen and nitrogen were detected. The gold peaks (Au) found in the spectra were due to the gold purposely settled to increase electrical conductive of the sample. After the adsorption of copper(II) ions (**Figure 10(b)**) surface morphology of SB changed drastically. The surface was found to be very uneven and many small coarser particles were formed on the surface. The EDX spectra showed the presences of copper on the surface of SB especially on these smaller particles. These observation proved that copper(II) ions were deposited on the surface of the adsorbent by means of physical adsorption. Physical adsorption occurs through the existences of van der Waals forces between the adsorbent and the adsorbate.

4. Conclusion

Schiff base derived from 4-aminoantipyrine and 2-methoxybenzaldehyde can act as a potential adsorbent to remove copper(II) ions from the aqueous solution. Present study showed that SB has a monolayer adsorption capacity of 5.64 mg/g for copper(II) ions. Chemisorption of copper(II) ions could be occurred through the interaction of copper(II) ions with the methoxy and imine groups on SB. Copper(II) ions also can be removed via physical adsorption. The kinetic studies showed that the removal of copper(II) ions by SB followed the pseudo second-order kinetic reaction.

5. Acknowledgements

The authors would like to express their sincere gratitude to Universiti Sains Malaysia for funding this study

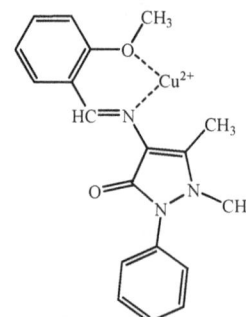

Figure 9. Proposed mechanism for the adsorption of copper(II) ions onto SB.

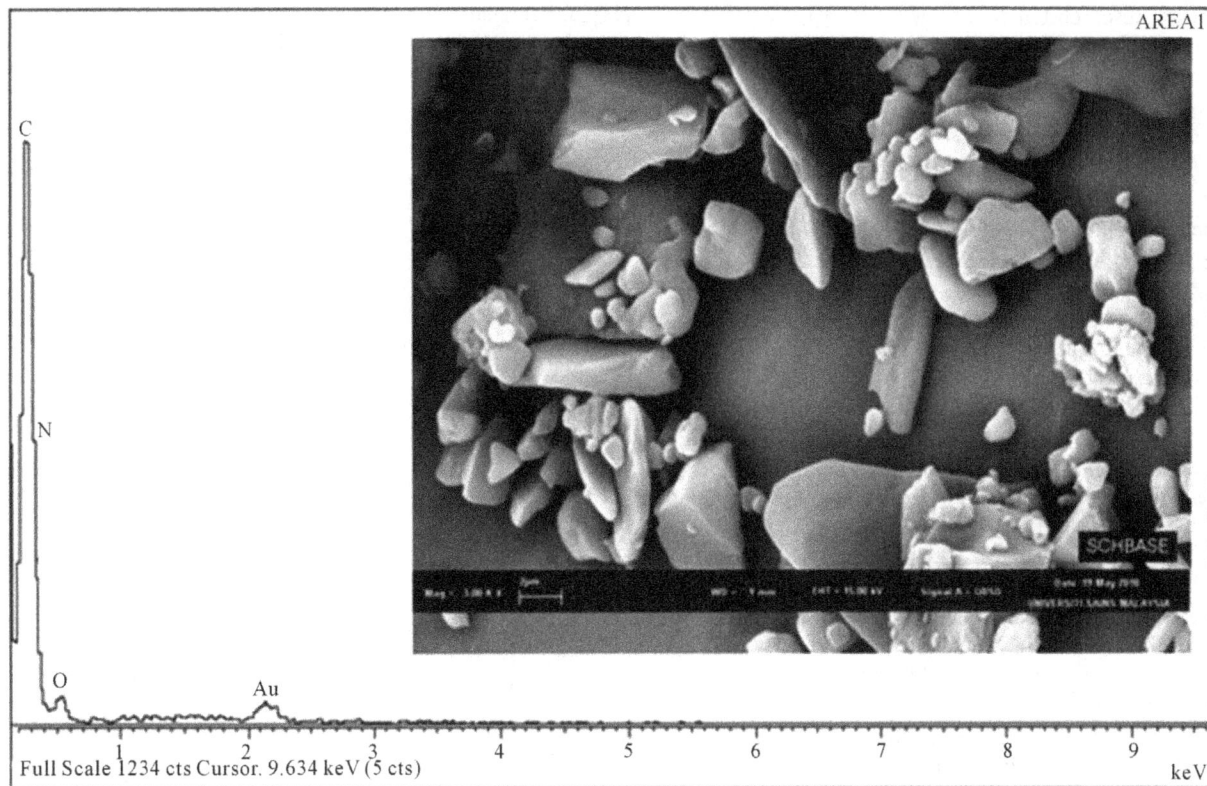

(a)

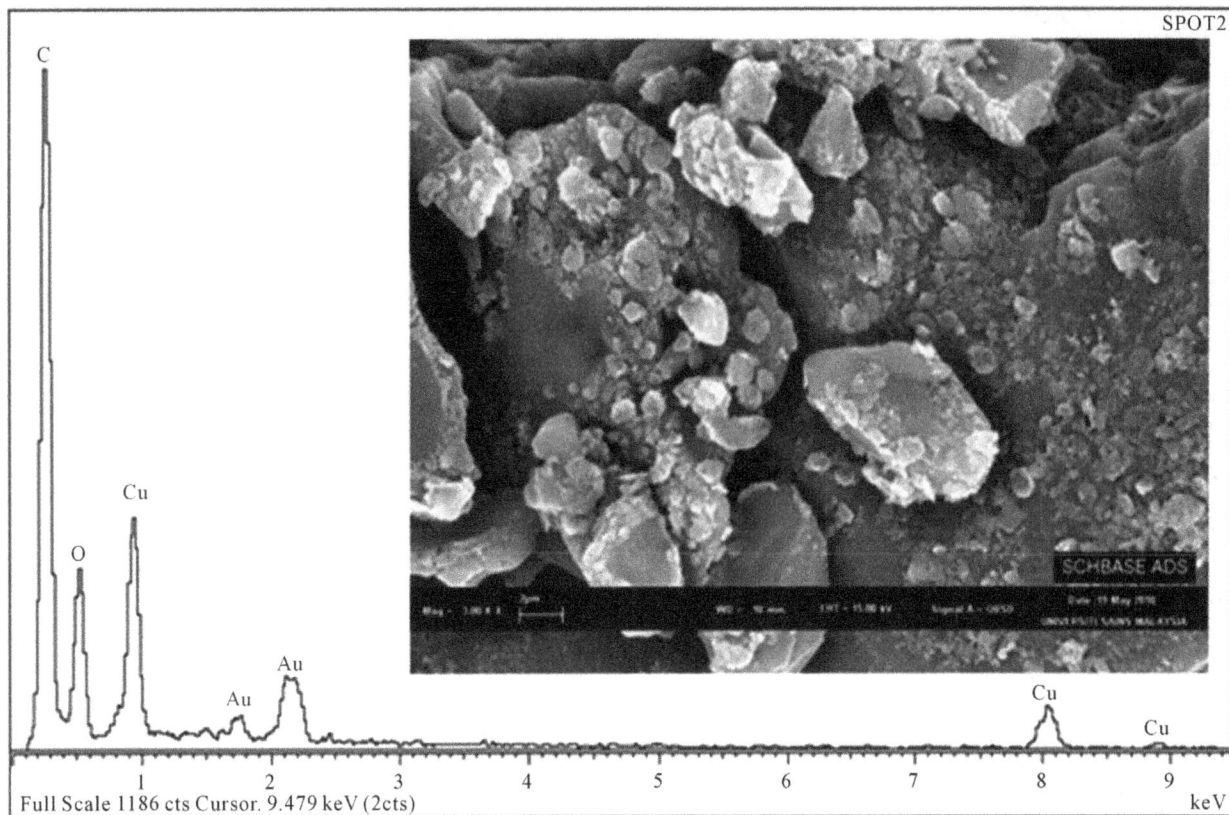

(b)

Figure 10. SEM-EDX spectra of SB (a) before and (b) after the adsorption of copper(II) ions.

through the Research University grant no. 1001/PKIMIA /811134.

REFERENCES

[1] E. A. Susan, J. O. Trudy, R. M. Bricka and D. D. Adrian, "A Review of Potentially Low-Cost Sorbents for Heavy Metals," *Water Research*, Vol. 33, No. 11, 1999, pp. 2469-2479. doi:10.1016/S0043-1354(98)00475-8

[2] N. Unlu and M. Ersoz, "Adsorption Characteristics of Heavy Metal Ions onto a Low Cost Biopolymeric Sorbents from Aqueous Solutions," *Journal of Hazardous Materials*, Vol. 136, No. 2, 2006, pp. 272-280. doi:10.1016/j.jhazmat.2005.12.013

[3] C. I. Lee, W. F. Yang and C. I. Hsieh, "Removal of Copper(II) by Manganese-Coated Sand in a Liquid Fluidized-Bed Reactor," *Journal of Hazardous Materials*, Vol. 114, No. 1-3, 2004, pp. 45-51. doi:10.1016/j.jhazmat.2004.06.033

[4] N. Boujelben, J. Bouzid and Z. Elouear, "Adsorption of Nickel and Copper onto Natural Iron Oxide-Coated Sand from Aqueous Solutions: Study in Single and Binary Systems," *Journal of Hazardous Materials*, Vol. 163, No. 1, 2009, pp. 376-382. doi:10.1016/j.jhazmat.2008.06.128

[5] G. Palma, J. Freer and J. Baeza, "Removal of Metal Ions by Modified *Pinus radiata* Bark and tannins from Water Solutions," *Water Research*, Vol. 37, No. 20, 2003, pp. 4974-4980. doi:10.1016/j.watres.2003.08.008

[6] Q. F. Yin, B. Z. Ju, S. F. Zhang, X. B. Wang and J. Z. Yang, "Preparation and Characteristics of Novel Dialdehyde Aminothiazhole Starch and Its Adsorption Properties for Cu(II) Ions from Aqueous Solution," *Carbohydrate Polymers*, Vol. 72, No. 2, 2008, pp. 326-333. doi:10.1016/j.carbpol.2007.08.019

[7] G. Vazquez, J. Gonzalez-Alvarez, S. Freire, M. Lopez-Lorenzo and G. Antorrena, "Removal of Cadmium and Mercury Ions from Aqueous Solution by Sorption on Treated *Pinus pinaster* Bark: Kinetics and Isotherms," *Bioresource Technology*, Vol. 82, No. 3, 2002, pp. 247-251. doi:10.1016/S0960-8524(01)00186-9

[8] M. H. Mashhadizadeh, M. Pesteh, M. Talakesh, I. Sheeikhshoaie, M. M. Ardakani and M. A. Karimi, "Solid Phase Extraction of Copper(II) by Sorption on Octadecyl Silica Membrane Disk Modified with a New Schiff Base and Determination with Atomic Absorption Spectrometry," *Spectrochimica Acta Part B*, Vol. 63, No. 8, 2008, pp. 885-888. doi:10.1016/j.sab.2008.03.018

[9] P. A. Amoyaw, M. Williams and X. R. Bu, "The Fast Removal of Low Concentration of Cadmium(II) from Aqueous Media by Chelating Polymers with Salicylaldehyde Units," *Journal of Hazardous Materials*, Vol. 170, No. 1, 2009, pp. 22-26. doi:10.1016/j.jhazmat.2009.05.028

[10] E. A. Oliveira, S. F. Montanher, A. D. Andrade, J. A. Nobrega and M. C. Rollemberg, "Equilibrium Studies for the Sorption of Chromium and Nickel from Aqueous Solutions Using Raw Rice Barn," *Process Biochemistry*, Vol. 40, No. 11, 2005, pp. 3485-3490. doi:10.1016/j.procbio.2005.02.026

[11] J. Febrianto, A. N. Kosasih, J. Sunarso, Y.-H. Ju, N. Indraswati and S. Ismadji, "Equilibrium and Kinetic Studies in Adsorption of Heavy Metals Using Biosorbent: A Summary of Recent Studies," *Journal of Hazardous Materials*, Vol. 162, No. 2-3, 2009, pp. 616-645. doi:10.1016/j.jhazmat.2008.06.042

[12] Y. S. Ho and G. Mckay, "The Sorption of Lead(II) Ions on Peat," *Water Research*, Vol. 33, No. 2, 1999, pp. 578-584. doi:10.1016/S0043-1354(98)00207-3

[13] F. Gode and E. Pehlivan, "Adsorption of Cr(III) Ions by Turkish Brown Coals," *Fuel Processing Technology*, Vol. 86, No. 8, 2005, pp. 875-884. doi:10.1016/j.fuproc.2004.10.006

[14] D. Singh, K. P. Singh and V. K. Singh, "Trivalent Chromium Removal from Wastewater Using Low Cost Activated Carbon Derived from Agricultural Waste Material and Activated Carbon Fabric Cloth," *Journal of Hazardous Materials*, Vol. 135, No. 1-3, 2006, pp. 280-295. doi:10.1016/j.jhazmat.2005.11.075

[15] V. C. Taty-Costodes, H. Fauduet, C. Porte and A. Delacroix, "Removal of Cd(II) and Pb(II) Ions from Aqueous Solutions by Adsorption onto Sawdust of *Pinus sylvestris*," *Journal of Hazardous Materials*, Vol. B105, 2003, pp. 121-142. doi:10.1016/j.jhazmat.2003.07.009

[16] G. McKay, Y. S. Ho and J. C. Y. Ng, "Biosorption of Copper from Waste Waters: A Review," *Separation & Purification Reviews*, Vol. 28, No. 1, 1999, pp. 87-125. doi:10.1080/03602549909351645

[17] Y. S. Ho, "Isotherms for the Sorption of Lead onto Peat: Comparison of Linear and Non-Linear Methods," *Polish Journal of Environmental Studies*, Vol. 15, No. 1, 2006, pp. 81-86.

[18] K. Vijayaraghavan, T. V. N. Padmesh, K. Palanivelu and M. Velan, "Biosorption of Nickel (II) Ions onto *Sargassum wightii*: Application of Two-Parameter and Three-Parameter Isotherm Models," *Journal of Hazardous Materials*, Vol. B133, 2006, pp. 304-308. doi:10.1016/j.jhazmat.2005.10.016

[19] B. Volesky, J. Weber and R. Vieira, "Biosorption of Cd and Cu by Different Types of *Sargassum biomass*," *Process Metallurgy*, Vol. 9, 1999, pp. 473-482. doi:10.1016/S1572-4409(99)80136-9

[20] A. Dąbrowski, "Adsorption—From Theory to Practice," *Advances in Colloid and Interface Science*, Vol. 93, No. 1-3, 2001, pp. 135-224. doi:10.1016/S0001-8686(00)00082-8

[21] S. Al-Asheh, F. Banat, R. Al-Omari and Z. Duvnjak, "Prediction of Binary Sorption Isothems for the Sorption of Heavy Metals by Pine Bark Using Single Isotherm Data," *Chemosphere*, Vol. 41, 2000, pp. 659-665. doi:10.1016/S0045-6535(99)00497-X

[22] T. Karthikeyan, S. Rajgopal and L. R. Miranda, "Chromium (VI) Adsorption from Aqueous Solution by *Hevea Brasilinesis* Sawdust Activated Carbon," *Journal of Hazardous Materials*, Vol. B124, 2005, pp. 192-199. doi:10.1016/j.jhazmat.2005.05.003

[23] S. P. Mishra, D. Tiwari, R. S. Dubey and M. Mishra, "Biosorptive Behaviour of Casein for Zn^{2+}, Hg^{2+} and Cr^{3+}: Effect of Physico-Chemical Treatment," *Bioresource Tech-*

nology, Vol. 63, 1998, pp. 1-5.
doi:10.1016/S0960-8524(97)00110-7

[24] W. Zheng, X.-M. Li, F. Wang, Q. Yang, P. Deng and G.-M. Zeng, "Adsorption Removal of Cadmium and Copper from Aqueous Solution by Areca—A Food Waste," *Journal of Hazardous Materials*, Vol. 157, No. 2-3, 2008, pp. 490-495. doi:10.1016/j.jhazmat.2008.01.029

[25] M. Mahramanlioglu, I. Kizilcikli and I. O. Bicer, "Ad-

sorption of Fluoride from Aqueous Solution by Acid Treated Spent Bleaching Earth," *Journal of Fluorine Chemistry*, Vol. 115, No. 1, 2002, pp. 41-47.
doi:10.1016/S0022-1139(02)00003-9

[26] O. A. Wahab, "Kinetic and Isotherm Studies of Copper (II) Removal from Wastewater Using Various Adsorbents," *Egyptian Journal of Aquatic Research*, Vol. 30, No. 1, 2007, pp. 125-143.

2

Extractive Metallurgy and National Policy

Fathi Habashi

Department of Mining, Metallurgical, and Materials Engineering, Laval University, Quebec City, Canada
Email: Fathi.Habashi@arul.ulaval.ca

ABSTRACT

Hydrometallurgical technology offers a unique possibility for developing countries to exploit their mineral resources locally instead of shipping them as concentrates. Production plants may start on a small scale with small capital investment then increase productivity later when the economy permits without financial penalty. This is in contract to smelting operations which necessitates large scale production from the start with high capital investment that may not be available locally.

Keywords: Autoclaves; Flash Tanks; Piston Pumps; Smelting; Sulfuric Acid Production; Environmental Problems

1. Introduction

Should a country export its mineral wealth in form of concentrates or should it locally treat its ores to produce metals as final products? In the past centuries pyrometallurgy was the only route to extract metals from ores and this required a large capital investment many countries could not afford to raise. This situation encouraged marketing of concentrates. Today, metallurgists have the option to use the hydrometallurgical route to process ores and concentrates at a reasonable capital investment. This opened the way to the possibility of processing ores locally. When need arises to increase production new units can be added economically. This is not possible with pyrometallurgical processes because one large furnace is more economical than a number of small furnaces with the same capacity due to heat radiation losses. However, another factor must also be considered in processing the ores locally: the availability of local experienced manpower. Hence, it is necessary to invest in metallurgical education and research. This is essential to ensure the availability of local engineers who are capable of designing, building, and operating metallurgical plants.

2. Mineral Wealth and Technology

Extracting metals from ores or the production of Industrial minerals requires a number of operations: some mechanical, physical, and physicochemical known as mineral beneficiation and some chemical known as extractive metallurgy (**Figure 1**). The chemical processes can be divided for convenience as pyro-, hydro-, and electrometallurgy. Pyrometallurgy is the most ancient technology, while hydro- and electrometallurgy are more recent. No wonder then that, in some cases, these new technologies are displacing the older ones gradually, either because of improved efficiency and better economics or to comply with environmental regulations [1-3].

Hydrometallurgy is the technology of extracting metals from ores by aqueous methods, pyrometallurgy by dry thermal methods, and electrometallurgy by electrolytic methods. In general, hydrometallurgy involves two distinct steps (**Figure 2**):

- Selective dissolution of the metal values from an ore— a process known as *leaching*.
- Selective recovery of the metal values from the solution, an operation that involves a *precipitation* method.

Sometimes a *purification/concentration* operation is conducted prior to precipitation. These processes are aimed at obtaining a pure and a concentrated solution from which the metal values can be precipitated effectively.

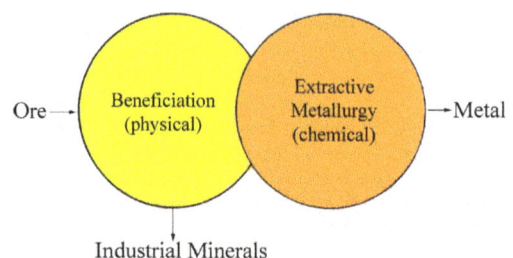

Figure 1. Extracting metals from ores.

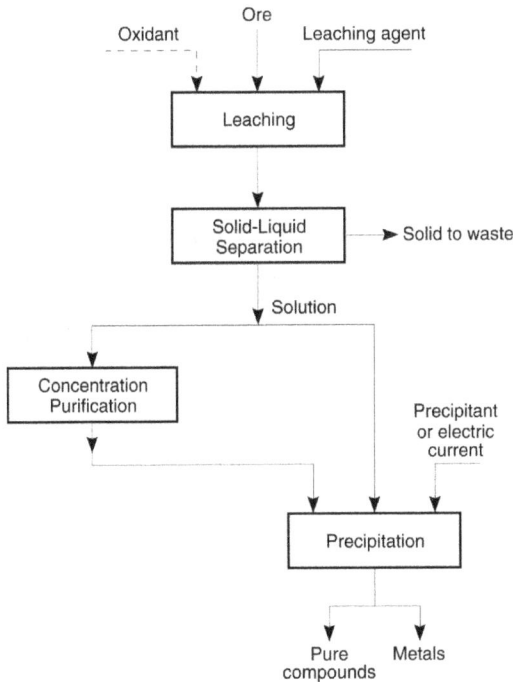

Figure 2. General outline of hydrometallurgical processes.

3. Hydro-Versus Pyrometallurgy

In the past fifty years, hydrometallurgy has been vigorously competing with thermal methods. Pyrometallurgy was most successful when high-grade massive sulfide ores were treated in a blast furnace, because such furnace has maximum heat economy being itself a heat exchanger: the cold charge descending from the top is preheated by the hot gases ascending in the furnace. Dust problems were also minimum because the ore was in form of large lumps. With the exhaustion of such raw material, metallurgists turned their attention towards the treatment of low-grade ores. This necessitated extensive grinding and flotation of the sulfides which resulted in a finely divided concentrates as raw material. These, naturally, could not be charged to a blast furnace because if charged, they would block the movement of the ascending gases—hence the birth of the fossil-fired horizontal reverberatory furnace. This was a turning point to the worst with respect to pollution of the environment, high energy consumption, and excessive dust formation.

The first success for hydrometallurgy was at the beginning of the twentieth century when a process for the production of alumina from bauxite was invented which rapidly displaced the existing pyrometallurgical method. Another success for hydrometallurgy was in the production of zinc when in 1980 a new process displaced the old pyromtallurgical process. At the beginning of the twenty first century a similar process was applied for treating copper concentrates thus demonstrating that the future is for hydrometallurgy and in particular pressure

hydrometallurgy. The advantages of hydrometallurgy can be outlined as follows.

3.1. Sulfur Dioxide Generation

During the pyrometallurgical treatment of sulfide ores, if SO_2 formed is in high enough concentration, it must be used for making acid and a nearby market for this acid must be found. Shipping sulfuric acid long distances is expensive and hazardous. If the SO_2 concentration is too low for making acid, disposal methods must be found. These are available but expensive. As a result, in many cases SO_2 is simply emitted to the atmosphere. On the other hand, sulfides can be treated by hydrometallurgical methods without generating SO_2, hence the independence of sulfuric acid manufacture. Elemental sulfur produced by this rout can be easily stockpiled, or transported safely for long distances at low cost.

3.2. Material Handling

In pyrometallurgical processes, the metallurgist is forced to transfer molten slags and matte from one furnace to the other in large, heavy, refractory-lined ladles. Beside the inconvenience and the cost of handling these materials, there is also the inevitable gas emission from them because they are usually saturated with SO_2 and during transfer they cool down a little resulting in decreased gas solubility hence the inconvenient working conditions. In hydrometallurgical plants, solutions and slurries are transferred by pipelines without any problem.

3.3. Energy Consumption

Because of the high temperatures involved in pyrometallurgical processes, which is usually around 1200°C, the reaction rates are high but much fuel will be needed. To make a process economical, heat recovery systems are essential. Heat can be readily recovered from hot gases, but rarely from molten slag or molten metal. Thus, a great deal of energy is lost. Further, the equipments needed for heat economy are bulky and expensive. In hydrometallurgical processes, on the other hand, less fuel is needed because of the low temperatures involved (usually below 100°C). Heat economy is usually no problem.

3.4. Dust Formation

Gases leaving furnaces carry over large amounts of fine dust. This must be recovered to abate pollution and because the dust itself is also a valuable material. The technology of dust recovery is well established but the equipment is bulky and expensive. In hydrometallurgical processes, this is no problem because wet material is usually handled.

3.5. Treatment of Complex Ores

Treatment of complex ores by pyrometallurgical method is unsuitable because separation is difficult; this is, however suitable by hydrometallurgy.

3.6. Treatment of Low-Grade Ores

Treatment of low-grade ores by pyrometallurgy is unsuitable because of the large amount of energy required to melt the gangue minerals. On the other hand it is especially suitable by hydrometallurgy if a selective leaching agent is used.

3.7. Economics

The economics of a pyrometallurgical process is usually suitable for large scale operations and this requires a large capital investment. On the other hand, hydrometallurgical processes are suitable for small scale operations and therefore low capital investment.

3.8. Waste Disposal

Many waste products of pyrometallurgical processes are usually coarse and harmless. For example, slags, which are a silicate phase, can be stored in piles exposed to air and rain without the danger of dissolution and contaminating the streams. They are, however, unacceptable from the aesthetic point of view. On the other hand most residues of hydrometallurgical processes are finely divided solids. If they are dry, they may create dust problems when the wind blows and when wet they will gradually release metal ions in solution which may contaminate the environment. However, disposal of waste solution in hydrometallurgical operations is less troublesome than disposal of gases in pyrometallurgy. Even in a pyrometallurgical process water-scrubbing of a gas before disposal will eventually have a disposal problem with that water.

4. Pressure Leaching

Pressure leaching is a relatively recent technology that can be effectively applied for treating a variety of concentrates. A pressure leaching plant (**Figure 3**) is essentially composed of a high pressure pump (**Figure 4**) that continuously forces the concentrate or ore slurry into an autoclave (**Figure 5**) heated at a temperature of 150°C - 200°C and operating at high pressure. The slurry leaving the autoclave is then introduced in flash tanks (**Figure 6**) where sudden expansion takes place thus releasing low-pressure steam that is used for pre-heating the feed slurry (**Figure 7**) and at the same time decreasing the pressure and temperature of the slurry to permit filtration at ambient conditions. In such system heat economy is maximum.

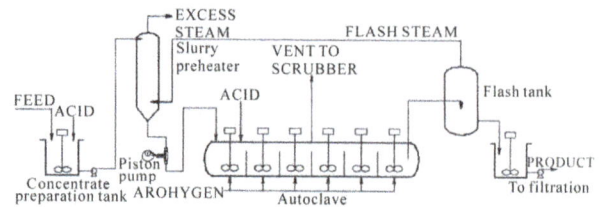

Figure 3. A pressure leaching plant [2].

Figure 4. High pressure piston pump [2].

Figure 5. A typical autoclave [2].

5. Abandoned but Not Forgotten

In the early 1970 many hydrometallurgical processes for

Figure 6. Flash tanks [2].

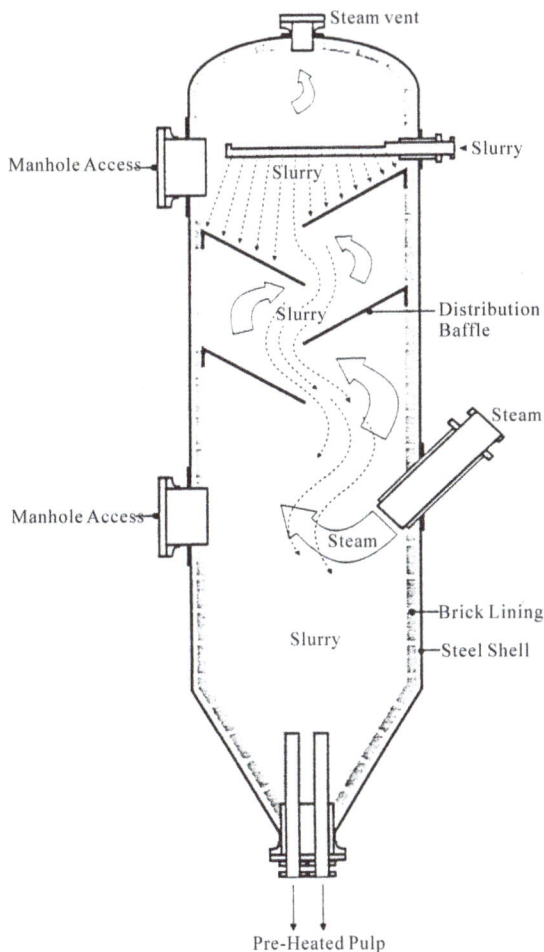

Figure 7. Slurry pre-heater [2].

copper production were developed but were abandoned few years later for various reasons. The situation how ever has changed and the industry is now mature enough [4].

6. Conclusion

In conclusion, the relatively new domain of hydrometallurgy and particularly pressure hydrometallurgy opens the way for many countries to exploit their mineral wealth without the need to rely on multinational companies to invest huge sums in building large smelters. Also, shipping concentrates is no longer a desired way to improve the economy of a country since local processing is now feasible using new technology. Investment in metallurgical education is essential.

REFERENCES

[1] F. Habashi, "Metals from Ores. An Introduction to Extractive Metallurgy," Métallurgie Extractive Québec, Québec City, 2003. www.zone.ul.ca

[2] F. Habashi, "A Textbook of Hydrometallurgy," 2nd Edition, Métallurgie Extractive Québec, Québec City, 1993, 1999. www.zone.ul.ca

[3] F. Habashi, "Handbook of Extractive Metallurgy," WILEY-VCH, Weinheim, 1997.

[4] F. Habashi, "Abandoned but Not Forgotten. The Recent History of Copper Hydrometallurgy," In: P. A. Riveras, *et al.*, Eds., *The John E. Dutrizac Symposium on Copper Hydrometallurgy*, CIM, Montreal, 2007, pp. 3-19.

Factors Research on the Influence of Leaching Rate of Nickel and Cobalt from Waste Superalloys with Sulfuric Acid

Xingxiang Fan[1], Weidong Xing[2], Haigang Dong[2,3], Jiachun Zhao[2],
Yuedong Wu[1], Bojie Li[2], Weifeng Tong[1], Xiaofeng Wu[1]

[1]Kunming Institute of Precious Metals, Kunming, China
[2]Sino-Platinum Metals Co. Ltd., Kunming, China
[3]State Key Laboratory of Advanced Technology of Comprehensive Utilization of Platinum Metals, Kunming, China
Email: fanxingxiang@tom.com, weizi314159@126.com

ABSTRACT

Unlike the reported leaching technologies of waste superalloys, the process of the "atomized spray-sulfuric acid leaching nickel and cobalt" technology was put forward in the present work according to the compositions of waste superalloys. The effects of sulfuric acid temperature, concentration, leaching time, stirring speed and size of superalloys on leaching of Ni and Co from waste superalloys have been mainly investigated, and the optimum leaching conditions were determined and reported. The leaching rates for nickel and cobalt were 96.68% and 96.63%, respectively, and the contents of nickel and cobalt in leaching slag were 6.77% and 0.96%, respectively. The obtained leaching solution containing Ni and Co could be used for production of Ni and Co products after removal.

Keywords: Waste Superalloys; Acid Leaching; Nickel; Cobalt; Leaching Rate

1. Introduction

Ni and Co have excellent physical and chemical properties and mechanical properties, such as high temperature resistance, corrosion resistance, high strength and strong magnetism etc. The production of new materials [1], especially superalloys, play a key role in aviation, aerospace, and other relevant departments of industry. With the development of aerospace career, the demand of these new materials is increasing rapidly, *i.e.*, the requirement of nickel and cobalt are also keeping increasing. However, the shortage of nickel and cobalt mineral resource is becoming more and more serious, secondary recovery of waste superalloys was thus put forward and studied in many countries to avoid wasting of the recyclable resources [2]. At present, the main extraction methods of cobalt and nickel include: acid leaching and high pressure acid leaching with sulfuric acid, hydrochloric acid, nitric acid, or ammonia leaching, chlorine leaching [3-12], and sulfide precipitation [13,14], electrolytic deposition method [15,16], and the carbothermal reduction [17], acid leaching following roasting [18] with hydrometallurgical process or pyrometallurgical process

or pyro-hydro-metallurgical process. These processes are mainly aimed at treating various kinds of raw ore, waste ion battery, waste catalyst materials, relatively rare to treat waste superalloys materials. In this paper, after melting and milling the waste superalloy scrap, the authors investigated leaching of nickel and cobalt from waste superalloys with sulfuric acid directly, the influence factors on the leaching rates of nickel and cobalt ware mainly discussed based one the experimental results.

2. Experimental

2.1. Experimental Materials

The materials used in this experiment were prepared by air-atomization. The average compositions of waste superalloys are listed in **Table 1**.which were determined with the methods of XRF, titration or FAAS. The SEM images in the **Figure 1** shows the micrograph of waste superalloys. **Figure 2** shows the waste superalloys mainly contain solid solution of Ni, Co, Al, Mo, Ta etc. as can be seen from the X-ray pattern shown in **Figure 2**. Acid insoluble components are mainly W, Mo, Ta etc.

(a)

(b)

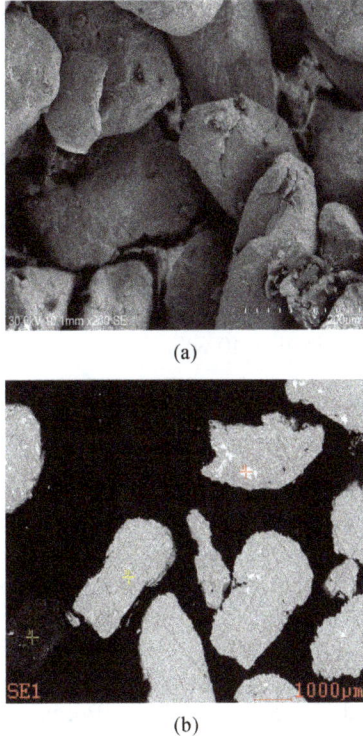

Figure 1. (a) The SEM images of waste superalloys powders; (b) Optical micrograph of waste superalloys (1. Ta, Nb, Ni; 2. Ni, Co, W, Al; 3. Ca, Si).

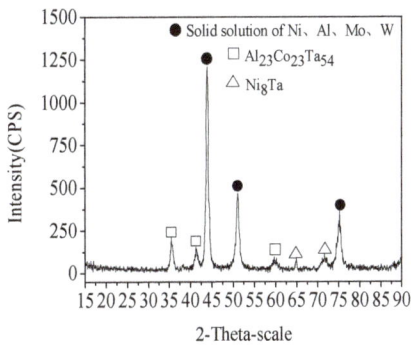

Figure 2. XRD pattern of waste superalloys.

Table 1. Chemical compositions of waste superalloys (wt/%).

Composition	Ni	Co	Mo	Re	W	Ta	Nb	Fe	Al	Cr	Ru
Content	62.72	8.54	1.65	2.42	7.64	6.12	0.51	0.64	5.87	4.77	42.3

Note: the value of Ru is g/t.

which were utilized in following process. The waste samples were crushed and screened under 80 - 120 mesh.

2.2. Experimental Methods

2.2.1. Experimental Process

The waste superalloys were melted at 1400°C, and then made into powders with gas atomization. The powders were milled into different sizes for leaching with sulfide

acid. The experimental flow chart of the leaching of Ni and Co from waste superalloys was shown in **Figure 3**.

2.2.2. Sulfide Leaching Tests

Sulfide leaching tests were carried out in 900 mL beakers. Add sulfide acid into the beaker and heat to a given temperature, then add the measured waste superalloys powders with mechanical stirring. During the reaction, constantly adding water to keep the volume unchanged. After the reaction, the volume of filtrate was measured, and the contents of Ni and Co in filter residue were analyzed.

2.2.3. Experimental Principle

Considering the waste superalloys contain Ni, Co and Fe, the tests were performed with sulfide acid as Ni, Co and Fe could be dissolved by sulfide acid, producing sulfate and H_2. During the leaching, the main reactions were considered as follows:

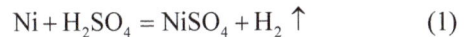

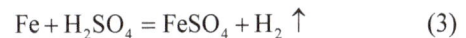

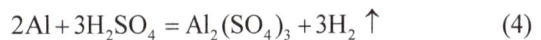

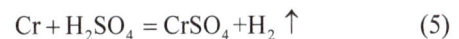

$$Ni + H_2SO_4 = NiSO_4 + H_2 \uparrow \qquad (1)$$

$$Co + H_2SO_4 = CoSO_4 + H_2 \uparrow \qquad (2)$$

$$Fe + H_2SO_4 = FeSO_4 + H_2 \uparrow \qquad (3)$$

$$2Al + 3H_2SO_4 = Al_2(SO_4)_3 + 3H_2 \uparrow \qquad (4)$$

$$Cr + H_2SO_4 = CrSO_4 + H_2 \uparrow \qquad (5)$$

3. Results and Discussion

3.1. Effect of the Temperature on Leaching Rate of Nickel and Cobalt

The effect of temperature on leaching rate of Ni and Co can be seen in **Figure 4** and the leaching time is fixed as 5 hours, the sulfide concentration is fixed as 40 wt%, the stirring speed is fixed as 250 r/min, the size of the powder is fixed as −80 + 120 mesh. It can be seen in **Figure 4** that the leaching rate of Ni and Co is influenced greatly by reaction temperature, the leaching rate of Ni and Co

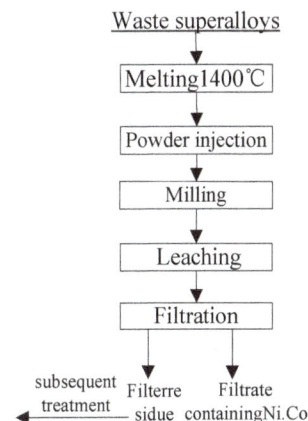

Figure 3. Flow chart of leaching Ni, Co from waste superalloys.

is as much as 96.68 wt% when temperature increased to 85°C. Although the leaching rate can be improved slightly with increasing temperature, it is difficult to keep the temperature increasing as the energy will increase simultaneously. Therefore, the optimum temperature is 85°C.

3.2. Effect of the Leaching Time on Leaching Rates of Nickel and Cobalt

Figure 5 shows the effect of leaching time on leaching rates of Ni and Co with other conditions are fixed. **Figure 5** shows that at the beginning of leaching, the leaching rate of Ni and Co increase very quickly when leaching time increased to 5 hours, the leaching rate of Ni and Co can achieve above 96%, and then although with the leaching time increasing, the leaching rate increased slowly. Considering the efficiency and energy, 5 hours was chosen as the optimum leaching time.

Figure 6 shows that the peaks intensity of solid solution of Ni (containing Co, Al etc.) are decreasing with the time increasing, and some new different complicate chemical phase began to appear.

3.3. Effect of the Sulfuric Acid Concentration on Leaching Rates of Nickel and Cobalt

The effects of sulfuric acid concentration on leaching rates of Ni and Co were shown in **Figure 7** with other conditions were fixed. **Figure 7** shows that with increasing of sulfuric acid concentration, the leaching rate of Ni and Co increase obviously. When the sulfuric acid concentration increased to 40 wt%, the leaching rate are above 96%, while it can be seen that the increasing is relatively slow. Therefore the optimum sulfuric acid concentration is determined as 40 wt%.

3.4. Effect of the Stirring Speed on Leaching Rates of Nickel and Cobalt

Figure 8 shows that the effects of stirring speed on leaching rates of Ni and Co with other conditions are fixed. It can be obviously seen that, the leaching rate increase greatly initially with increasing the stirring speed. When the stirring speed increased to 250 r/min, the leaching rate of Ni and Co can achieve a satisfied result. But when the speed is increased again, the leaching rate of Ni and Co decreases greatly. Because the materials will be rotating with the solution, the effect of stirring is decreased.

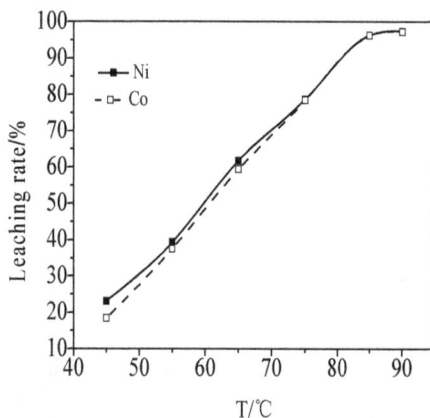

Figure 4. Effect of the temperature on leaching rate of nickel and cobalt.

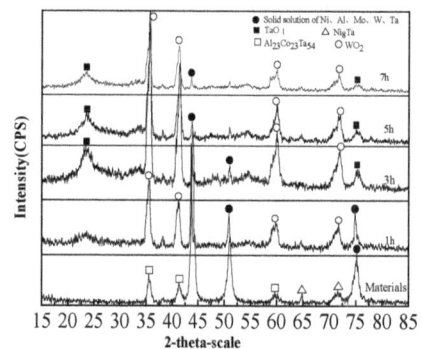

Figure 6. XRD pattern of material and leaching residue in different time.

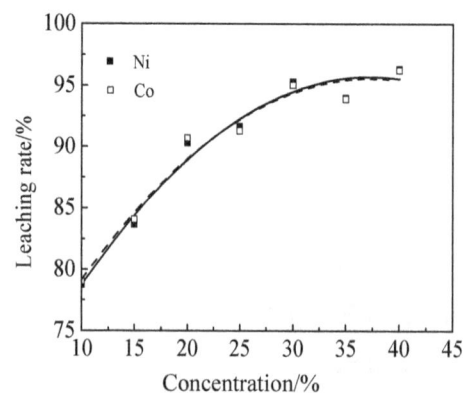

Figure 5. Effect of the leaching time on leaching rate of nickel and cobalt.

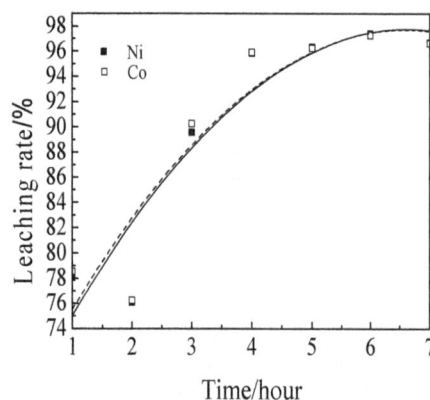

Figure 7. Effect of the sulfuric acid concentration on leaching rate of nickel and cobalt.

3.5. Effect of Particle Size on Leaching Rate of Nickel and Cobalt

It can be seen that the effect of particle size on leaching rate of Ni and Co in **Figure 9**. Theoretically, the smaller size of the materials, the easier to be heat and mass transfer, and the higher leaching rate. **Figure 9** shows that the leaching rate of Ni and Co is increasing slowly when the particle size is increased from 40 - 80 mesh to 160 - 200 mesh. When the size is increased to 200 - 250 mesh, the leaching rate will increase slowly. But it is hard to prepare small particle size, and also time-consuming, because the leaching rate of Ni and Co has achieved 96% when the particle size is 80 - 120 mesh. Therefore the optimum size of materials is determined as 80 - 120 mesh.

4. Results Analysis of Residues of Leaching Ni and Co with Sulfuric Acid

The chemical compositions of leaching residues are listed in **Table 2**. It can be seen that the contents of Ni

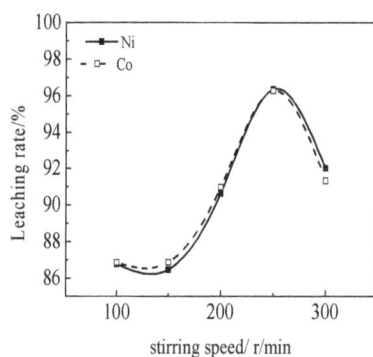

Figure 8. Effect of stirring speed on leaching rate of nickel and cobalt

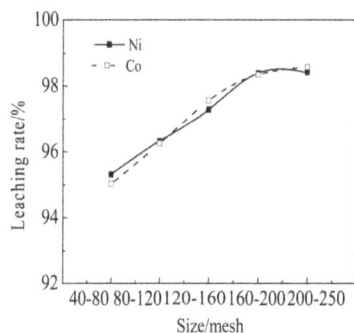

Figure 9. Effect of size on leaching rate of nickel and cobalt.

Table 2. Chemical compositions of leaching residues (wt/%).

Composition	Ni	Co	Mo	Re	W	Ta	Cr	Nb	Ru
Amount	6.77	0.96	4.86	7.31	36.74	23.58	2.89	1.69	110

Note: the value of Ru is g/t.

and Co were obviously decreased, and the contents of W and Ta etc. were enriched for recycling in the subsequent process. In comparison with **Figures 1(a)** and **10**, it can been seen that the particle size becomes much smaller.

The XRD analysis in **Figure 11** indicates that the peaks of Ni, Co and Al etc. decreased greatly compare to the **Figure 1(b)**. The peaks of W and Ta etc. increased, it shows that W and Ta etc. were enriched well in the residues of leaching.

5. Conclusions

1) The optimum leaching conditions for leaching Ni and Co from superalloys with sulfuric acid are reported as T = 85°C, sulfuric acid concentration = 40 wt%, leaching time = 5 h, stirring speed = 250 r/min, and materials particle size = 80 - 120 mesh. The leaching rate of Ni and Co can be at least 96% under this optimum leaching conditions. The contents of Ni and Co in leaching slags were measured as 6.77% and 0.96%, respectively. The leaching solution containing nickel and cobalt could be used for production of Ni and Co products.

2) This study investigated the optimum leaching conditions of Ni and Co from waste superalloys and supplies the important reference data for recycle of other waste superalloys.

Figure 10. The SEM graph of leaching residues.

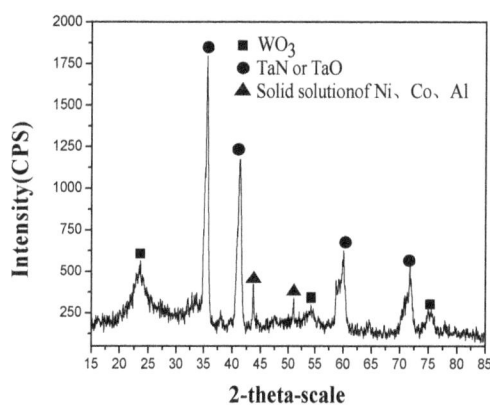

Figure 11. XRD pattern of leaching residues.

6. Acknowledgements

The project was sponsored by National High Technology Research and Development Program of China (863 Program, 2012AA063204).

REFERENCES

[1] A. L. Li, X. R. Zeng and L. M. Cao, "Current Status of Research on High Temperature Materials for Advanced Aircraft Engines," *Materials Review*, Vol. 17, No. 2, 2003, pp. 26-28.

[2] Q. L. Su, Z. W. Shi and J. Hunsaker, "Status and Consideration of Recycling of Superalloys in Our Country," *Non-Ferrous Metals Recycling and Utilization*, Vol. 4, 2006, pp. 19-20.

[3] Y.-F. Shen, W.-Y. Xue and W.-Y. Niu, "Recovery of Co(II) and Ni(II) from Hydrochloric Acid Solution of Alloy Scrap," *Transaction of Nonferrous Metals Society of China*, Vol. 18, No. 5, 2008, pp. 1262-1268.

[4] J. H. Li, X. H. Li and Q. Y. Hu, "Study of Extraction and Purification of Ni, Co and Mn from Spent Battery Material," *Hydrometallurgy*, Vol. 99, No. 1-2, 2009, pp. 7-12.

[5] G. J. van Tonder and P. J. Cilliers, "Cobalt and Nickel Removal from Zincor Impure Electrolyte by Molecular Recognition Technology (MRT)-Pilot Plant Demonstration," *The Journal of The South African Institute of Ming an Metalleurgy*, 2002, pp. 11-18.

[6] Z. J. Yu, Q. M. Feng and L. M. Ou, "Leaching of Cobalt Bearing Metallic Matte in Sulfhuric Acid at Normal Pressure," *Journal of Central South University (Science and Technology)*, Vol. 37, No. 4, 2006, pp. 675-679.

[7] M. V. Rane, V. H. Bafna and R. Sadanandam, "Recovery of High Purity Cobalt from Spentammonia Cracker Catalyst," *Hydrometallurgy*, Vol. 77, No. 3-4, 2005, pp. 247-251. doi:10.1016/j.hydromet.2004.12.004

[8] Y. J. Zheng, H. Teng and H. Q. Yan, "Nitric Acid Oxidation Leaching of Cobalt from Refractory High-Arenic Cobalt Ores," *The Chinese Journal of Nonferrous Metals*, Vol. 20, No. 7, 2010, pp. 1418-1423.

[9] N. J. Kang, "Development of Application of Hot Pressure Leaching Technology in Recovery of Nickel and Cobalt in China," *China Nonferrous Metallurgy*, Vol. 24, No. 2, 1995, pp. 1-7.

[10] J. H. Liu, H. R. Zhang and R. X. Wang, "Process of Ammonium Leaching Oxidation Ore of Cobalt and Copper at High Pressure," *Chinese Journal of Rare Metals*, Vol. 36, No. 1, 2012, pp. 149-153.

[11] Y. F. Shen, W. Y. Xue and W. Li, "Selective Recovery of Nickel and Cobalt from Cobalt-Enriched Ni-Cu Matte by Two-Stage Counter-Current Leaching," *Separation and Purification Technology*, Vol. 60, No. 2, 2008, pp. 113-119. doi:10.1016/j.seppur.2007.08.010

[12] X. C. Hou, L. S. Xiao and C. J. Gao, "Experimental Study on Leaching of Nickel and Cobalt from Waste High-Temperature Ni-Co Alloys," *Hydrrometallurgy of China*, Vol. 28, No. 3, 2009, pp. 164-169.

[13] Y. M. Zhou and B. L. Hu, "Nickel and Cobalt Recovered from Cobalt-Nickel Matte Leaching Lixivium," *Nonferrous Metals (Extractive Metallurgy)*, Vol. 6, No. 4, 2012, pp. 11-13.

[14] Y. B. Xu, Y. T. Xie and J. S. Liu, "Enrichment of Valuable Metals from the Sulfuric Acid Leach Liquors of Nickeliferous Oxide Ores," *Hydrometallurgy*, Vol. 95, No. 1-2, 2009, pp. 28-32.

[15] N. Pradhan, P. Singh and B. C. Tripathy, "Electrowining of Cobalt from Acidic Sulphate Solutions-Effect of Chloride Ion," *Minerals Engineering*, Vol. 14, No. 7, 2001, pp. 775-783. doi:10.1016/S0892-6875(01)00072-3

[16] W. P. Zhang, "Study on the Processing of Low-Co Cemented Carbide Scraps by Electrochemical Method," *Cenmented Carbide*, Vol. 23, No. 2, 2006, pp. 107-109.

[17] R. T. Jones, G. M. Denton and Q. G. Reynolds, "Recovery of Cobalt from Slag in a DC Arc Furnace at Chambishi, Zambia," *The Journal of the South African Institute of Mining and Metallurgy*, 2002, pp. 5-10.

[18] C. Arslan and F. Arslan, "Recvery of Copper, Cobalt and Zinc from Copper Smelter a Converter Slags," Vol. 67, No. 1-3, 2002, pp. 1-7.

4

Liquid-Liquid Extraction of V(IV) from Sulphate Medium by Cyanex 301 Dissolved in Kerosene

Ranjit Kumar Biswas[*]**, Aneek Krishna Karmakar**

Department of Applied Chemistry and Chemical Engineering, Rajshahi University, Rajshahi, Bangladesh

Email: [*]rkbiswas53@yahoo.com

ABSTRACT

The equilibrium of extraction of V(IV) in the $V(IV)$-SO_4^{2-} (H^+, Na^+)-Cyanex 301 (HA)-kerosene system has been studied. Significant extraction occurs above pH 1 within 10 min. $^C D$ (extraction ratio at constant $pH_{(eq)}$ and $[HA]_{(o,eq)}$) value is slightly decreased with increasing $[V(IV)]_{(ini)}$. $^C D$ is found to be directly proportional to $[H^+]^{-n}$ ($n \leq 2$), $[HA]^2$ and $(1 + 1.58 \left[SO_4^{2-} \right])$. The process is endothermic ($\Delta H = 16$ kJ/mol). Apparent K_{ex} values at 303 K are $10^{-1.419}$ and $10^{-0.94}$ in 0.10 and 1.50 mol/L SO_4^{2-} medium, respectively. The loading capacity is calculated to be 7.87 g V(IV) per 100 g Cyanex 301. Kerosene appears as the best diluent. Stripping to the extents of 100%, 94% and 97.7% are possible in single stage by 1 mol/L H_2SO_4, HCl and HNO_3, respectively. Separations of V(IV) from Cu(II) (at pH 0), Zn(II) (at pH 0.5) and Fe(III) (at pH 1.0) by Cyanex 301 are possible.

Keywords: Extraction Equilibrium; Vanadium(IV); Cyanex 301; Kerosene; Sulphate

1. Introduction

Vanadium is widely used to prepare ferro-vanadium and the oxidative catalyst, V_2O_5. Almost 90% vanadium is used to prepare ferro-vanadium (0.1% - 0.3% V) for automobile parts and high speed tools. Besides its use as oxidative catalyst, vanadium compounds are used in ink, dye, paint and varnish, insecticide, photographic chemicals, medicine and glass industries [1]. From carnotite ($K_2U_2V_2O_{11} \cdot 3H_2O$), it is extracted with uranium by the DAPEX process. It is also extracted from vanadinite ($3Pb_3(VO_4)_2 \cdot PbCl_2$). But in the present world, these ores together with patronite (V_2S_3) are rare on the earth's crust. So to meet up the demand, the processing of low grade ores and waste materials (tar sand, waste desulphurization catalyst, waste SO_2-oxidation catalyst of contact process, slag, ash, rock etc.) are desirable. Prior roas- ting and leaching [2,3] followed by solvent extraction is convenient for such purpose. It can build up concentra- tion by using low organic to aqueous phase ratio (O/A) in ex- traction and high O/A in stripping.

Works published before 1976 on solvent extraction of V(IV) by various extractants have been summarized by Sekine and Hasegawa [4]. D2EHPA is a promising extractant for V(IV) and V(V) [5-11]. Saji and Reddy [12] have reported the extractions of V(IV) and V(V) by

EHEHPA. In the field of solvent extraction, a delayed development is the use of organo-phosphinic compounds (Cyanex reagents) introduced by American Cyanamid and Cytec Canada Inc. Cyanex 302 and Cyanex 301 are mono-and di-sulphide analogues of Cyanex 272 (bis-2,4,4-trimethylpentylphosphinic acid). The sulphur substitution decreases the pK_a values (6.4, 5.6 and 2.6 for Cyanex 272 [13], Cyanex 302 [14] and Cyanex 301 [15], respectively) permitting to work at lower pH [16]. Cyanex reagents differ from other commercial organo-phosphorous reagents (e.g. D2EHPA, DDPA, TBP, EHEHPA etc.) in that the former reagents contain P-C bonding, whereas the latter reagents contain P-O-C bonding. The presence of P-C bonding in Cyanex reagents renders them to be less susceptible to hydrolysis and less soluble in water [17].

In recent past, the extraction behavior of V(IV) from SO_4^{2-} medium by Cyanex 272 [18,19] and by Cyanex 302 [20] and of V(IV) and V(V) in presence of Mo(VI), W(VI), U(VI), Ti(IV), Al(III), Cr(III), Fe(III) from Cl⁻ medium by Cyanex 272 and HA [17] had been reported. There is no report on the equilibrium of the extraction of V(IV) from SO_4^{2-} medium by HA. The present paper reports the extraction characteristics of V(IV) from SO_4^{2-} medium by HA dissolved in kerosene in order to determine the effects of various parameters on extraction ratio and the value of K_{ex}; and also to propose mecha-

---

[*]Corresponding author.

nism. The loading of HA by V(IV) and the stripping ability of various mineral acids are also investigated. Finally, the possibilities of separation of V(IV) from some 3*d*-block cations in binary mixtures have been examined.

2. Materials and Methods

2.1. Materials

Cyanex 301 is a product of Cytec Canada Inc. and received as a gift. It contains 75% - 83% R_2PS_2H, 5% - 8% R_3PS, 3% - 6% R_2PSOH and ~2% unknown compound [21] and used without further purification since R_3PS and R_2PSOH have the extracting powers too. Kerosene purchased from the local market is distilled (200°C - 260°C) to collect the colorless aliphatic fraction. As-received NH_4VO_3 (99%, Riedel-deHaen) and $VOSO_4 \cdot xH_2O$ (99.9%, Alfa Aesar-Johnson-Mathey), H_2O_2 (30% (w/v), Merck-Germany) are used. Analytically pure diluents (besides kerosene) are the products of Riedel-deHaen and E. Merck-India.

2.2. Analytical

The aqueous [V(IV)] has been measured by the HNO_3 oxidative-H_2O_2 method [22] at 450 nm using a UV-visible Spectrophotometer (UV-1650 PC, Shimadzu, Japan). For standard and test solution preparations, NH_4VO_3 and $VOSO_4 \cdot xH_2O$, respectively, are used. A Mettler Toledo pH meter (model MP 220) is used for pH measurement.

2.3. Procedure for Extraction

The procedure for extraction is given elsewhere [20,23]. Two phases are agitated at O/A = 1 (O = 20 mL) and 303 K (otherwise stated) for a predetermined time (15 min). The phase separation is quick; and the aqueous phase after disengagement is analyzed for its $pH_{(eq)}$ and V(IV)-content. Then "D" is calculated as usual [20,23].

2.4. Procedure for Loading

The loading procedure is given elsewhere [24]. An aliquot of 100 mL 0.20 mol/L HA-kerosene solution has been used for V(IV)-loading from an aqueous solution containing 1.0 g/L V(IV) and 0.10 mol/L SO_4^{2-} at $pH_{(ini)}$ = 2.60. After each stage of contact, cumulative $[V(IV)]_{(o,eq)}$ is calculated to monitor the progress of stripping.

2.5. Procedure for Stripping

The stripping procedure is similar to the extraction procedure. The fully loaded organic solution obtained in the loading study is diluted to contain 200 mg/L V(IV) and 0.10 mol/L HA in kerosene. Vanadium(IV) in this solution has been stripped by (0.1, 0.3 or 1.0) mol/L (H_2SO_4, HCl or HNO_3) solution. A shaking time of 1 h is allowed

arbitrarily. After equilibration and phase separation, the aqueous phase is analyzed for its [V(IV)] in order to calculate % V(IV) stripped by:

$$\left(\left[V(IV) \right]_{(aq,eq)} / \left[V(IV) \right]_{(o,ini)} \right) \times 100$$

2.6. Treatment of Extraction Equilibrium Data

R_2PS_2H is the principal constituent of Cyanex 301. This species is monomeric in non-polar diluents (owing to low electronegativity of sulphur) [21,25]. Consequently, Cyanex 301 has been abbreviated as HA. In aqueous solution, VO^{2+} can form complexes with co-existing OH^- and HSO_4^- or SO_4^{2-}. On considering the existence of $[VO(OH)_jL_k]$ (L is HSO_4^- or SO_4^{2-} and the charge of the complex is neglected for simplicity) in the aqueous phase and the monomeric charge-less extracted complex does not contain any OH^- and HSO_4^- or SO_4^{2-}, the equilibrium for its extraction by HA can be represented as ("x", "2 − j" and "k" are experimental extractant, pH and co-existing ligand dependences, respectively):

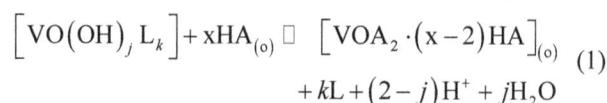

$$\left[VO(OH)_j L_k \right] + xHA_{(o)} \quad \left[VOA_2 \cdot (x-2)HA \right]_{(o)} \quad (1)$$
$$+ kL + (2-j)H^+ + jH_2O$$

On defining "D" as

$$\left[\left[VOA_2 \cdot (x-2)HA \right]_{(o)} \right] / \left[\left[VO(OH)_j L_k \right] \right]$$

K_{ex} of Equation (1) can be expressed as:

$$\log D = \log K_{ex} + (2-j)pH_{(eq)}$$
$$+ x\log[HA]_{(o,eq)} - k\log[L]_{(eq)} \quad (2)$$

The Equation (2) represents the basic equation for solvated chelate formation by reaction of a metal ion with an acidic extractant. All concentration terms and pH in Equation (2) refer to the equilibrium values. Although it is difficult to collect D-values experimentally at a constant set of $pH_{(eq)}$, $[HA]_{(o,eq)}$ and $[SO_4^{2-}]_{(eq)}$, it is possible to modify the experimental D-values to ^CD values at a chosen set of constant $pH_{(eq)}$, $[HA]_{(o,eq)}$ and $[SO_4^{2-}]_{(eq)}$. Since in almost all experiments, 0.10 mol/L SO_4^{2-} (~25 times greater than [V(IV)]) has been used; it can be assumed that $[SO_4^{2-}]_{(eq)}$ will not be significantly changed from $[SO_4^{2-}]_{(ini)}$. Consequently, after determining the approximate pH and extractant dependences and rounding up these values, $\log{^CD}$ can be calculated by:

$$\log{^CD}$$
$$= \log D + (2-j)\left(\text{constant } pH_{(eq)}\text{chosen} - pH_{(eq)}\text{exptl.} \right)$$
$$+ x\left\{ \log \left(\text{constant}[HA]_{(o,eq)} \text{ chosen} \right) \right.$$
$$\left. - \log \left([HA]_{(o,ini)} - x[V(IV)]_{(o,eq)} \right) \right\} \quad (3)$$

Moreover, $[H_2A_2]_{(o,eq)}$ is equal to

$$[H_2A_2]_{(o,ini)} - x[V(IV)]_{(o,eq)}$$

Consequently, according to Equation (2), $\log {}^C D$ should be independent of $[V(IV)]$ if the solutions behave ideally; while it should depend on $pH_{(eq)}$, $[HA]_{(eq)}$ or $[L]$ at a constant set of other parameters. Moreover, as K_{ex} is related to temperature by Van't Hoff equation, $\log {}^C D$ will also depend on temperature.

3. Results and Discussion

3.1. Extraction Equilibrium

Some preliminary experiments indicate that V(IV) is extractable by HA at pH ~ 0.50. When 0.20 g/L V(IV) containing 0.10 mol/L SO_4^{2-} at $pH_{(ini)}$ of 2.30 is extracted by 0.10 mol/L HA in kerosene at 303 K and O/A = 1, then it is found that $[V(IV)]_{(o)}$ is increased up to phase contact of 10 min. It is therefore concluded that the equilibration time is 10 min (15 min is allowed in subsequent experiments).

Variation of "D" with $[V(IV)]_{(ini)}$ is found out at four different set of experimental parameters. It is found in all cases that $[V(IV)]_{(o)}$ is increased, but 'D' is decreased continuously with increasing $[V(IV)]_{(ini)}$ in the aqueous phase. This is contrary to Equation (2) which is valid at constant $[HA]_{(o,eq)}$ and $pH_{(eq)}$. The observed decreasing behavior might be due to the non-constancy of $[HA]_{(o,eq)}$ and $pH_{(eq)}$ for various extents of V(IV) extraction. On calculating $\log {}^C D$ (by Equation (3) on considering "2 − j" = 1.50 at constant $pH_{(eq)}$ of 1.40 or 1.67 at constant $pH_{(eq)}$ of 1.80 and x = 2.00 for all systems), $\log {}^C D$ vs. $\log ([V(IV)]_{(ini)}, mol/L)$ plots are drawn in **Figure 1**. Plots indicate the independency of $^C D$ on $[V(IV)]$ at least up to 300 mg/L. Systems with higher $[V(IV)]$ behave non-ideally.

At a constant $[HA]_{(o, eq)}$, the plot of $\log D$ vs $pH_{(eq)}$ should be a straight line with slope equaling to "2 − j" (cf. Equation (2)). For low $[SO_4^{2-}]$ of 0.10 mol/L, **Figure 2** represents $\log {}^C D$ vs $pH_{(eq)}$ plots at constant $[HA]_{(o,eq)}$ of 0.20 and 0.30 mol/L. In both cases, straight lines are not obtained. Three or four points at *lpHr* produce straight lines of slope 2. At *hpHr*, the slope is decreased gradually and at pH 2.15, it becomes 1.6. It is concluded that single type of extractable species is formed from two different types of aqueous V(IV) species, viz. VO^{2+} and $VO(OH)^+$. In the first case, two H^+ are liberated; whereas, in the second case, one H^+ is liberated, per V(IV) being extracted into the organic phase. As an evidence to this statement, the formation of same extractable species from VO^{2+}, $VO(OH)^+$ and $VO(OH)_2$ by D2EHPA may be cited [5]. The effect of pH on extraction at *hcr* of $[SO_4^{2-}]$ (1.50 mol/L) has also been investigated (cf. **Figure 2**). A

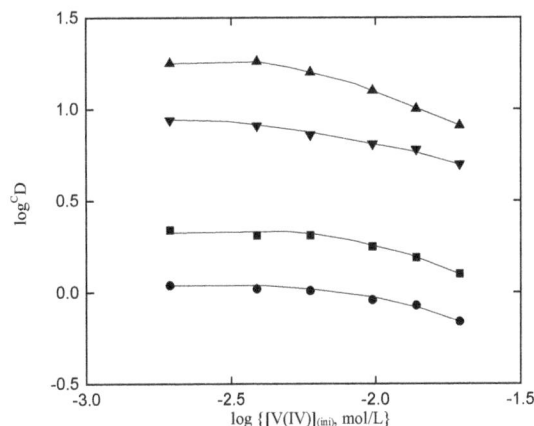

Figure 1. Effect of $[V(IV)]$ on its extraction. Temp. = 303 K, $[SO_4^{2-}]$ = 0.10 mol/L, Equilibration time = 15 min, O/A = 1 (O = 20 mL). (■), $pH_{(ini)}$ = 1.52, constant $pH_{(eq)}$ chosen = 1.4, $[HA]_{(o,ini)}$ = 0.30 mol/L = $[HA]_{(o,eq)}$; (●), $pH_{(ini)}$ = 1.52, , constant $pH_{(eq)}$ chosen = 1.42, $[HA]_{(o,ini)}$ = 0.20 mol/L = $[HA]_{(o,eq)}$; (▲), $pH_{(ini)}$ = 1.90, constant $pH_{(eq)}$ chosen = 1.78, $[HA]_{(o,ini)}$ = 0.30 mol/L = $[HA]_{(o,eq)}$; and (▼), $pH_{(ini)}$ = 1.90, constant $pH_{(eq)}$ chosen = 1.78 $[HA]_{(o,ini)}$ = 0.20 mol/L = $[HA]_{(o,eq)}$.

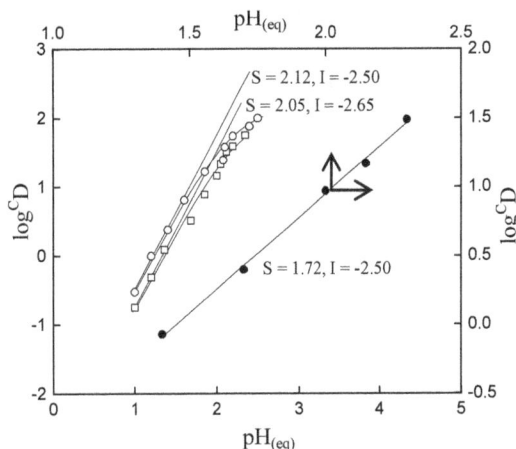

Figure 2. Effect of $pH_{(eq)}$ on extraction. $[V(IV)]_{(ini)}$ = 200 mg/L. (○), $[HA]_{(o,ini)}$ = 0.30 mol/L = $[HA]_{(o,eq)}$; (□), $[HA]_{(o,ini)}$ = 0.20 mol/L = $[HA]_{(o,eq)}$; (●), $[SO_4^{2-}]$ = 1.50 mol/L, $[HA]_{(o,ini)}$ = 0.15 mol/L= $[HA]_{(o,eq)}$. Other parameters are as in Figure 1.

straight line having slope of 1.72 is obtained. It is therefore concluded that the pH dependency in the present system depends on $[SO_4^{2-}]$. As $[SO_4^{2-}]$ is increased, pH dependency is decreased.

According to Equation (2), the plot of $\log {}^C D$ vs. $\log [HA]_{(o, eq)}$ at a constant $pH_{(eq)}$ should be a straight line with slope giving moles of HA(x) associated with 1 mol of V(IV) in extracted species. The $\log D$ vs $\log [HA]_{(o, ini)}$ plots (as there will be a very little variation between $pH_{(ini)}$ and $pH_{(eq)}$ and between $[HA]_{(o, ini)}$ and $[HA]_{(o, eq)}$) are shown in **Figure 3**. Straight lines having slope of 2 are indeed obtained at both *lcr* and *hcr* of SO_4^{2-}. It is

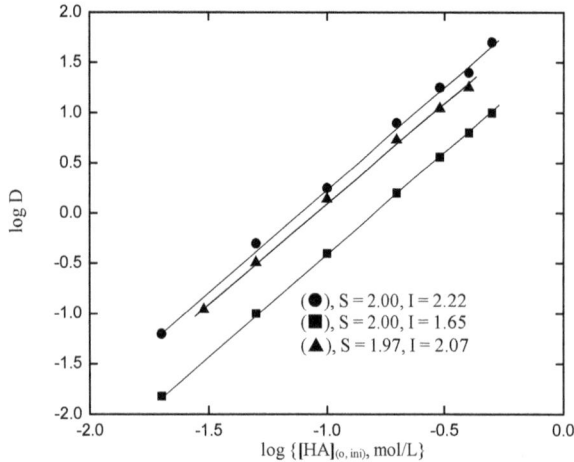

Figure 3. Effect of $[HA]_{(o,eq)}$ on extraction. (●), $pH_{(ini)} = 2.00$; (■), $pH_{(ini)} = 1.50$; (▲), $pH_{(ini)} = 1.40$, $[SO_4^{2-}] = 1.50$ mol/L. Other parameters are as in Figure 1.

therefore concluded that the HA-dependency is always 2 irrespective of $[SO_4^{2-}]$, though the pH-dependency is dependent of $[SO_4^{2-}]$.

The log ^CD vs log ($[SO_4^{2-}]$, mol/L) plot is displayed in **Figure 4**. Experimental points fall on a curve. In *lcr* of SO_4^{2-}, ^CD is seldom changed; whilst in the *hcr*, it is considerably increased with increasing $[SO_4^{2-}]$. The curve in the figure is theoretical and represented by:

$$\log {}^CD = 0.95 + \log \left(1 + 1.58 \left[SO_4^{2-}\right]\right) \qquad (4)$$

which is obtained to fit experimental points as described in the caption of **Figure 4**. It is seen that $[SO_4^{2-}]$ has little effect on extraction when its concentration is kept ~0.10 mol/L; but at *hcr* of $[SO_4^{2-}]$, log ^CD is almost directly proportional to log $[SO_4^{2-}]$.

The Van't Hoff plots are shown in **Figure 5**. It is found that ^CD is increased with increasing temperature and the straight line relationship holds. Slopes of the lines are −870 ($\Delta H = 16.70$ kJ/mol) and −830 ($\Delta H = 15.95$ kJ/mol) for pH = 1.35 and 1.25 systems, respectively. Therefore, the process is endothermic with low ΔH value of ~16 kJ/mol.

It is evident from these studies that the value of "x" is 2 irrespective of the experimental parameter but the value of "k" is 0 at low $[SO_4^{2-}]$ and −1 at high $[SO_4^{2-}]$. The value of "$2 - j$" is 2 in low $pH_{(eq)}$ and <2 in high $pH_{(eq)}$. At *lcr* of SO_4^{2-} and at $lpH_{(eq)}r$, "$2 - j$" = 2 implies that "j" = 0; but at $hpH_{(eq)}r$, "$2 - j$" < 2 (but >1) implies that $1 < j < 2$. On the other hand, at *hcr* of SO_4^{2-} and at both $lpH_{(eq)}r$ and $hpH_{(eq)}r$, "$2 - j$" < 2 implies $j > 0$.

3.2. Evaluation of Extraction Equilibrium Constant

The foregoing results give the equation for log^CD at 303

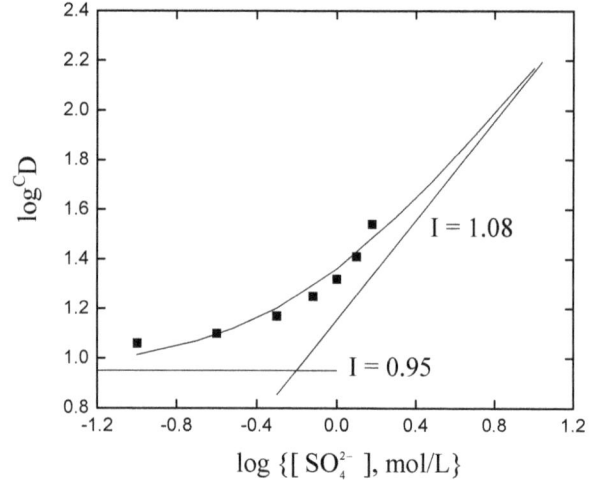

Figure 4. Effect of $[SO_4^{2-}]$ on extraction. $pH_{(ini)} = 2.00 = pH_{(eq)}$, $[HA]_{(o,ini)} = 0.225$ mol/L $= [HA]_{(o,eq)}$. Other parameters are as in Figure 1. Experimental points fall on a curve represented by: $\log {}^CD = $ constant $+ \log(1 + K_{SO_4} [SO_4^{2-}])$. The curve possesses two asymptotes: at *lcr* of SO_4^{2-}: $\log {}^CD = $ constant (represented by horizontal line) and at *hcr* of SO_4^{2-}: $\log {}^CD = $ constant $+ \log K_{SO_4} + \log[SO_4^{2-}]$ (represented by inclined line). At the point of intersection of two asymptotes: $\log K_{SO_4} + \log[SO_4^{2-}] = 0$; or, $\log K_{SO_4} = -\log[SO_4^{2-}] = -(-0.20) = 0.20$; or, $\log K_{SO_4} = 1.58$. Intercept of asymptote at *hcr* is 1.08 and that at *lcr* is 0.95.

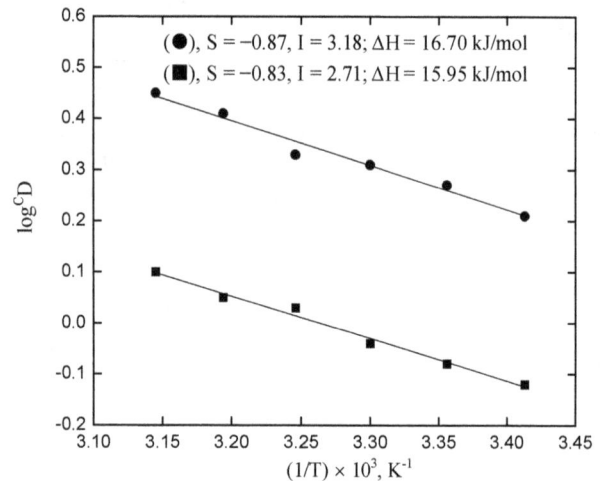

Figure 5. Effect of temperature on extraction. (●), $pH_{(ini)} = 1.35$, $[HA]_{(o,ini)} = 0.30$ mol/L $= [HA]_{(o,ini)}$; (■), $pH_{(ini)} = 1.25$, $[HA]_{(ini)} = 0.20$ mol/L $= [HA]_{(o,ini)}$. Other parameters are as in Figure 1.

K in *lcr* of sulphate as:

$$\log {}^CD = \log K_{ex} + 2 \, pH_{(eq)} + 2 \log[HA]_{(o,eq)} + \log \left(1 + 1.58 \left[SO_4^{2-}\right]\right) \qquad (5)$$

Based on Equation (5), value of log K_{ex} has been evaluated from intercepts of the straight lines or asymptotic lines in **Figures (2)-(4)** and tabulated (**Table 1**). The average log K_{ex} is -1.419 at *lcr* of SO_4^{2-} with *stand. dev.* of 0.105. Besides, log ^CD at 303 K in *hcr* of SO_4^{2-} can be put as:

$$\log{}^C D = \log K_{ex} + 1.72\, pH_{(eq)} + 2\,\log[HA]_{(o,\,eq)} + \log 1.58 + \log\left[SO_4^{2-}\right] \quad (6)$$

On using Equation (6), log K_{ex} is evaluated as -0.94 with *stand. dev.* of 0.026.

3.3. Extraction Mechanism

The empirical equation for K_{ex} at 303 K is:

$$K_{ex} = 10^{-1.42}$$

$$= \frac{\left[V(IV)\right]_{(o)}\left[H^+\right]^{\leq 2}\left(1 + 1.58\left[SO_4^{2-}\right]\right)^{-1}}{\left[V(IV)\right]\left[HA\right]_{(o)}^2} \quad (7)$$

The 1st and 2nd ionization constants of H_2SO_4 are 10^3 [26] and 10^{-2} [27], respectively. These values suggest that SO_4^{2-} will be more available than HSO_4^- in the working pH region. So, L in Equation (1) represents SO_4^{2-}. As the values of "x", "k", "l" and "(2 − j)" are known at different experimental conditions, Equation (1) will provide extraction mechanisms. Although in Equation (1), "L" is presented as a product (liberated during complex formation); experimental results indicate that it is associated with V(IV) during complex formation. As "x" is always 2, non-solvated chelate (VOA$_2$) is formed at *lcr* of SO_4^{2-}, whereas, solvated complex (VOSO$_4$·2HA) is formed at *hcr* of SO_4^{2-}. Typical equilibria are suggested as:

1) in *lcr* of SO_4^{2-} and *lpHr*:

$$VO^{2+} + 2\,HA_{(o)} \,\square\, \left[VOA_2\right]_{(o)} + 2\,H^+ \quad (8)$$

2) in *lcr* of SO_4^{2-} and *hpHr* (limiting):

$$VO \cdot OH^+ + 2\,HA_{(o)} \,\square\, \left[VOA_2\right]_{(o)} + H_2O + H^+ \quad (9)$$

3) in *hcr* of SO_4^{2-} (limiting):

$$VO^{2+} + SO_4^{2-} + 2\,HA_{(o)} \,\square\, \left[VO \cdot SO_4 \cdot 2HA\right]_{(o)} \quad (10)$$

An alternative option of the formation of $[VO \cdot SO_4 \cdot 2HA]_{(o)}$ may be the formation of $[VO(HSO_4)(A) \cdot HA]_{(o)}$ with the simultaneous liberation of a proton. These are only presumptions from the experimental results and not proven by other means.

3.4. Effect of Diluent

In order to determine the effect of diluent on V(IV)-distribution, D-values have been measured when the same aqueous phase has been extracted separately by 0.10 mol/L HA in different diluents keeping all other parametric conditions ([V(IV)] = 200 mg/L, pH$_{(ini)}$ = 2.00 and $[SO_4^{2-}]$ = 0.01 mol/L) identical. It is observed that the extraction ratio increases in the following order with the variation of diluent: CHCl$_3$ (ε = 4.807; D = 0.42) < 1,2-C$_2$H$_4$Cl$_2$ (ε = 10.42; D = 0.54) < C$_6$H$_4$-(CH$_3$)$_2$ (ε = 2.26; D = 0.68) < cyclo-C$_6$H$_{12}$ (ε = 2.02; D = 1.31) = C$_6$H$_5$Cl (ε = 5.69; D = 1.31) < C$_6$H$_5$-CH$_3$ (ε = 2.385; D = 1.64) < n-C$_7$H$_{16}$ (ε = 1.921; D = 2.08) = C$_6$H$_6$ (ε = 2.274; D = 2.08) < CCl$_4$ (ε = 2.228; D = 2.34) < petroleum benzin (D = 3.62) < kerosene (ε = 2.00; D = 3.93). The study helps draw the conclusion that kerosene is a very good diluent followed by petroleum benzin and CCl$_4$ for the extraction of V(IV) by Cyanex 301. CHCl$_3$, 1,2-C$_2$H$_4$Cl$_2$ and C$_6$H$_4$ (CH$_3$)$_2$ are not recommended. 79.72% V(IV) extraction in kerosene phase is decreased to only 29.70% V(IV) extraction in CHCl$_3$ phase.

3.5. Loading of Cyanex 301 with V(IV)

The cumulative [V(IV)]$_{(o)}$ (g/L) has been plotted against the number of phase contact in **Figure 6**. It is observed that the loading of V(IV) in the organic phase is ended up at the 13th contact. An aliquot of 1 L 0.20 mol/L HA is saturated with 5.07 g V(IV) and so the loading capacity is calculated as 7.87 g V(IV) per 100 g HA. The loading capacity is considerably high, and so it can be recommended for a large scale separation of V(IV) from an aqueous solution. The extraction of 5.07 g V(IV)/L by 1 L 0.20 molar HA at saturated loading implies the HA/V(IV) mole ratio of 2.01 which is identical to that obtained from the extractant dependence study. The loading results indicate that the mechanism of extraction at high loading is not changed from that suggested at low loading.

3.6. Stripping of V(IV)-Loaded Organic Phase by Mineral Acids

The maximum V(IV) loaded organic phase containing 5.07 g/L V(IV) with theoretically no free extractant, after proper dilution and adjustment of free [HA], has been subjected for stripping by 0.1, 0.3 and 1.0 mol/L H$_2$SO$_4$, HNO$_3$ and HCl solutions 303 K and O/A = 1. The stripping results are given in **Table 2**. It is found that stripping percentage is more or less acceptable in all three acids used alone. In all cases, stripping percentage is increased with increasing concentration of acid. It is seen that 71.50% stripping by 0.10 mol/L H$_2$SO$_4$ is increased to 100% stripping by 1 mol/L H$_2$SO$_4$. Similarly, 45% stripping by 0.10 mol/L HCl is increased to 94% stripping with 1 mol/L HCl; whereas, 78% stripping by 0.10 mol/L HNO$_3$ is increased to ~98% stripping by 1 mol/L HNO$_3$. Sulphuric acid (1 mol/L) is sufficient to strip off V(IV) quantitatively. Nitric acid and hydrochloric acid

Table 1. Evaluation of the values of K_{ex} at 303 K.

Fig. No.	$pH_{(eq)}$	[Cyanex 301], mol/L	$[SO_4^{2-}]$, mol/L	Intercept, I	log K_{ex}	Avg. log K_{ex}	*Stand. dev.*
i) At *lcr* of SO_4^{2-}							
2	variable	0.200	0.10	−2.65	-1.313		
		0.300	0.10	-2.50	−1.517		
3	1.42	variable	0.10	1.65	−1.253	−1.419	0.105
	1.78		0.10	2.22	−1.403		
4	1.85	0.225	variable	0.90 (*lcr*)	−1.504		
				1.08 (*hcr*)	−1.523		
ii) At *hcr* of SO_4^{2-}							
2	variable	0.150	1.5	−2.43	−0.958	−0.940	0.026
3	1.65	variable	1.5	2.06	−0.921		

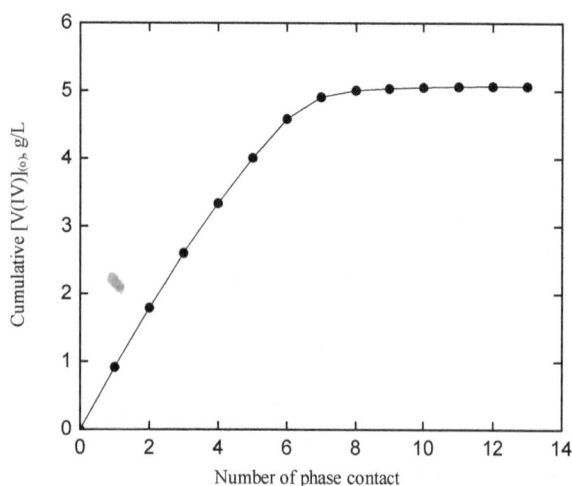

Figure 6. Loading of HA with V(IV). $[V(IV)]_{(ini)}$ = 1000 mg/L, $[SO_4^{2-}]$ = 0.10 mol/L, $pH_{(ini)}$ = 2.60, [HA] = 0.20 mol/L, Stage equilibration time = 15 min, Temp. = 303 K, O/A = 1 (O = 100 mL).

Table 2. Stripping of V(IV) loaded organic phase using different acid solutions. $[V(IV)]_{(o)}$ = 200 mg/L, [Cyanex 301] = 0.10 mol/L, Equilibration time = 1 h, Temp. = (303 ± 0.5) K, O/A = 1 (O = 20 mL).

Stripping agent	Acid concentration, mol/L	$[V(IV)]_{(aq)}$, mg/L	% of V(IV) stripped
H_2SO_4	0.10	140.0	71.5
	0.30	190.0	95.0
	1.00	200.0	100.0
HCl	0.10	90.0	45.0
	0.30	170.0	85.0
	1.00	180.0	94.0
HNO_3	0.10	156.0	78.0
	0.30	165.6	82.8
	1.00	195.3	97.7

can also be used in stripping if two stage stripping is practiced.

It is reported that Cyanex 302 and Cyanex 301 undergo oxidation in oxidizing environment [28-31] (oxidation products being $R_2(P=S)$-S-S-$(S=P)R_2$, $R_2(P=S)OH$ and $R_2(P=O)OH$). It can be demonstrated, however, that when V(IV) is extracted repeatedly from fresh aqueous solutions ([V(IV)] = 0.20 g/L, $[SO_4^{2-}]$ = 0.10 mol/L, $pH_{(ini)}$ = 1.50) by 0.30 mol/L HA in kerosene (fresh in the first step and regenerated afterwards) and stripped subsequently with 1 mol/L H_2SO_4, then (79 ± 2)% extraction and 100% stripping are observed from the 1st - 25th extraction-stripping steps. It is therefore concluded that HA does not undergo any sort of oxidation as also reported by Sole *et al.* [32].

3.7. Separation Ability of V(IV) from Some Other Metal Ions

In order to examine the effectiveness of HA towards the mutual separations of some *3d*-block metal ions *viz.* Ti(IV), V(IV), Fe(III), Co(II), Ni(II), Cu(II) and Zn(II), the extraction percentages of these metal ions have been estimated. For this purpose, 0.20 g/L metal ion is extracted from 0.10 mol/L SO_4^{2-} (or, $[SO_4^{2-}]$ = H_2SO_4 when $[H_2SO_4]$ > 0.10 mol/L) medium at different $pH_{(eq)}$ values by 0.10 mol/L HA in kerosene at 303 K and O/A = 1 (O = 20 mL) on equilibration for 1 h. The extraction results given in **Table 3** predict the following:

1) V(IV) can be separated from Cu(II) at pH 0 in single step (0% V(IV) extraction and 100% Cu(II) extraction).

2) On using counter-current extraction stages, V(IV)

Table 3. Solvent extraction data of some *3d*-block elements by Cyanex 301 dissolved in kerosene. [Cyanex 301] = 0.10 mol/L (in kerosene); [Metal ion] = 0.20 g/L; [SO_4^{2-}] = [H_2SO_4] or 0.10 mol/L, Temp = 303 K, O/A = 1 (O = 20 mL), Equilibration time = 1 h.

pH$_{(eq)}$	V(IV)	Ti(IV)	Fe(III)	Co(II)	Ni(II)	Cu(II)	Zn(II)
0.0	NE	NE	11.0	8.0	8.0	CE	72.0
0.5	1.0	NE	82.0	30.0	35.0	CE	98.0
1.0	4.3	0.9	99.0	67.0	76.0		CE
1.5	38.7	6.2	CE	91.0	95.0		CE
2.0	86.3	35.5	CE	98.0	99.0		
2.5	96.2	70.6		99.5	CE		
3.0	98.4	90.5*		CE	CE		
3.5	99.1	96.8		CE			
4.0	CE*	98.9		CE			
4.5		CE					

NE: non-extractable, CE: complete extraction; *Aqueous solution becomes cloudy before extraction but becomes clear after extraction.

can be separated from:

- Zn(II) at pH ~ 0.5 (1% V(IV) extraction and 98% Zn(II) extraction).
- Fe(III) at pH 1.0 (4.3% V(IV)-extraction and 99% Fe(III)-extraction),
- Co(II) at pH 1.5 (38.7% V(IV)-extraction and 91% Co(II) extraction), and
- Ni(II) at pH 1.5 (38.7% V(IV)-extraction and 95% Ni(II) extraction).

3) Separation of V(IV) from Ti(IV) is difficult but not impossible. Separation can be achieved at pH 2.0 on using counter-current multistage extraction.

4. Conclusions

The following conclusions are drawn:

1) Vanadium(IV) can be extracted by HA at pH above 1. The equilibration time 10 min. Up to at least 0.30 g/L V(IV), the extraction ratio (D) is independent of V(IV) concentration in the aqueous phase.

2) The corrected extraction ratio (^CD) is proportional to $[H^+]^{-2}$ at its *lcr*, $[HA]^2$ and the factor $(1 + 1.58 [SO_4^{2-}])$. The K$_{ex}$ values at 303 K are $10^{-1.419}$ and $10^{-0.94}$ in 0.10 and 1.50 mol/L SO_4^{2-} medium, respectively.

3) The extraction process is endothermic with a ΔH value of 16 kJ/mol.

4) The extracted species are VOA$_2$ at *lcr* and VO·HSO$_4$·A·HA or VOSO$_4$ at *hcr* of SO_4^{2-}.

5) 100 g HA can extract as much as 7.87 g V(IV). This gives HA/V(IV) mole ratio of 2.01 indicating that the mechanism of extraction is not changed with loading.

6) Among the diluents used, kerosene is the best. The least effective diluent is CHCl$_3$ followed by 1,2-C$_2$H$_4$Cl$_2$ and xylene.

7) 1 mol/L H$_2$SO$_4$, HCl and HNO$_3$ can strip off 100%, 94% and 97.7% V(IV), respectively, in single step.

8) Using HA as extractant, V(IV) can be separated from Cu(II). It can be separated from Zn(II) at pH 0.5 but for clear-cut separation counter-current 2-3 stage extractions may be required. In a similar way, V(IV) can be separated from Fe(III) at pH 1.

Separations V(IV) from Ti(IV), Co(II) and Ni(II) by HA appear to be difficult.

REFERENCES

[1] S. Prakash, G. D. Tuli, S. K. Banu and R. D. Madan, "Advanced Inorganic Chemistry," S. Chand and Company Ltd., New Delhi, 1990.

[2] R. K. Biswas, M. Wakihara and M. Taniguchi, "Recovery of Vanadium and Molybdenum from Heavy Oil Desulphurization Waste Catalyst," *Hydrometallurgy*, Vol. 14, No. 2, 1985, pp. 219-230. doi:10.1016/0304-386X(85)90034-9

[3] C. O. Gomez-Bueno, D. R. Spink and G. L. Rempel, "Extraction of Vanadium from Athabasca Tar Sands Fly Ash," *Metallurgical and Materials Transactions B*, Vol. 12, No. 2, 1981, pp. 341-352.

[4] T. Sekine and Y. Hasegawa, "Solvent Extraction Chemistry: Fundamentals and Applications," Marcel Dekker, Inc., New York, 1977, pp. 564-567.

[5] F. Islam and R. K. Biswas, "The Solvent Extraction of Vanadium(IV) with HDEHP in Benzene and Kerosene: The Solvent Extraction of Vanadium(IV) from Sulphuric Acid Solution with Bis-(2-Ehylhexyl) Phosphoric Acid in Benzene and Kerosene," *Journal of Inorganic and Nuclear Chemistry*, Vol. 42, 1980, pp. 415-420. doi:10.1016/0022-1902(80)80018-2

[6] F. Islam and R. K. Biswas, "Kinetics of Solvent Extraction of Metal Ions with HDEHP-II: Kinetics and Mechanism of Solvent Extraction of V(IV) from Acidic Aqueous Solutions with Bis-(2-Ethylhexyl)Phosphoric Acid in Benzene" *Journal of Inorganic and Nuclear Chemistry*, Vol. 42, No. 3, 1980, pp. 421-429. doi:10.1016/0022-1902(80)80019-4

[7] T. Sato, T. Nakamura and M. Kawamura, "The Extraction of Vanadium(IV) from Hydrochloric Acid Solutions by Di-(2-Ethylhexyl)-Phosphoric Acid," *Journal of Inorganic and Nuclear Chemistry*, Vol. 40, No. 5, 1978, pp. 853-856. doi:10.1016/0022-1902(78)80164-X

[8] J. P. Brunette, F. Rastegar and M. J. F. Leroy, "Solvent Extraction of Vanadium(V) by Di-(2-Ethylhexyl)-Phosphoric Acid from Nitric Acid Solutions," *Journal of Inorganic and Nuclear Chemistry*, Vol. 41, No. 5, 1979, pp. 735-737. doi:10.1016/0022-1902(79)80364-4

[9] M. A. Hughes and R. K. Biswas, "The Kinetics of Vanadium(IV) Extraction in the Acidic Sulphate-D2EHPA-n-Heptane System Using the Rotating Diffusion Cell

Technique," *Hydrometallurgy*, Vol. 26, No. 3, 1991, pp. 281-297. doi:10.1016/0304-386X(91)90005-7

[10] R. S. Juang and R. H. Lo, "Stoichiometry of Vanadium(IV) Extraction from Sulfate Solutions with Di(2-Ethylhexyl) Phosphoric Acid Dissolved in Kerosene," *Journal of Chemical Engineering of Japan*, Vol. 26, No. 2, 1993, pp. 219-222. doi:10.1252/jcej.26.219

[11] R. K. Biswas and M. G. K. Mondal, "Kinetics of VO^{2+} Extraction by D2EHPA," *Hydrometallurgy*, Vol. 69, No. 1-3, 2003, pp. 117-133. doi:10.1016/S0304-386X(02)00208-6

[12] J. Saji and M. L. P. Reddy, "Solvent Extraction Separation of Vanadium(V) from Multivalent Metal Chloride Solution Using 2-Ethylhexyl Phosphonic Acid Mono-2-Ethylhexyl Ester," *Journal of Chemical Technology and Biotechnology*, Vol. 77, No. 10, 2002, pp. 1149-1156. doi:10.1002/jctb.690

[13] J. Saji and M. L. P. Reddy, "Selective Extraction and Separation of Titanium(IV) from Multivalent Metal Chloride Solutions Using 2-Ethylhexyl Phosphonic Acid Mono 2-Ethylhexyl Ester," *Separation Science and Technology*, Vol. 38, No. 2, 2003, pp. 427-441. doi:10.1081/SS-120016583

[14] J. Saji, J. K. Saji and M. L. P. Reddy, "Liquid-Liquid Extraction of Tetravalent Titanium from Acidic Chloride Solutions by Bis(2,4,4-Trimethylpentyl)Phosphinic Acid," *Solvent Extraction and Ion Exchange*, Vol. 18, No. 5, 2000, pp. 877-894. doi:10.1080/07366290008934712

[15] M. Ulewicz and W. Walkowiak, "Selective Removal of Transition Metal Ions in Transport through Polymer Inclusion Memberances with Organophosphorus Acid," *Environment Protection Engineering*, Vol. 31, No. 3-4, 2005, pp. 74-81.

[16] W. A. Rickelton, "Novel Uses for Thiophosphinic Acids in Solvent Extraction," *Journal of Metals*, Vol. 44, No. 5, 1992, pp. 52-54.

[17] A. Saily and S. N. Tandon, "Liquid-Liquid Extraction Behavior of V(IV) Using Phosphinic Acids as Extractants," *Fresenius' Journal of Analytical Chemistry*, Vol. 360, No. 2, 1998, pp. 266-270. doi:10.1007/s002160050688

[18] P. Zhang, K. Inoue and H. Tsuyama, "Recovery of Molybdenum and Vanadium from Spent Hydrodesulfurization Catalysts by Means of Liquid-Liquid Extraction," *Kagaku KogakuRonbunshu*, Vol. 21, No. 1, 1995, pp. 451-456. doi:10.1252/kakoronbunshu.21.451

[19] P. Zhang, K. Inoue, K. Yoshizuka and H. Tsuyama, "Solvent Extraction of Vanadium(IV) from Sulfuric Acid Solution by Bis(2,4,4-Trimethylpentyl)Phosphinic Acid in Exxsol D80," *Journal of Chemical Engineering of Japan*, Vol. 29, No. 1, 1996, pp. 82-87. doi:10.1252/jcej.29.82

[20] R. K. Biswas and A. K. Karmakar, "Equilibrium of the Extraction of V(IV) in the V(IV)-SO_4^{2-}(H^+, Na^+)- Cyanex 302-Kerosene System," *International Journal of Nonfer-*

rous Metallurgy, Vol. 1, 2012, pp. 23-31.

[21] K. C. Sole and J. B. Hiskey, "Solvent Extraction Characteristics of Thio Substituted Organophosphinic Acid Extractants," *Hydrometallurgy*, Vol. 30, 1992, pp. 345-365. doi:10.1016/0304-386X(92)90093-F

[22] J. Bassett, R. C. Denney, G. H. Jeffery and J. Mendham, "Vogel's Textbook of Quantitative Inorganic Analysis including Elementary Instrumental Analysis," 4th Edition, ELBS and Longman, London, 1979, pp. 752-753.

[23] M. R. Ali, R. K. Biswas, S. M. A. Salam, A. Akhter, A. K. Karmakar and M. H. Ullah, "Cyanex 302: An Extractant for Fe^{3+} from Chloride Medium," *Bangladesh Journal of Scientific and Industrial Research*, Vol. 46, No. 4, 2011, pp. 407-414.

[24] R. K. Biswas and D. A. Begum, "Solvent Extraction of Fe(III) from Chloride Solution by D2EHPA in Kerosene," *Hydrometallurgy*, Vol. 50, No. 2, 1998, pp. 153-168. doi:10.1016/S0304-386X(98)00048-6

[25] E. Paatero, T. Lantto and P. Ernola, "The Effect of Trioctylphosphine Oxide on Phase and Extraction Equilibria in Systems Containing Bis(2,4,4-Trimethylpentyl) Phosphinic Acid," *Solvent Extraction and Ion Exchange*, Vol. 8, No. 3, 1990, pp. 371-388. doi:10.1080/07366299008918006

[26] E. C. Potterr, "Electrochemical Principles and Applications," Cleaver House Press, London, 1961, p. 51.

[27] R. M. Smith and A. E. Martell, "Critical Stability Constants," *Inorganic Complexes*, Vol. 4, Plenum Press, New York and London, 1976.

[28] N. E. El-Hefny, "Kinetics and Mechanism of Extraction of Cu(II) by Cyanex 302 from Nitrate Medium and Oxidative Stripping of Cu(I) using Lewis Cell," *Chemical Engineering and Processing*, Vol. 49, No. 12, 2010, pp. 84-90. doi:10.1016/j.cep.2009.11.012

[29] K. C. Sole and J. B. Hiskey, "Solvent Extraction of Copper by Cyanex 272, Cyanex 302 and Cyanex 301," *Hydrometallurgy*, Vol. 37, No. 2, 1995, pp. 129-147. doi:10.1016/0304-386X(94)00023-V

[30] A. Bhattacharyya, P. K. Mohapatra and V. K. Manchanda, "Seperation of Americium(III) and Europium(III) from Nitrate Medium Using a Binary Mixture of Cyanex 301 with N-Donor Ligands," *Solvent Extraction and Ion Exchange*, Vol. 24, 2006, pp. 1-17. doi:10.1080/07366290500388459

[31] B. Menoya, M. P. Elizalde and A. Almela, "Determination of the Degradation Compounds Formed by the Oxidation of Thiophosphinic Acids and Phosphine Sulphides with Nitric Acid," *Analytical Sciences*, Vol. 18, 2002, pp. 799-804. doi:10.2116/analsci.18.799

[32] K. C. Sole, J. B. Hiskey and T. L. Ferguson, "An Assessment of the Long Term Stabilities of Cyanex 302 and Cyanex 301 in Sulfuric and Nitric Acids," *Solvent Extraction and Ion Exchange*, Vol. 11, 1993, pp. 783-796. doi:10.1080/07366299308918186

Notations and Abbreviations

ΔH	Apparent enthalpy change
ε	Dielectric constant
^cD	D at a constant $pH_{(eq)}$ and $[HA]_{(o, eq)}$
D	Extraction or distribution ratio
HA	Cyanex 301 (monomeric)
hcr	Higher concentration region
hpHr	Higher pH region
K_{ex}	Extraction equilibrium constant
K_{SO_4}	A proportionality constant in sulphate dependence study
L	Co-existing anion except OH⁻ in the aqueous phase
lcr	Lower concentration region
lpHr	Lower pH region
R	$CH_3\text{-}C(CH_3)_2\text{-}CH_2\text{-}CH(CH_3)\text{-}CH_2\text{-}$
[ ]	Sign of complex species
[ ]	Sign of concentration
Suffix (o)	Organic phase
(ini)	Initial
(eq)	Equilibrium
(aq)	Aqueous phase

5

Comparative Study of Gold Concentration by Elutriation from Different Precious Metal Bearing Ores

Martín A. Encinas-Romero[*], Guillermo Tiburcio-Munive, Jesús L. Valenzuela-García

Departamento de Ingeniería Química y Metalurgia, Universidad de Sonora, Hermosillo, México
Email: [*]maencinas@iq.uson.mx

ABSTRACT

Conventional methods for precious metals gravimetric concentration involve equipment such as shaking tables, centrifuging concentrators, jigs, trommels, or a combination of those. A less commonly used technique is elutriation, which represents an efficient, safe and low-cost method of separation. The goal of the present investigation was to make a comparative study of gold concentration by elutriation from different precious metal bearing ores: an oxide ore, a mineral consisting of a sulfide matrix, a mineral in which the precious metals are free and disseminated and a slimy and clayey black sand material. The best recoveries of precious metals by elutriation were attained for the free disseminated ore and for the black sands, obtaining gold recoveries of 70% and 96% respectively, with appreciable ratios of concentration as well.

Keywords: Elutriation; Precious Metals; Oxides, Sulfides; Free Gold; Black Sands

1. Introduction

Elutriation is a particulate separation process in which an upward fluid stream generally air or water is used. The classification is made through a series of tubular or conical vessels of increasing size, so that the flow rate decreases successively from one of vessel to the next.

Generally, an elutriator consists of one or more "sorting columns" in which the fluid flows upwards at a constant velocity. Feed particles introduced into the sorting column will be separated into two fractions, according to their terminal velocities calculated from Stoke's Law. Particles with a terminal velocity smaller than the fluid mean flow rate will overflow, whereas those with a larger velocity than the mean flow rate will sink toward the underflow. Elutriation is carried out until either no visual signs of a further separation are observed, or there is no change in the weight proportions of the products [1].

Elutriation with air is faster than elutriation with a liquid. It also tends to be more efficient due to a lower air resistance to the particles fall, and to a less tendency for agglomeration of particles [1,2].

Elutriation with liquids is a process of separation or sub-separation of particulate of different sizes within a fluid stream, such as water. If the specific gravity of the feed material is uniform, the resulting grades of the product streams can be significantly uniform even for very narrow size ranges. On the other hand, a considerable size variation of the product streams occurs when there are large differences in specific gravity or particle size of the feed even for narrow size ranges [3,4].

The main advantage of elutriation is the absence of moving parts. It represents an economic alternative as a method for precious metal concentration. The main disadvantage of elutriation is related to the velocity profile, originating across the fluid stream due to the resistance imposed by the vessel walls. Thus, the particles are exposed to a fluid velocity field that varies with the radial position in the vessel. In fact, these particles are carried toward the region of high-speed flow due to the pressure differences on their surfaces. Consequently, the high-speed fluid captures these particles, reducing thereafter the efficiency of the elutriator [5].

There are no recent reports on the use of the elutriation with liquids as applied to upgrade precious metal minerals. Therefore, as a part of a general project on the use of the non-conventional methods for precious metal gravity concentration, the following study was undertaken. The goal of this study is to compare the precious metals concentration by elutriation as applied to different ores. Gold

---

[*]Corresponding author.

recoveries and concentration ratios for those materials were measured and used as comparison parameters.

2. Materials and Methods

2.1. Feed Materials

The study involved the treatment of four different types of feed materials from regional mining deposits:

1) Oxide ore
2) Sulfide ore
3) Free gold ore, and
4) Black sands

The as-received materials were first crushed in a 170 × 135 jaw crusher. Afterwards, they were secondary crushed in a 222 mm cone crusher. Then, they were pulverized in a 222 mm ring pulverizer. Finally, the materials were wet-screened and dried to prepare close-sized feeding for the elutriation tests. A representative simple for every size range was taken for gold analysis and the results are presented in **Table 1**.

With the exception of black sands, chemical and X-ray analysis were made for bulk samples of the different ores. Also, specific gravities for every mineral studied were determined by a picnometer. **Table 2** presents the results. The main mineral species present in the oxide ore are quartz and Fe-bearing minerals (hematite, limonite and jarosite). For the sulfide ore, pyrite is the main component, whereas silicates (quartz, orthoclase, albite and muscovite) are the most abundant species in the free-gold ore.

2.2. Elutriation Equipment

The elutriation equipment consisted of two sorting columns of PVC as shown in **Figure 1**. Both columns have diameters of 50.8 mm (2") and the lengths for the elutriation section are 1054.1 mm (41.5") and 482.6 mm (19"), respectively. Details of equipment facilities are given elsewhere [6].

A typical run involved the following steps. Firstly, the elutriator is assembled as shown in **Figure 1** and the concentrate 1, concentrate 2 and tailings containers are positioned tightly in place. Afterwards, the water feed valve is opened to fill up completely the first column and partially the second one. The sample was mixed with water aiming at preparing a slurry containing 30% of solids (weight basis). This pulp is put into the feed tank and is kept continuously stirred to maintain the solids in suspension.

A test started by opening the pulp feed valve and regulating its flow rate within the range 6 to 13 l/h, and the feed water flow rate, within the range 30 to 121 l/h. The precise velocity ratio for every particle size is adjusted when a clear and constant separation of the two fractions is obtained. Typically, when the particle size decreases much slower feed and washing flow rates are used than with coarser particles. The experiment was stopped until no visible signs of particle separation were observed. The products: concentrates 1 and 2 and tailings were dried at 100°C and analyzed by fire assay for its gold contents.

3. Results and Discussion

3.1. Oxide Ore

Figures 2 and **3** show the gold distribution and ratio of concentration for concentrates 1 and 2, respectively, for the oxide ore. For small particles sized the greatest recoveries are found in the concentrate 2 (>90%). However, as **Figure 3** shows, the concentration ratios in this fraction have an average value of 1:1, which means that most of the feed ore stays in the concentrate 2 without any appreciable upgrading. On the other hand, as the particle

Table 1. Gold head essay for different particle size of the minerals tested.

Size Distribution		Gold Head Essay (g/ton)			
Mesh Size	Average Particle Size (μm)	Oxide Mineral	Sulfide Mineral	Free Gold Mineral	Black Sand
−40; +60	302.5	na	na	na	0.72
−60; +100	200	na	na	na	0.39
−80; +100	165	0.3	7.07	38.2	na
−100; +150	128	0.26	66.75	36.6	0.65
−150; +200	90.5	0.20	74.94	26.65	na
−200; +270	64	0.33	102.7	23.12	na
0.270; +325	49	0.33	74.98	15.73	na
−325	45	0.46	106.6	na	0.40

na: not applicable.

Table 2. Chemical composition and X-ray difraction analysis for the different minerals tested.

Species	Composition (%w/w)			
	Oxide Mineral	Sulfide Mineral	Fee Gold Mineral	Black Sand
SiO_2	>30	>30	>30	8.4
Fe	14.2	19.8	3.92	58.83
Cu	0.031	0.21	0.30	
Pb		1.96	0.020	
Zn	0.20	0.26	0.020	
Cr_2O_3				0.04
ZrO_2				0.054
Fe_2O_3 (hematite)	<5	<5	<5	
$Fe_2O_3Fe_2O_3 \cdot H_2O$ (limonite)	<5	<5		
$KFe(SO_4)_2(OH)_6$ (jarosite)	<5	<5		
$(NaK)Al_3(SO_4)_2(OH)_6$ (natrualunite)	<5			
$PbFe_6(SO_4)_4(OH)_{12}$ (plumbojarosite)	<5			
FeS_2 (pyrite)	<5	<5	10 - 15	
$KAlSi_2O_8$ (orthoclase)	<5	<5	10 - 15	
$NaAlSiO_3$ (albite)	<5			
PbS (galena)	<5			
$KAl_2Si_2AlO_{10}(OH)_2$ (muscovite)		<5	5 - 10	
K-Na-Mg-Fe-Al-Si-O-H_2O (iccite)		<5		
$Al_2SiO_5(OH)_4$ (kaolinite)		<5		
FeO(OH) (goethite)		<5		
$PbCO_3$ (cerusite)		<5		
Ca-Na-Mg-Fe-Al-Si-O-OH-H_2O (montmorillonite)	0.20	74.49	39.43	0.36
Au-80 mesh (g/ton) 0.2				
Bulk Specific Gravity (g/cm^3)	3.62	3.56	3.64	5.34

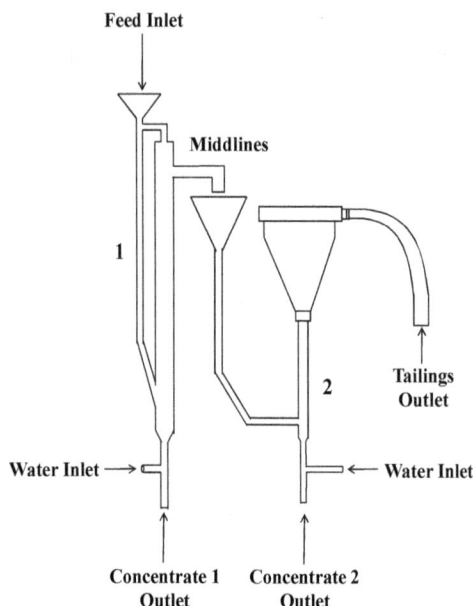

Figure 1. Schematic diagram of experimental elutriator. Numbers 1 and 2 refer to first and second elutriators.

size increases recovery in the concentrate 1 increases. Gold recoveries of 74% were obtained for the particle sizes of 165 μm (−80 + 100 mesh), and gold concentration ratios of the order of 5.5:1. As an important %w/w in this ore corresponds to iron oxides, others researchers had reported that the smaller iron oxides particles are not easily elutriated from fluidized beds of mixed size particles [7]. These results are consistent with the results of present study.

3.2. Sulfide Ore

Regarding the sulfide ore, **Figures 4** and **5** indicate a poor gold recovery for both concentrates (<8%). For coarser sizes a slight increase in recovery was observed for concentrate 1 whereas for concentrate 2 the increase is more significant (from 10% to 60%), similar to the behavior observed with the oxide ore. However, the increase in gold distribution is accompanied by a drastic decrease in the ratio of concentration indicating that most of the gold values are lost in the tailings.

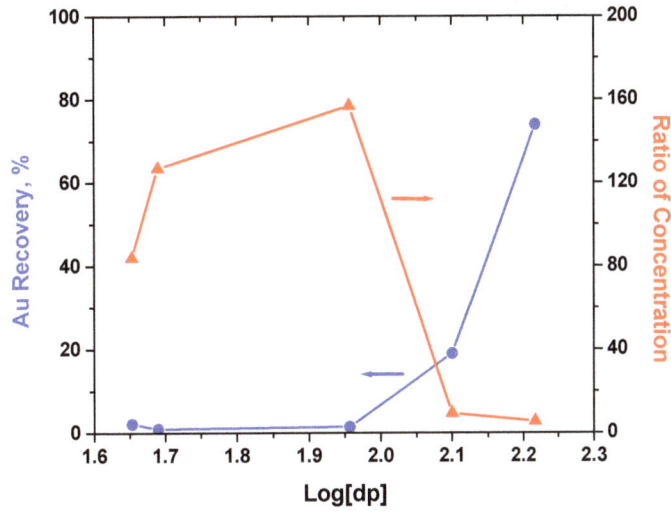

Figure 2. Gold recovery and ratio of concentration, as a function of a particle size, for concentrate 1 of oxide mineral.

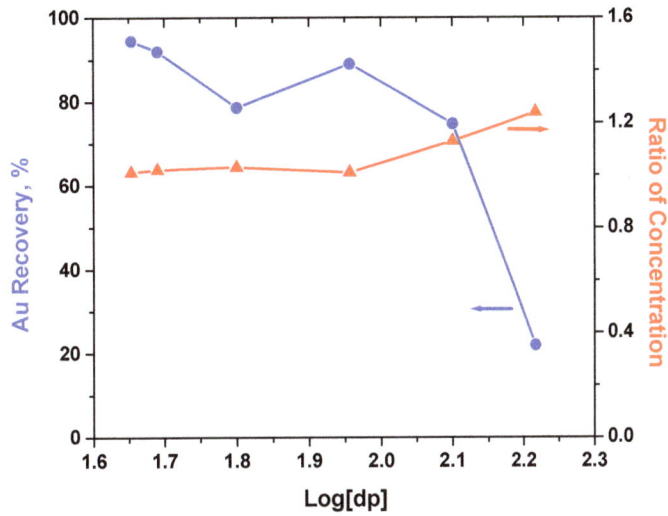

Figure 3. Gold recovery and ratio of concentration, as a function of a particle size, for concentrate 2 of oxide mineral.

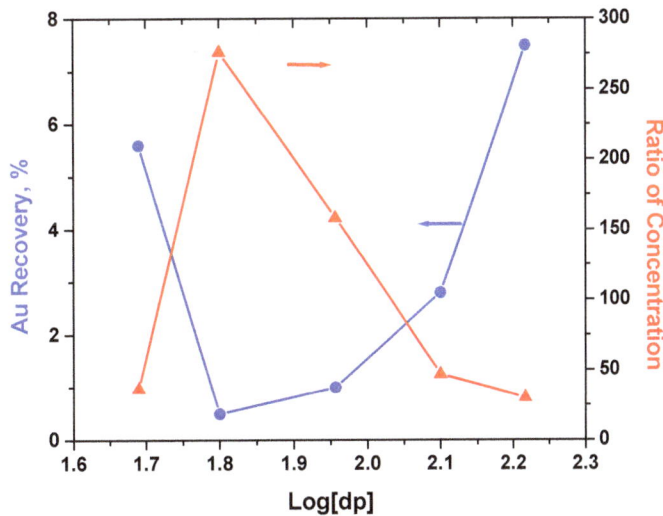

Figure 4. Gold recovery and ratio of concentration, as a function of a particle size, for concentrate 1 of sulfide mineral.

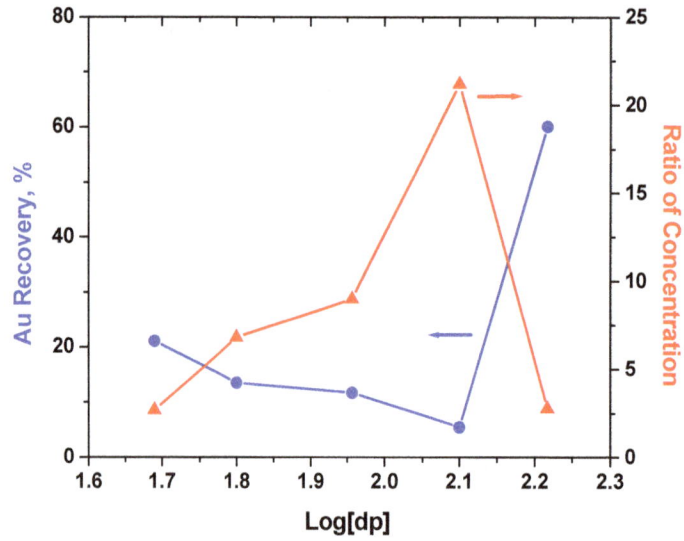

Figure 5. Gold recovery and ratio of concentration, as a function of a particle size, for concentrate 2 of sulfide mineral.

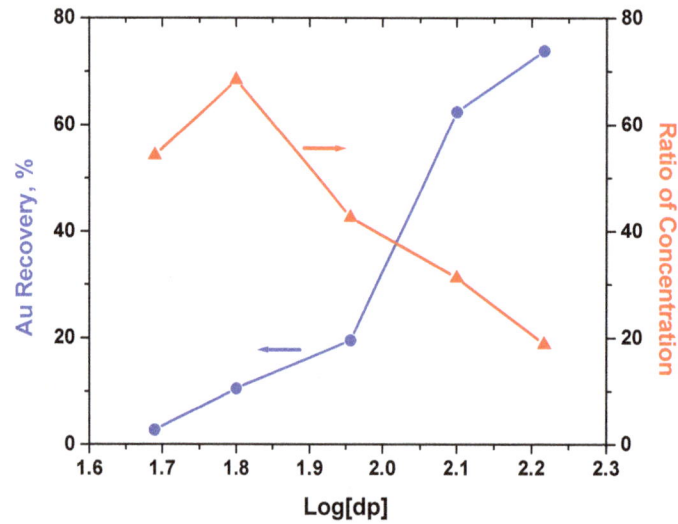

Figure 6. Gold recovery and ratio of concentration, as a function of a particle size, for concentrate 1 of free gold mineral.

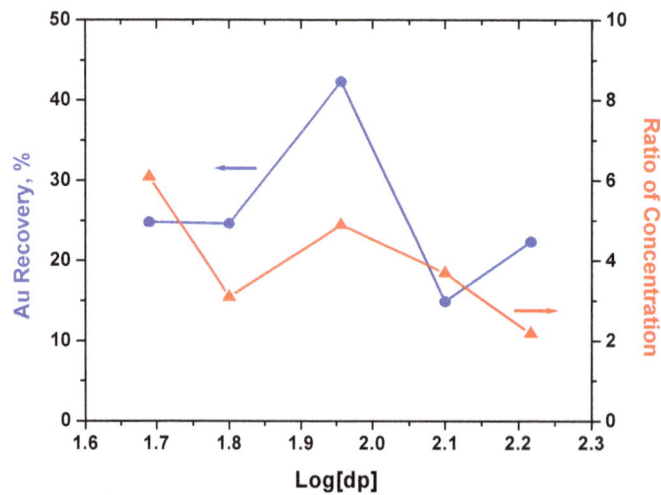

Figure 7. Gold recovery and ratio of concentration, as a function of a particle size, for concentrate 2 of free gold mineral.

3.3. Free Gold

For the free gold ore, **Figures 6** and **7** depict the variation of gold recovery and ratios of concentration against particle size for concentrates 1 and 2, respectively. Compared to previous materials, here a significant difference is observed. For coarser sizes concentrate 1 has gold contents above 70% and reasonable ratios of concentration average 70:1. For smaller particle sized, the greatest proportion of gold is present in concentrate 2, with appreciable ratios of concentration as well. This kind of materials showed a better response to elutriation as compared to the other minerals.

3.4. Black Sands

In the case of black sands, a great susceptibility to this treatment was observed. **Figures 8** and **9** indicate that for almost the whole particle size range studied, excellent recoveries were obtained, e.g. 96% gold recovery in concentrate 1 fraction with adequate concentration ratios, as **Figure 8** shows. For coarser particle sizes, most of the gold reported in concentrate 2 as shown in **Figure 9**. Because this kind of ores contain relatively pure, well-sorted heavy mineral concentrates, they are separated from larger, less dense particles limiting the settling of larger, less dense particles, but also allows smaller, denser particles to settle unhindered [8].

Tables 3 and **4** summarize the complete comparison of gold distribution and ratio of concentration in concentrates 1 and 2, from the different minerals tested in this study.

4. Conclusions

Elutriation is not suitable for gravimetric concentration of precious metals if they are embedded in complex oxide or sulfide matrix.

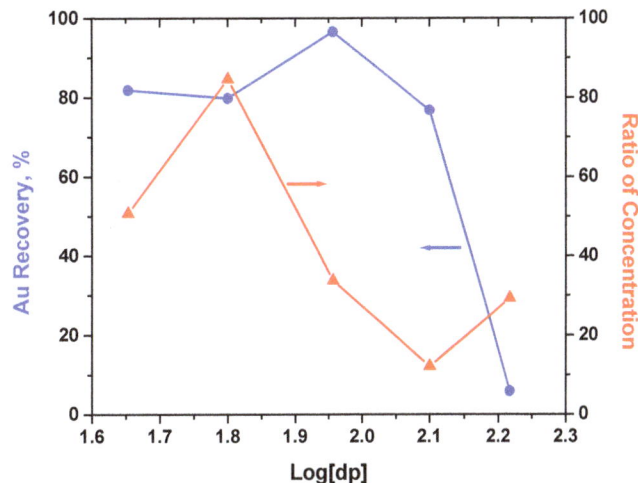

Figure 8. Gold recovery and ratio of concentration, as a function of a particle size, for concentrate 1 of black sand mineral.

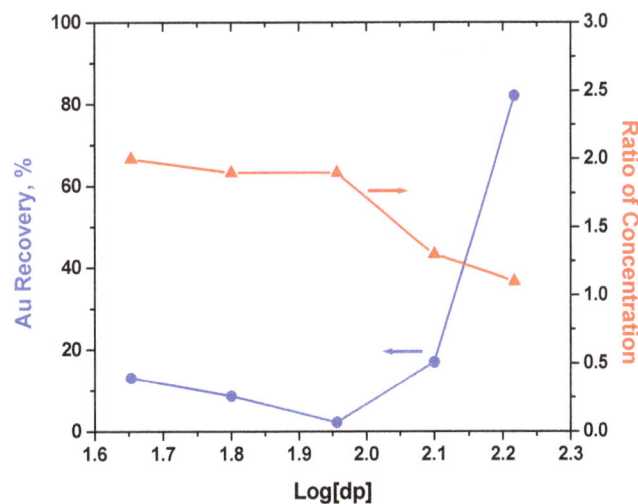

Figure 9. Gold recovery and ratio of concentration as a function of a particle size, for concentrate 2 of black sand mineral.

Table 3. Gold recovery as a function of a particle size, from different minerals tested. (C-1): Concentrate 1, (C-2): Concentrate 2.

Average Particle Size, [dp] (μm)	Gold Recovery Oxide Mineral (%)		Gold Recovery Sulfide Mineral (%)		Gold Recovery Free Gold Mineral (%)		Gold Recovery Black Sand Mineral (%)	
	C-1	C-2	C-1	C-2	C-1	C-2	C-1	C-2
302.5	na	na	na	na	na	na	6.0	82.22
200	na	na	na	na	na	na	76.93	17.08
165	74	22	7.5	60.21	73.9	22.4	na	na
128	19	74.8	2.8	5.6	62.5	15	96.70	2.35
90.5	1.46	89	1.0	11.81	19.6	42	na	na
64	-	78.8	0.5	13.61	10.6	24.7	na	na
49	1	92	5.6	21.16	2.8	24.9	na	na
45	2.2	94.5	-	-	na	na	82	13.2

na: not applicable.

Table 4. Ratio of concentration as a function of a particle size, from different minerals tested. (C-1): Concentrate 1, (C-2): Concentrate 2.

Average Particle Size [dp] (μm)	Ratio of Concentration Oxide Mineral (%)		Ratio of Concentration Sulfide Mineral (%)		Ratio of Concentration Free Gold Mineral (%)		Ratio of Concentration Black Sand Mineral (%)	
	C-1	C-2	C-1	C-2	C-1	C-2	C-1	C-2
302.5	na	na	na	na	na	na	29.4	1.1
200	na	na	na	na	na	na	12.3	1.3
165	5.5	1.24	30.2	2.79	18.9	2.2	na	na
128	9.3	1.13	46.8	21.21	31.3	3.7	33.1	1.9
90.5	157.5	1.01	158.2	9.02	42.7	4.9	na	na
64	-	1.03	276.2	6.82	68.5	3.1	na	na
49	127	1.02	36.4	2.70	54.3	6.1	na	na
45	84	1.01	-	-	na	na	50.9	2.0

na: not applicable.

cious metals if they are present in slimly and clayey black sands or as free gold.

For the free gold ores the best gold recoveries (70%) and concentration ratios (70:1) were obtained for coarser particles sizes, from 90.5 to 165 μm.

For the black sands the best results (96% of gold recovery) were for smaller sizes, *i.e.* 45 to 128 μm.

REFERENCES

[1] B. A. Wills, "Mineral Processing Technology," Pergamon Press, Oxford, England, 1979.

[2] M. G. Rasul, V. Rudolph and F. Y. Wang, "Particle Separation Using Fluidization Techniques," *International Journal of Mineral Processing*, Vol. 60, No. 3-4, 2000, pp. 163-179. http://dx.doi.org/10.1016/S0301-7516(00)00016-8

[3] A. F. Taggart, "Handbook of Mineral Dressing," John Wiley & Sons, New York, 1945.

[4] J. Elder, W. Kow, J. Domenico and D. Wyatt, "Gravity Concentration—A Better Way," *Proceedings of the International Heavy Minerals Conference*, 18-19 June 2001, Australasian Institute of Mining and Metallurgy, pp. 115-118.

[5] E. G. Kelly and D. J. Spottiswood, "Introduction to Mineral Processing," John Wiley & Sons, Inc., New York, 1992.

[6] A. Escamilla Ruvalcaba, "Tratamiento por Elutriación para la Concentración de Minerales de Oro y Plata," B.S. Thesis, Universidad de Sonora, Hermosillo, Sonora, 1999.

[7] J. M. Rodríguez, J. R. Sánchez, A. Alvaro, D. F. Florea and A. M. Estévez, "Fluidization and Elutriation of Iron Oxide Particles. A Study of Attrition and Agglomeration Processes in Fluidized Beds," *Powder Technology*, Vol. 111, No. 3, 2000, pp. 218-230. http://dx.doi.org/10.1016/S0032-5910(99)00292-2

[8] L. J. Cabri, "New Developments in Process Mineralogy of Platinum-bearing Ores," *Proceedings of the Canadian Mineral Processors*, *36th Annual Meeting*, Ottawa, 2004, pp. 189-198.

6

Axisymmetric Strain Stability in Sheet Metal Deformation

Tatjana V. Brovman
Tver State Technical University, Tver, Russia
Email: brovman@mail.ru

ABSTRACT

Production of axisymmetric pieces by technology of sheet metal drawing is widespread nowadays. So the calculation analysis of capacity and forces necessary for deformation is of special interest. The length of cylindrical pieces with axisymmetric deformation is limited by loss of stability and buckling due to the development of side strains. A new technological process is based on making considerable number of folds—18 - 26 with the amplitude of 0.8 - 0.9 mm—before the deformation or immediately after the partial one. That reduces the stiffness of billets and prevents from development of large size buckles. A new technological process is developed for producing a long run of high-quality products.

Keywords: Plastic Axisymmetric Deformation; Loss of Stability; Side Strain Development

1. Introduction

Axisymmetric deformation of sheet metal blanks are widely used in the manufacture of cylindrical pieces made of flat metal sheets.

Such products as various vessel bodies, nozzles, connecting pipes are made of low carbon steels as well as of copper, brass, magnesium alloys and other metals.

This paper examines only the deformation of the products with axial symmetry, which are increasingly used in industry.

Regularities of plastic deformation are studied in research papers [1-4]. They demonstrate that under the pressure of a cylindrical die on the center of the workpiece which was initially shaped like a disk, the plane stress is developed, with only two of the six tensor components being nonzero. Due to tensile radial stress and compressive tangential stress [1,2] the buckling of deformable workpiece accompanied with folds is possible. Usually this leads to waste metal and therefore it is necessary to limit the length of a product, to use drawing with subsequent additional deformation (secondary drawing operation, rolling).

To prevent the occurrence of folds the blank holders are used, but this significantly increases the deformation strain to an extent of the occasional overdrawing occurrence in a deformable work piece. Sometimes drawing with the billet or its edge parts heating (near the maximum diameter) is used. For some metals billet heating can increase the allowed length of a product by 1.5 - 2.0 times.

Magnesium alloy sheets are thought to be used for deep drawing only with the billet heating. But as the thickness of a sheet metal to be deformed is small (up to 1 - 3 mm) and is exposed to chilling it has to be heated directly on a die. That complicates the design of the equipment, enhances significantly power consumption and makes it difficult to ensure safety conditions.

In this paper we discuss some features of the sheet blanks axsymmetric strain. A new drawing technique which allows to improve the technology and to expand significantly the range of products is proposed and investigated.

2. Kinematically Admissible Velocity Fields by Drawing Deformation

The graphic pattern of drawing deformation is illustrated in **Figure 1**, where one can see matrix 1, deformed billet 2 with outer radius R_0 and inner one R_1, thickness h and punch 3 which moves along axis Z with constant velocity $V_z = -V_0$.

Kinematically admissible velocity field determines the upper limit of deformation ca capacity N and the associated moving die force P. Such a velocity field is to satisfy: a) the incompressibility condition, b) velocity boundary conditions, c) the condition under which the deformation capacity is positive [1]. The volume of a deformable billet can be divided into two parts. One of them having a cylindrical shape with inner radius R_1

moves with velocity $V_z = -V_0$ together with moving die 3 [2,3]. In this zone V_r is equal to 0 (the coordinate axes location is shown in **Figure 1**). As this part of a billet moves like a rigid body, its capacity of plastic deformation is equal to zero [3]. The second zone includes the part of a billet which has a shape of a ring with thickness h limited by radius $r = R_1 + h$ and $r = R_0$. Some features of this deformation are considered in paper [4]. In this area the kinematically admissible velocity field could be

$$V_r = -V_0 \frac{R_1}{r}; \quad V_z = 0 \qquad (1)$$

with the strain-rate tensor components

$$\varepsilon_r = V_0 \frac{R_1}{r^2}; \quad \varepsilon_\theta = -V_0 \frac{R_1}{r^2}; \quad \varepsilon_z = 0$$

(the other tensor components are equal to zero). Value V_0 is velocity under $r = R_1$ and the second invariant of strain-rate tensor is equal to

$$H = 2\frac{V_0 R_1}{r^2}$$

This velocity field can be assumed as kinematically admissible only if there exists surface $z(r)$ separating the two selected deformation zones. On this surface the normal rate components are to be equal, (tangential components may differ and their difference determines the capacity of the shear). Such a surface exists and is a paraboloid [5]. As a result, it is possible to determine the deformation capacity

$$N_1 = \int_{R_1}^{R_0} 2\pi r h H dr = 4\pi k V_0 R_1 h \ln \frac{R_0}{R_1}$$

where k is yield point in shear, and shear capacity over the surface of paraboloid is

$$N_2 = \int_{R_1}^{R_1+h} k 2\pi r k \Delta V \sqrt{1 + \left(\frac{dz}{dr}\right)^2} \, dr$$

with paraboloid equation:

$$z(r) = \frac{(R_1 + h)^2 - r^2}{2R_1 + h} \qquad (2)$$

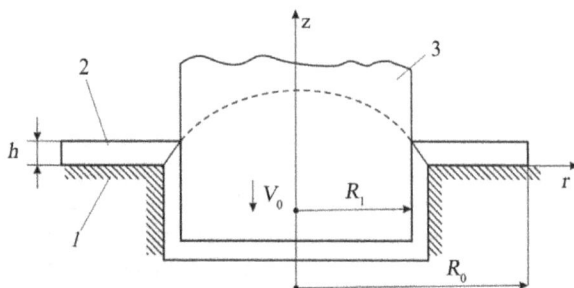

Figure 1. Drawing Scheme.

$$\frac{dz}{dr} = -\frac{2r}{2R_1 + h}$$

tangential components break at the boundary of two zones

$$\Delta V = V_0 \sqrt{1 + \left(\frac{dz}{dr}\right)^2} \left(\frac{dz}{dr}\right)^{-1}$$

with shear capacity

$$N_2 = 2\pi k V_0 R_1 h \left(1 + \frac{h}{2R_1}\right),$$

Full capacity with provision for friction load at the surface of lower die is

$$N = PV_0 = N_1 + N_2 + \mu V_0 P,$$

where μ is coefficient of sliding friction. Moving die force is

$$P = \frac{N_1 + N_2}{V_0(1 - \mu)} \qquad (3)$$

If the deformation is carried out with the use of blankholder, with force T, the friction load increases and force P is equal to

$$P = \frac{N_1 + N_2}{V_0(1 - \mu)} + \frac{2T\mu}{1 - \mu} \qquad (4)$$

Usually $\frac{h}{2R_1} \ll 1$ and the force is

$$P = \frac{2\pi k R_1 h}{1 - \mu}\left(1 + 2\ln\frac{R_0}{R_1}\right) + \frac{2T\mu}{1 - \mu} \qquad (5)$$

$$k = \frac{\sigma_T}{\sqrt{3}}$$

If the steel sheet deformation is carried out without the use of blankholder, as shown in **Figure 1** (with $T = 0$, $R_1 = 80$ mm, $R_0 = 96$ mm, $\frac{R_0}{R_1} = 1.2$; $h = 1$ mm, $\mu = 0.10$, $\sigma_T = 300\frac{MH}{m^2}$, $k = 173 \times \frac{MH}{M^2}$), the value of force according to (5) is $P = 0.13$ MN.

As usual, the kinematically admissible velocity field determines the upper limit of forces (*i.e.*, overstated values of P). According to experimental data, in case of drawing this overstatement amounts to 20% - 25%. This is acceptable for the choice of equipment in most cases.

3. Workpiece Buckling in the Drawing Deformation

In the process of deep drawing a sheet shaped initially

like a disk undergoes tangential compressive stress that can lead to loss of stability. Thereby folds occur on the part of a surface, as shown in **Figure 2**. Paper [1] indicates that the location and number of folds depends on the metal anisotropy, for example for some alloys the number of waves is four, with the location being at an angle of 45° to the rolling direction of a sheet. But the number may come up to six [1].

Usually, the fold occurrence, *i.e.* wavy surface with amplitude of up to 10 - 20 mm, leads to inability to output products and to rejected material. From Saint-Venan condition end equation of equilibrium

$$\frac{\mathrm{d}\sigma_r}{\mathrm{d}r} + \frac{1}{r}(\sigma_r - \sigma_\theta) = 0 \qquad (6)$$

and condition of plasticity

$$\sigma_r - \sigma_\theta = 2k$$

it is clear that stress is $\sigma_r = 2k \ln \dfrac{R_0}{r}$

$$\sigma_\theta = 2k \ln \frac{R_0}{r} - 2k \qquad (7)$$

where the integration constant is determined by the condition of $\sigma_r = 0$ with $r = R_0$. Tangential stress with $r = R_0$ is equal to $\sigma_\theta = -2k$.

The minimum allowable value $\dfrac{R_0}{R_1}$, (under which folds occur) is determined by parameters: R_0, R_1, h, σ_T, E_T with the dimensions: m (for the first three) и MH/m^2 for yield limit σ_T and plastic deformation hardening modulus $E_T = \dfrac{\mathrm{d}\sigma}{\mathrm{d}\varepsilon}$. Here E_T is a value characterizing metal hardening in plastic deformation.

Out of these five variables it is possible to form three dimensionless parameters:

$$A_1 = \frac{R_1}{R_0}, \quad A_2 = \frac{h}{R_0} \quad \text{and} \quad A_3 = \frac{E_T}{\sigma_T}$$

As well as to determine approximately strain-hardening characteristic $E_T = \sigma_B - \sigma_T$, where σ_B is ultimate strength.

According to π theorem the relationship between the above-mentioned parameters must take the following form $f(A_1, A_2, A_3) = 0$. Having solved it for A_1, we get

$$\frac{R_1}{R_0} = f(A_2, A_3) \qquad (8)$$

Based on the data of many deformation experiments with carbon steel sheets of 0.5 mm, 1.00 mm, 1.5 mm and 2.0 mm thicknesses the graphs $\dfrac{R_1}{R_0} = f(A_2, A_3)$ were plotted as given on **Figure 3**.

The relationship (8) can be determined with the empirical formula for a maximum permissible (minimum) value

$$\left(\frac{R_1}{R_0}\right)_{\min} = 1 - 0.5 \frac{E_T}{\sigma_T} \sqrt{\frac{h}{R_0}} \qquad (9)$$

To provide the sheet stability the following condition should be fulfilled:

$$\frac{R_1}{R_0} > \left(\frac{R_1}{R_0}\right)_{\min} \qquad (10)$$

Since the maximum possible upset length is equal to $l = R_0 - R_1 = R_0\left(1 - \dfrac{R_1}{R_0}\right)$, it follows from (9) that the maximum possible length of work is

$$l_{\max} = 0.9 R_0 \frac{E_T}{\sigma_T} \sqrt{\frac{h}{R_0}} \qquad (11)$$

For example, for a steel sheet of $h = 1$ mm thickness

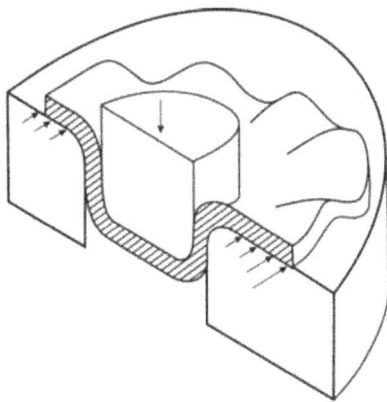

Figure 2. Development of Folds.

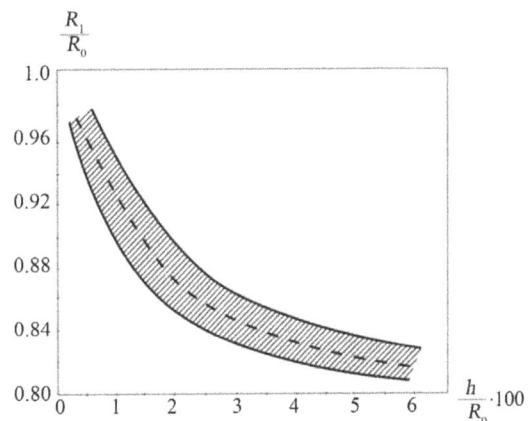

Figure 3. Graph of Function $\dfrac{R_1}{R_0}\left(\dfrac{h}{R_0}\right)$.

with $\sigma_T = 250 \dfrac{MN}{m^2}$, $\sigma_B = 490 \dfrac{MN}{m^2}$,

$\dfrac{E_T}{\sigma_T} = \dfrac{490 - 250}{250} = 0.96$; $R_0 = 100$ mm. $\dfrac{h}{R_0} = 0.01$ according to Formula (9) we get $\left(\dfrac{R_1}{R_0}\right)_{min} = 0.914$,

$R_1 = 91.4$ mm; $l = 8.6$ mm.

Formula (9) can be applied if $\dfrac{E_T}{\sigma_T}\sqrt{\dfrac{h}{R_0}} \geq 0.4$ and the experimental data spread is considerable which is seen from **Figure 3**. The dotted line here divides the data span into two parts. Below the dottedline there is a zone for specially selected samples with thickness differences of no more than 0.02 mm. The higher the gage interference, the more often the buckling failure and buckle development happen. Therefore it is important to increase the dimensional accuracy of sheets used for deep drawing. Empirical formula (9) gives that if

$\dfrac{E_T}{\sigma_T} \to 0, \left(\dfrac{R_1}{R_0}\right)_{min} \to 1$. It conforms to the test which

shows what should be applied for metal (alloys) drawing with high hardening: the higher $\dfrac{\sigma_B}{\sigma_T}$ is, the easier the

deformation can be made, and the larger work length can be produced when drawing. Metals with minor hardening with graph of function $\sigma(\varepsilon)$ being close to a perfectly plastic body graph (*i.e.* when $\sigma = \text{const.} = \sigma_T$) are unfit for deep drawing.

Steel sheets with up to 0.05% - 0.15% temper and not more than 20 - 30 µm grain size with extension strain of no less than 40% are often used fro drawing.

According to the above-mentioned the holder allows to increase l value and decrease parameter $\left(\dfrac{R_1}{R_0}\right)$. However, the ring crack wastage increases as shown on **Figure 4**. There can be seen ring cracks being the reasons of up to 10% and more increase of waster number which aggravates the manufactures' economic indexes significantly.

4. New Method of Drawing

Based on the research conducted a new technological drawing process has been suggested (see Patent Application of Russian Federation №2011123174/02, MPK 7B1D 22/02 from 08 June 2011).

The method consists in forming a series of waves (ridges) along a ring on a disk-form billet, see **Figure 5**. The amplitude of the waves is small (usually not more than (1/2)h) but their quantity is large up to 16 - 28

waves located along the ring as can be seen on **Figure 5**. All the deepenings (waves) are reasonable to be made simultaneously with one die then the billet should be upturned and deformed by drawing, the clearance between the sheet being deformed and the punch should exceed the "waves" amplitude 1.5 - 2.0 times in order to avoid sheet jamming. With drawing deformation the buckled sheet compressive deformation relieves (its hardness is reduced).

Experiments were made with samples of diameters 70 - 120 mm and height 80 - 140 mm of low carbon steels.

The waves may be located not along the billet edge as shown on **Figure 5(a)** but in its middle as on **Figure 5(b)**. Even making 0.5 - 1 mm amplitude waves on a billet parallel portion gives a positive result after partial drawing deformation.

Various wave variants are also shown on **Figures 5(c)** and **(d)**, e. The depth for 0.5 mm sheets of carbon steel was 0.5 mm, the number of waves being 18 - 26. "The loss of stability" with large wave formation, e.g. shown on **Figure 2**, does not occur and there is a possibility of manufacturing products with a length 2 - 3 times higher the admissible length of a conventional drawing.

The deformation is sure to result in increasing the wave amplitude, sometimes 1.5 times (there occurs their compression with "wave" shortening sometimes in 2 - 3 times), but the new waves of large amplitude do not emerge (e.g. as those shown on **Figure 2**).

Figure 4. Blanks with Cracks.

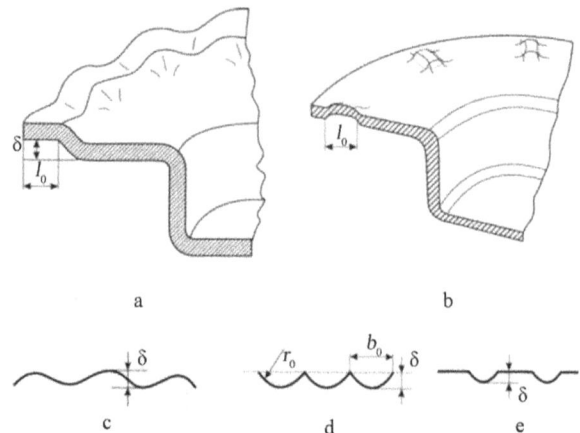

Figure 5. New Method of Drawing.

The main idea of the method consists in making a large number (18 - 26) of waves of small amplitude on a sheet (disk) in advance in order to avoid the potential of emerging 4 - 8 "waves" (buckles) of large amplitude (up to 10 - 20 mm). It decreases the rigidity of a deformed billet, with the amplitude being able to reach 1.5 mm after drawing, which does not lead to sharp changes of the whole product pattern (it usually happens with the stability loss) and allows to manufacture the required product of high quality.

Figure 6 shows the samples manufactured at OAO TsBPR (the town of Tver, Russia) with the new method used.

These Filter Casing Blanks were made from low carbon steel (st.08 according Russian standart 1050 - 88). Quality of Carbon was 0.05% - 0.25%. Part of filter Casings was made of semi-killed steel (st.3). Diameters of these Casing Blanks were 96 mm and their height was 92 mm. Artificial waves were made with length 28 mm and amplitudes 1 mm. Quanlity of Filters was good and reject was decreased twice till 0.75%.

The artificial waves (ridges) are certain to remain on the product. The method is sure not to be used in cases when even small waves on the product cannot be allowed. But in many cases such waves with amplitude of less than 1mm do not at all make the product quality worse and therefore are possible. So the above-mentioned Tver plant manufactured more than 10 thousand filter casings (**Figure 6**) of high quality. An analogy may be drawn to the case of bar pressing (see **Figure 7**). When a straight cantilever bar is compressed with stress P (**Figure 7(a)**), according to Euler formula

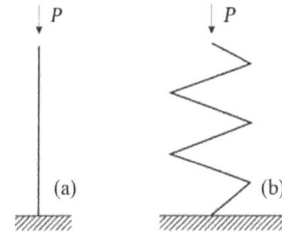

(a)

(b)

Figure 6. Filter Casing Blanks Produced with New Method.

Figure 7. Scheme of Stability under Deformation.

$$\varepsilon_0 = 7.84 \frac{J}{Fl^2}$$

it loses stability when being subject to compressive deformation, (where F is an area of bar section, J is its moment of inertia, l is a bar length).

With further compression a bar loses its configuration, with the deflection increasing rapidly up to 0.5 l and more.

If a bar can be manufactured as shown on **Figure 7(b)**, it keeps its whole configuration in higher deformations $2\varepsilon_0, 3\varepsilon_0$. Its rigidity is a considerably less than that of a straight bar but no sharp changes of the whole form occur.

(Both the rings and deformable bodies of **Figure 7(b)** can certainly "buckle" but, as a rule, in cases when the deformation is several times more than value ε_0 for a straight bar).

In some cases the manufactured rigidity reduction of the billet being drawn can be rational.

5. Conclusions

The following conclusions can be deduced from the above-said study:

1) The axially symmetric sheet blank deformation under drawing has been examined.

The kinematically admissible velocity field for force upper limits has been built.

2) A new method of drawing which included making ridges-waves of small amplitude on a blank before drawing has been proposed and studied. But the number of such waves should be considerable and it gives the possibility to avoid the formation of more significant form distortions of a billet.

3) The method was used for producing oil filter casings and getting a high-quality product. More than 10 thousand filter casings of good qualities of carbon steel were produced with use of new method.

REFERENCES

[1]   R. Hill, "Mathematical Theory of Plasticity," Clarendon Press, Oxford University Press, 1950, 407 pages.

[2]   L. A. Shofman, "Theory and Calculation of Cold Form-

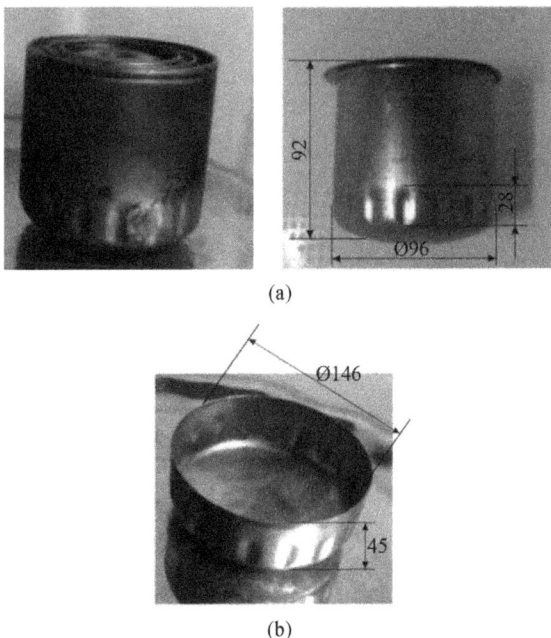

ing," Machine-Building, Moscow, 1964, 375 pages. [in Russian]

[3] M. Ja. Brovman, "Application of the Theory of Plasticity in Rolling," Metallurgiya, Moscow, 1991, 265 pages. [in Russian]

[4] H. Vergus and R. Sowerby, "The Pure Plastic Bending of Laminated Sheet Metals," *Journal of Mechanical Sciences*, Vol. 17, No. 1, 1975, pp. 31-51.

[5] T. V. Brovman, "Sheet Stamping Forces Calculation," *Vestnik Mashinostroyenia*, No. 3, 2004, pp. 61-63. [in Russian]

7

Solvent Extraction of Lanthanum Ion from Chloride Medium by Di-(2-ethylhexyl) Phosphoric Acid with a Complexing Method

Shaohua Yin, Wenyuan Wu[*], Xue Bian, Yao Luo, Fengyun Zhang

School of Materials and Metallurgy, Northeastern University, Shenyang, China

Email: [*]wuwy@smm.neu.edu.cn

ABSTRACT

Solvent extraction experiments of La(III) with di-(2-ethylhexyl)phosphoric acid (P204) from chloride solution in the presence of a complexing agent (lactic acid) have been performed. The effective separation factors can be achieved when the complexing agent is added to the aqueous phase of the extraction system. The complexing agent lactic acid can be effectively recycled using tributyl phosphate (TBP) as extractant, by the use of a countercurrent extraction process, and the chemical oxygen demand (COD) value in the raffinate is 57.7 mg/L, which meets the emission standards of pollutants from rare earths industry. Thus, the simple and environment-friendly complexing method has been proved to be an effective strategy for separating light rare earths, and provides a positive influence on the purification of La(III).

Keywords: La(III); Extraction and Separation; P204 Extractant; Lactic Acid; TBP

1. Introduction

Rare earths are important elements, from an industrial point of view. They are extensively used in the metallurgy, lasers, magnets, and batteries. Lanthanum (La), for example, one of the most abundant rare earths, is of current commercial interest as it is used in the hydrogen storage materials, various alloy materials, etc. With the increasing demand for it, the separation and purification of La(III) have gained considerable importance.

Actually, solvent extraction is a classical chemical analytical method, playing an important role as a separation technique [1]. Saponified acidic extractant such as di-2-ethylhexyl phosphoric acid (P204) has been used to separate La(III) and rare earths under some conditions, which can enhance the extraction capacity of acidic extractants, but the resulting loss of ammonium ion to aqueous phase causes serious pollution [2,3], which becomes an important issue in the current rare-earth industry. Therefore, it is essential to find a new path for separating La(III) from the adjacent REs.

Up to date, much effort has been devoted to exploring some new extraction systems or extractants superior to the existing saponified extraction system, such as synergistic extraction systems [4-6], ionic liquids (ILs) [7-9] containing functional groups used in the rare-earth ions extraction and separation. One of the most effective methods for improving the separation is to add a water-soluble complexing agent into the aqueous phase [10]. Sujatha studied the separation of Ce(III) and Nd(III) using the glycine as complexing agent, and found the average separation factor between these lanthanides was improved from 3.2 to 3.8 [11]. Similar trends have been found in the separation of the lanthanides using other complexing agents such as EDTA, DTPA and HEDTA [12-16]. However, the complexing agents have high cost and are difficult to be recycled.

We investigated the extraction of La(III) using a complexing agent citric acid with unsaponified P204, and found the extraction effect was as good as the saponified system [17]. However, some disadvantages still exist in these processes, such as very high extraction acidity and a difficult stripping. Hence, there is a growing interest in the development of new systems using a complexing agent hydroxy carboxylic acid for effective separation of REs. As a kind of hydroxy carboxylic acid, lactic acid (abbreviated as HLac) is similar to the citric acid, and is currently a promising option for the hydrometallurgical separation of the trivalent lanthanides. The separation of Pr(III)/Ce(III) using the HLac has been investigated, and the extraction performance is better than that without the complexing agent [18].

---

[*]Corresponding author.

In this paper, the extraction of La(III) using the un-saponified P204 in the presence of the HLac in the laboratory was investigated. The various extraction effects on different La(III) were reported and considered for the separation of La(III) from the adjacent rare earths.

2. Materials and Methods

2.1. Materials

P204 and TBP supplied by Tianjin Kermel Chemical Reagent Co. Ltd. were used without purification and diluted in sulphonating kerosene. A $LaCl_3$ solution with a concentration of 0.2 mol/L was prepared by dissolving lanthanum carbonate with a certain proportion of HLac and a small amount of HCl. Lactic acid was supplied by Sinopharm Chemical Reagent Co. Ltd. All other chemicals used were of analytical reagent grade.

Digital pH meter (pHs-3C, Shanghai Rex Instruments Factory), calibrated daily with 4.01 and 6.86 standard buffer solutions, was employed to measure pH values of the aqueous phase. An Agilent 1100 model HPLC was employed to measure the concentration of lactic acid in the recycling experiments.

2.2. Solvent Extraction Procedure

For the equilibrium experiments, equal volumes (20 mL) of the aqueous and P204 extractant were mixed and shaken for 30 min at 298 ± 1 K using a mechanical shaker. After phase separation, the concentration of La^{3+} left in the aqueous phase was analyzed by titration with a standard solution of EDTA at pH 5.5 using xylenol orange as an indicator, and that in the organic phase was obtained by mass balance. Distribution ratio was obtained by $D = [La^{3+}]_o/[La^{3+}]_a$, where "$a$" and "$o$" denote aqueous phase and organic phase. pH value was determined after extraction and phase separation.

For the recycling experiments, equal volumes (10 mL) of the raffinate after the above extraction and TBP extractant were mixed and shaken for 30 min at 298 ± 1 K using a mechanical shaker. The HLac concentration was analyzed by a HPLC method after entire stripping, and the extraction efficiency E is defined as follows:

$$E\% = \frac{[HLac]_t - [HLac]_a}{[HLac]_t} \qquad (1)$$

where $[HLac]_t$ and $[HLac]_a$ represent initial and strip liquor concentrations of HLac in aqueous phase, respectively.

3. Results and Discussion

3.1. The Effect of Acidity on Extraction of La(III)

It should be noted that the pH value in the aqueous solu-

tion plays an important role in the extraction system with acidic extractants. The plots of distribution ratio (D) versus aqueous pH are shown in **Figure 1**. It can be seen that the D shows an increasing trend with pH increasing at a constant concentration of lactic acid.

3.2. The Effect of HLac Concentration on the Extraction

The effect of HLac concentration on the extraction of La^{3+} with P204 is studied and the results are shown in **Figure 2**. From **Figure 2**, it can be found that the distribution ratios of La(III) increase with the increasing of lactic acid concentration, and D reaches largest at the 0.6 mol/L HLac concentration and initial aqueous pH 3.5. This may be explained in terms of the buffering action of HLac, which slows down the effect of higher acidity in the raffinate on the distribution ratios.

In practical application, the effect of initial aqueous pH is considered to be an important parameter, primarily because the acidity control of the aqueous phase is one of

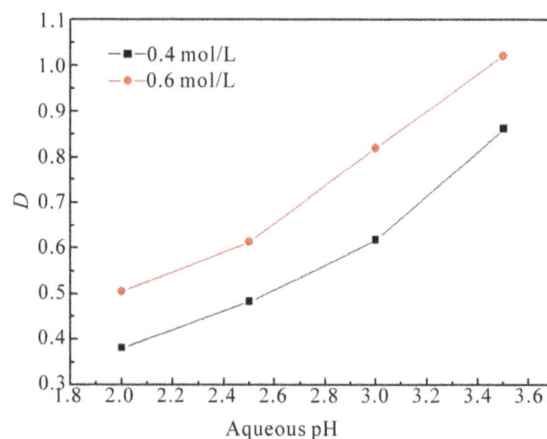

Figure 1. The effect of aqueous pH on the extraction of La(III).

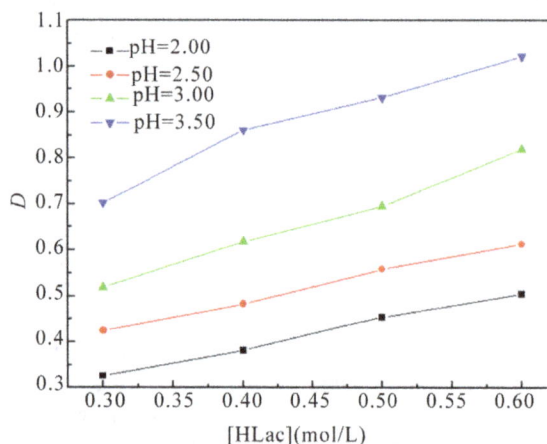

Figure 2. The effect of lactic acid concentration on the extraction of La(III).

the key links in industrial processes. To our knowledge, extraction abilities of acidic organophosphorous extractants are commonly effected by the hydrogen ion from the hydroxyl group in P204 [19]. As a rule, the extraction of the trivalent rare-earth ions with P204 is an ion-exchange mechanism (Equation (2)). The rare earths extraction efficiency can be decreased due to the increase of the acidity in the raffinate, which is caused by the hydrogen ions exchanged with rare earth ions. To overcome the disadvantages, the acidic extractants saponified with aqueous ammonia have been widely applied for the extraction (Equation (3)), but saponification will result in serious pollution, emulsification and the third phase formation during extraction. Therefore, the careful control of the acidity in the raffinate is a critical component to the successful operation of the extraction process. The application of lactic acid is an effective strategy for eliminating this phenomenon, as shown in Equation (4), where "x" is the number of the ligand, and equals to 1, 2 or 3, and Lac is the lactate ion. The rare-earth ions in the aqueous phase can form a variety of complexes in the presence of lactic acid. However, a particular complex is prominent in the feed solution under the experimental conditions. After extraction, some hydrogen ions in the raffinate can combine with the lactate, which reduces the acidity in the aqueous phase and slows down the effect of higher acidity in the raffinate on the distribution ratios. On the other hand, no ammonium ion is released in the extraction process, indicating that it is a better environment-friendly process than the saponification extraction method.

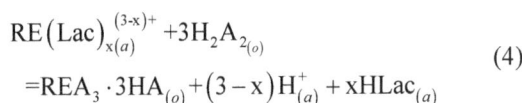

$$RE^{3+} + 3H_2A_2 = RE(HA_2)_3 + 3H^+ \quad (2)$$

$$H_2A_2 + 2NH_3 \cdot H_2O = 2NH_4^+A^- + 2H_2O \quad (3)$$

$$RE(Lac)_{x(a)}^{(3-x)+} + 3H_2A_{2(o)}$$
$$= REA_3 \cdot 3HA_{(o)} + (3-x)H_{(a)}^+ + xHLac_{(a)} \quad (4)$$

3.3. Separation Performance

It is of considerable interest to quantitatively compare the separation ability. Under equilibrium conditions, the separation factors of the adjacent rare earth elements, β, can be defined as:

$$\beta = \frac{D_2}{D_1} \quad (5)$$

where the D_1 and D_2 refer to the distribution ratios of metal ion 1 and metal ion 2, respectively. Generally, the value of D_2 is larger than that of D_1.

As can be seen from **Table 1**, the separation factors for Ce/La, Pr/La and Nd/La increase with the aqueous pH value increasing at HLac concentration of 0.6 mol/L. For example, at pH 3.5 and HLac concentration of 0.6 mol/L in the range studied, the separation factors for Ce/La, Pr/La and Nd/La become values of 3.42, 6.98 and 11.16. From the separation factors, we can come to the conclusion that this system should be a more effective method to separate La^{3+} from the other rare earths.

3.4. Recycling Experiments of the Complexing Agent HLac

As for the above experimental results, this system could be considered an efficient potential method for separating REs. However, there is a standard about the waste-water containing high organic loadings in the rare earths industry, otherwise, it will affect the chemical oxygen demand (COD). So the treatment of HLac from the extraction system becomes an important issue. In this study, we determine the optimum conditions by orthogonal test firstly when the lactic acid concentration is 0.6 mol/L. Using a countercurrent extraction process at a phase ratio $V_o:V_w = 1:1$, t = 20°C, pH = 0.6, and 75 vol% TBP in kerosene, the HLac recovery reaches more than 99% for 10 extraction stages as shown in **Table 2**.

In order to evaluate whether the lactic acid concentration achieves the emission standards of pollutants from rare earths industry or not, we calculate the COD in the raffinate after the 11 extraction stages. As can be seen from the **Table 3**, the COD value after the 11 extraction stages can be up to 57.7 mg/L, which meets the emission standards of pollutants from rare earths industry (COD ≤ 80 mg/L), where theory of chemical oxygen demand of HLac is 1.07 g/g [20]. The results indicate that the HLac can be effectively recycled.

Table 1. Separation factors (D_i/D_{La}) under the experimental conditions.

pH	2	2.5	3	3.5
$\beta_{Ce/La}$	2.42	2.8	3	3.42
$\beta_{Pr/La}$	3.92	4.82	5.58	6.98
$\beta_{Nd/La}$	5.1	6.5	8.04	11.16

Table 2. Extraction efficiencies (E%) of each row.

Parameter	Stages	1	2	3	4	5
Row 1 (E%)	HLac	0	0.06	0.21	1.01	1.91
Row 2 (E%)	HLac	0.05	0.13	0.65	1.57	3.8
Row 3 (E%)	HLac	0.67	1.46	1.98	7.98	9.85
Parameter	Stages	6	7	8	9	10
Row 1 (E%)	HLac	22.22	59.65	84.81	97.31	99.19
Row 2 (E%)	HLac	24.2	49.09	77.85	91.93	99.18
Row 3 (E%)	HLac	15.23	41.78	67.53	88.53	99.19

Table 3. The experimental results of multistage counter-current extraction.

Parameter		10 stages countercurrent extraction		
Mass concentration	Feed (g/L)	E%	Raffinate	COD (mg/L)
HLac	54.048	99.2	0.4	428
Parameter		11 stages countercurrent extraction		
		E%	Raffinate	COD (mg/L)
		99.9	0.054	57.7

4. Conclusions

The following conclusions are drawn:

1) The distribution ratios of the extraction of La(III) by P204 increase with the increase of the pH value in the feed solution and lactic acid concentration. The maximum separation factors of Ce/La, Pr/La and Nd/La become values of 3.42, 6.98 and 11.16 at pH 3.5 and HLac concentration of 0.6 mol/L.

2) The recycling experiments show that the complexing agent lactic acid could be efficiently recycled, and the COD in the raffinate meets the emission standards of pollutants from rare earths industry.

5. Acknowledgements

Financial aid from the following programs is gratefully acknowledged: the National Natural Science Foundation of China (50974042, 51104040 and 51274060), the National Program on Key Basic Research Project of China (973 Program) (2012CBA01205), the National Key Technology Research and Development Program of the Ministry of Science and Technology of China (2012BAE01B00) and the Scientific Research special Foundation of Doctor subject of Chinese Universities (20100042110008).

REFERENCES

[1] D. H. Cheng, X. W. Chen, Y. Shu and J. H. Wang, "Selective Extraction/Isolation of Hemoglobin with Ionic Liquid 1-Butyl-3-trimethylsilylimidazolium Hexafluorophosphate (BtmsimPF6)," *Talanta*, Vol. 75, No. 5, 2008, pp. 1270-1278. doi:10.1016/j.talanta.2008.01.044

[2] X. W. Huang, H. W. Li, X. X. Xue and G. C. Zhang, "Development Status and Research Progress in Rare Earth Hydrometallurgy in China," *Journal of the Chinese Rare Earth Society*, Vol. 24, No. 2, 2006, pp. 129-133.

[3] Z. G. Zhu and J. G. Cheng, "The Research on Waste Water's Processing Method by Rare Earth Hydrometallurgy," *China's Manganese Industy*, Vol. 28, No. 3, 2010, pp. 34-36.

[4] Z. F. Zhang, H. F. Li, F. Q. Guo, S. L. Meng and D. Q. Li, "Synergistic Extraction and Recovery of Cerium(IV) and Fluorin from Sulfuric Solutions with Cyanex 923 and Di-2-ethylhexyl Phosphoric Acid," *Separation and Puri-*

fication Technology, Vol. 63, No. 2, 2008, pp. 348-352. doi:10.1016/j.seppur.2008.05.023

[5] N. Z. Song, S. S. Tong, W. Liu, Q. Jia, W. H. Zhou and W. P. Liao, "Extraction and Separation of Rare Earths from Chloride Medium with Mixtures of 2-Ethylhexylphosphonic Acid Mono-(2-ethylhexyl) Ester and Sec-Nonylphenoxy Acetic Acid," *Journal of Chemical Technology and Biotechnology*, Vol. 63, No. 12, 2009, pp. 1798-1802. doi:10.1002/jctb.2248

[6] X. H. Luo, X. W. Huang, Z. W. Zhu, Z. Q. Long and Y. J. Liu, "Synergistic Extraction of Cerium From Sulfuric Acid Medium Using Mixture of 2-Ethylhexyl Phosphonic Acid Mono 2-Ethylhexyl Ester and Di-(2-ethyl hexyl) Phosphoric Acid as Extractant," *Journal Rare Earths*, Vol. 27, No. 1, 2009, pp. 119-122. doi:10.1016/S1002-0721(08)60204-5

[7] X. Q. Sun, Y. Ji, F. C. Hu, B. He, J. Chen and D. Q. Li, "The Inner Synergistic Effect of Bifunctional Ionic Liquid Extractant for Solvent Extraction," *Talanta*, Vol. 81, No. 4-5, 2010, pp. 1877-1883. doi:10.1016/j.talanta.2010.03.041

[8] V. M. Egorov, D. I. Djigailo, D. S. Momotenko, D. V. Chernyshov, I. I. Torocheshnikova, S. V. Smirnova and I. V. Pletnev, "Task-Specific Ionic Liquid Trioctylmethylammonium Salicylate as Extraction Solvent for Transition Metal Ions," *Talanta*, Vol. 80, No. 3, 2010, pp. 1177-1182. doi:10.1016/j.talanta.2009.09.003

[9] W. Wang, H. L.Yang, H. M. Cui, D. L. Zhang, Y. Liu and J. Chen, "Application of Bifunctional Ionic Liquid Extractants [A336][CA-12] and [A336][CA-100] to the Lanthanum Extraction and Separation from Rare Earths in the Chloride Medium," *Industrial and Engineering Chemistry Research*, Vol. 50, No. 12, 2011, pp. 7534-7541. doi:10.1021/ie2001633

[10] X. B. Sun, Y. G. Wang and D. Q. Li, "Selective Separation of Yttrium by CA-100 in the Presence of a Complexing Agent," *Rare Earths* 2004 *Conference*, Nara, 7-12 November 2004, pp. 999-1002.

[11] S. Sujatha, M. L. P. Reddy, T. R. Ramamohan and A. D. Damodaran, "Effect of Water Soluble Complexing Agent on the Extraction of Ce(III) and Nd(III) by 2-Ethylhexylphosphonic Acid Mono-2-ethylhexyl Ester," *Journal of Radioanalytical and Nuclear Chemistry*, Vol. 174, No. 2, 1993, pp. 271-278. doi:10.1007/BF02037914

[12] S. Nishihama, T. Hirai and I. Komasawa, "Advanced Liquid-Liquid Extraction Systems for the Separation of Rare Earth Ions by Combination of Conversion of the Metal Species with Chemical Reaction," *Journal of Solid State Chemistry*, Vol. 171, No. 1-2, 2003, pp. 101-108. doi:10.1016/S0022-4596(02)00198-6

[13] S. Nishihama, T. Hirai and I. Komasawa, "Selective Extraction of Y from a Ho/Y/Er Mixture by Liquid-Liquid Extraction in the Presence of a Water-Soluble Complexing Agent," *Industrial and Engineering Chemistry Research*, Vol. 39, No. 10, 2000, pp. 3907-3911. doi:10.1021/ie000030a

[14] Y. G. Wang, Y. Xiong, S. L. Meng and D. Q. Li, "Separation of Yttrium from Heavy Lanthanide by CA-100

Using the Complexing Agent," *Talanta*, Vol. 63, No. 2, 2004, pp. 239-243. doi:10.1016/j.talanta.2003.09.034

[15] S. Nishihama, T. Hirai and I. Komasawa, "Review of Advanced Liquid Liquid Extraction Systems for the Separation of Metal Ions by a Combination of Conversion of the Metal Species with Chemical Reaction," *Industrial and Engineering Chemistry Research*, Vol. 40, No. 14, 2001, pp. 3085-3091. doi:10.1021/ie010022+

[16] X. Q. Sun, B. Peng, J. Chen, D. Q. Li and F. Luo, "An Effective Method for Enhancing Metal-Ions' Selectivity of Ionic Liquid-Based Extraction System: Adding Water-Soluble Complexing Agent," *Talanta*, Vol. 74, No. 4, 2008, pp. 1071-1074. doi:10.1016/j.talanta.2007.07.031

[17] H. T. Chang, W. Y. Wu, G. F. Tu, Y. H. Zhao, D. Li and X. Y. Sang, "Extraction of Light Rare Earth in the System of P204-HCl-H$_3$cit," *Chinese Rare Earths*, Vol. 28, No. 3, 2008, pp. 18-21.

[18] S. H. Yin, W. Y. Wu, X. Bian and F. Y. Zhang, "Effect of Complexing Agent Lactic Acid on the Extraction and Separation of Pr(III)/Ce(III) with Di-(2-ethylhexyl) Phosphoric Acid," *Hydrometallurgy*, Vol. 131-132, 2013, pp. 133-137. doi:10.1016/j.hydromet.2012.11.005

[19] N. V. Thakur, D. V. Jayawant, N. S. Iyer and K. S. Koppiker, "Separation of Neodymium from Lighter Rare Earths Using Alkyl Phosphonic Acid, PC88A," *Hydrometallurgy*, Vol. 34, No. 1, 1993, pp. 99-108. doi:10.1016/0304-386X(93)90084-Q

[20] X. K. Wu and Q. P. Jin, "Treatment Technology of Organic Wastewater," East China Institute of Chemical Press, Shanghai, 1989.

8

Developing a Thermodynamical Method for Prediction of Activity Coefficient of TBP Dissolved in Kerosene

Eskandar Keshavarz Alamdari[1,2], Sayed Khatiboleslam Sadrnezhaad[3]
[1]Department of Mining and Metallurgical Engineering, Amirkabir University of Technology, Tehran, Iran
[2]Research Center for Materials and Mining Industries Technology, Amirkabir University of Technology, Tehran, Iran
[3]Department of Materials Science and Engineering, Sharif University of Technology, Tehran, Iran
Email: alamdari@aut.ac.ir, sadrnezh@sharif.ac.ir

ABSTRACT

Results of the experimental measurements on the partial molar volume of kerosene used as a medium for dissolving TBP are utilized to determine the activity of TBP in the binary kerosene-TBP solution through the application of Gibbs-Duhem equation. The treatment is based on combination of the experimental data with the thermodynamic values available on the compressibility factor of pure kerosene at room temperature. It is shown that the activity of TBP in kerosene has a positive deviation from ideality with an activity coefficient derived as follows: 1) at $X_{TBP} \leq 0.01$:

$\gamma_{TBP} = 42.530$, 2) at the $0.01 < X_{TBP} < 0.2$: $\ln \gamma_{TBP} = \dfrac{0.2913 - 0.2843 X_{kerosene}}{1 - 1.8674 X_{kerosene} + 0.8687 X_{kerosene}^2}$ 3) at higher TBP concen-

trations $0.2 < X_{TBP} < 0.97$: $\ln \gamma_{TBP} = \dfrac{-0.0146 + 1.2826 X_{kerosene}}{1 + 1.5595 X_{kerosene} - 1.9594 X_{kerosene}^2}$ and 4) at TBP Raoultian concentrations

$0.97 \leq X_{TBP}$: $\gamma_{TBP} = 1$. These quantities can be utilized at temperature closed to 298 K.

Keywords: Thermodynamics; Activity; Activity Coefficient; Kerosene; TBP; Organic Solution

1. Introduction

Activities of the spices dissolved in organic aromatic solutions are of the important information required for understanding of the thermodynamics of the solvent extraction regimes usually utilized in production of the nonferrous metals. It has been reported that the activity coefficients of involved components in the extraction reaction of metals during extraction processes are usually equal to one [1-10]. However, the activities in the real component values are significantly different from the ideal state. The activity coefficient of components (especially components in aqueous media) was estimated by using some conventional thermodynamic models such as Debye-Hückel or Pitzr Equation [11]. On the other hand, due to the physicochemical interaction of organic components, the mathematical models could be used in some special cases. By applying the correct value of the activity coefficients in the extraction equations, a correct mathematical model can predict an acceptable value for extracted metals.

The thermodynamic evaluation of the distribution ratio of metals, for instance, becomes much easier if the activity of coefficient tri-n-butyl phosphate (TBP) dissolved in kerosene becomes precisely known. There is, however, no data available in the literature on the activity coefficient of different spices dissolved in such aromatic or aliphatic solutions as kerosene.

TBP is a common organic material which uses as extractant and/or modifier in the presence of some aliphatic diluents such as kerosene. Therefore, developing an analytical method for the prediction of the activity coefficients of organic component could be useful for future investigations. In this paper, an analytical method for determination of the activity and the activity coefficient of TBP dissolved in kerosene is developed and presented.

2. Thermodynamical Parameters and Prediction of the Activity and the Activity Coefficients

It is shown that the excess partial molar Gibbs free

energy of the spices i depends on the composition of the solution. The difference between the partial molar Gibbs free energy of the spices i and the molar Gibbs free energy of pure i is the change in the Gibbs free energy accompanying the formation of one mole of i dissolved in the solution; $\Delta \overline{G}_i^S$ [12-15]. Thus:

$$\Delta \overline{G}_i^S = RT \ln a_i = RT \ln\left(\gamma_i X_i\right) = RT\left(\ln \gamma_i + \ln X_i\right) \quad (1)$$

on the other hand:

$$d\left(\Delta \overline{G}_i^S\right) = -\Delta \overline{S}_i^S dT + \Delta \overline{V}_i^S dP \quad (2)$$

where $\Delta \overline{S}_i^S$ and $\Delta \overline{V}_i^S$ are the partial molar entropy and the partial volume change of the dissolution reaction, respectively. In the isothermal condition, Equation (2) is rewritten as:

$$d\left(\Delta \overline{G}_i^S\right) = \Delta \overline{V}_i^S dP \quad (3)$$

The molar volume of a multi component solution is defined by:

$$\overline{V}^S = \sum_{i=1}^n X_i \overline{V}_i^S \quad (4)$$

The molar volume of the mechanical mixture can similarly be defined by:

$$V^\circ = \sum_{i=1}^n X_i V_i^\circ \quad (5)$$

The volume change due to the formation of the solution is, thus, given by:

$$\Delta \overline{V}^S = \overline{V}^S - V^\circ = \sum_{i=1}^n X_i\left(\overline{V}_i^S - V_i^\circ\right) \quad (6)$$

The value of ΔV^S for a binary solution, which exhibits negative deviation from ideality, is less than zero. Based on known thermodynamic relationships available [12-15], the volume change of the species A in a binary A-B system can be obtained from:

$$\Delta \overline{V}_A^S = \Delta \overline{V}^S + \left[\frac{\partial\left(\Delta \overline{V}^S\right)}{\partial X_A}\right]_{T,P} X_B \quad (7)$$

Also, the isothermal compressibility of a substance, or a system, is defined as:

$$\beta = -\frac{1}{V}\left(\frac{\partial V}{\partial P}\right)_T \quad (8)$$

This is the fractional decrease in the volume of the system for unit increase in pressure at constant temperature. For pure A, the isothermal compressibility is defined as:

$$\beta_{A^\circ} = -\frac{1}{V_A^\circ}\left(\frac{\partial V_A^\circ}{\partial P}\right)_T \quad (9)$$

and for species A of the binary solution:

$$\beta_{\overline{A}} = -\frac{1}{\overline{V}_A}\left(\frac{\partial \overline{V}_A}{\partial P}\right)_T \quad (10)$$

if we assume that:

$$\beta_{A^\circ} = \beta_{\overline{A}} \quad (11)$$

then from Equations (9) and (10):

$$-\frac{1}{V_A^\circ}\left(\frac{\partial V_A^\circ}{\partial P}\right)_T = -\frac{1}{\overline{V}_A}\left(\frac{\partial \overline{V}_A}{\partial P}\right)_T \quad (12)$$

hence at a constant temperature, we have:

$$\frac{\partial V_A^\circ}{V_A^\circ} = \frac{\partial \overline{V}_A}{\overline{V}_A} \quad (13)$$

or:

$$\frac{\partial V_A^\circ}{\partial \overline{V}_A} = \frac{V_A^\circ}{\overline{V}_A} \quad (14)$$

and also:

$$\frac{\partial V_A^\circ}{\partial \overline{V}_A - \partial V_A^\circ} = \frac{V_A^\circ}{\overline{V}_A - V_A^\circ} \quad (15)$$

or:

$$\frac{\partial V_A^\circ}{\partial\left(\Delta \overline{V}_A^S\right)} = \frac{V_A^\circ}{\Delta \overline{V}_A^S} \quad (16)$$

then:

$$-\frac{1}{V_A^\circ}\left(\frac{\partial V_A^\circ}{\partial P}\right)_T = -\frac{1}{\Delta \overline{V}_A^S}\left(\frac{\partial\left(\Delta \overline{V}_A^S\right)}{\partial P}\right)_T \quad (17)$$

hence:

$$\beta_A = -\frac{1}{\Delta \overline{V}_A^S}\left[\frac{\partial\left(\Delta \overline{V}_A^S\right)}{\partial P}\right]_T \quad (18)$$

At a constant temperature, Equation (18) can be rearranged as:

$$dP = -\frac{d\left(\Delta \overline{V}_A^S\right)}{\beta_A \Delta \overline{V}_A^S} \quad (19)$$

by substituting the value of (dP) from Equation (19) into Equation (3), we have:

$$d\left(\Delta \overline{G}_A^S\right) = -\frac{d\left(\Delta \overline{V}_A^S\right)}{\beta_A} \quad (20)$$

and from Equation (1), we obtain:

$$d\left(\ln a_A\right) = -\frac{d\left(\Delta \overline{V}_A^S\right)}{RT\beta_A} \quad (21)$$

Integrating Equation (21) from the initial condition where $X_A = 1$ and $\Delta \overline{V}_A^S = 0$, one can write:

$$\int_{\ln a_A \text{ at } X_A=1}^{\ln a_A \text{ at } X_A} d\left(\ln a_A\right) = \int_{\Delta \bar{V}_A^S \text{ at } X_A=1}^{\Delta \bar{V}_A^S \text{ at } X_A} - \frac{d\left(\Delta \bar{V}_A^S\right)}{RT\beta_A} \qquad (22)$$

or:

$$\ln a_A = \int_0^{\Delta \bar{V}_A^S} - \frac{d\left(\Delta \bar{V}_A^S\right)}{RT\beta_A} \qquad (23)$$

The value of the right hand side at Equation (23) can graphically be obtained by plotting the quantity of $\frac{1}{RT\beta_A}$ vs $\Delta \bar{V}_A^S$, and determining the area under the curve. The activity coefficient of the species A of binary solution can thus be obtained from:

$$\gamma_A = \frac{a_A}{X_A} \qquad (24)$$

The activity coefficient of the second component of the solution can be determined by integration the Gibbs-Duhem equation [12-15]:

$$\int_{\ln \gamma_B \text{ at } X_B=1}^{\ln \gamma_B \text{ at } X_B} d\left(\ln \gamma_B\right) = \int_{\ln \gamma_A \text{ at } X_B=1}^{\ln \gamma_A \text{ at } X_B} - \frac{X_A}{X_B} d\left(\ln \gamma_A\right) \qquad (25)$$

At the boundary condition where X_B and γ_B are equal to one, γ_A equals to γ_A°, thus:

$$\ln \gamma_B = \int_{\ln \gamma_A^\circ}^{\ln \gamma_A} - \frac{X_A}{X_B} d\left(\ln \gamma_A\right) \qquad (26)$$

The activity of species B is, thus, determined from:

$$a_B = X_B \gamma_B \qquad (27)$$

3. Material and Methods

Both TBP and kerosene, which were used, were of analytical grade from Fluka AB., Switzerland. Small quantity of TBP weighted with a mettler 240 balance system was added to a calibrated 100-ml. Kerosene was instilled to the 100-ml flask and the solution was mixed thoroughly. The solution was then retained for five to ten minutes to absorb the required heat for reaching to chemical equilibrium. The total weight of the solution was then measured.

The experiments were carried out at constant room temperature (298 K). The molecular weight of the pure TBP was equal to 263.32 gr/mole. The molecular weight of the used kerosene was determined by the gas chromatography and the change of the melting point methods. In the latter, the melting temperature change was measured by analytical grade phenol with a molecular weight of 94.11 gr/mole .The molecular weight of kerosene was the determined from [16,17]:

$$\ln X_{\text{phenol}} = -\frac{1000\Delta H_{f,\text{phenol}}^\circ \Delta T}{M_{\text{phenol}} RT_{m,\text{phenol}} m} \qquad (28)$$

where X_{phenol} is mole fraction of phenol in dilute phenol-kerosene solution, $\Delta H_{f,\text{phenol}}^\circ$ is the latent heat of melting of phenol in its melting point $T_{m,\text{phenol}}$, ΔT is the change of melting point of solution when the molality of kerosene in dilute solution is m and M_{phenol} is the molecular weight of phenol. The result was equal to 173.3 gr/mole.

With the experimental data, the integral molar volume of the solution $\left(\bar{V}^S\right)$ and the integral molar volume change of the solution $\left(\Delta \bar{V}^S\right)$ were determined from Equations (4) and (6). a_{kerosene} (Equation (23)), γ_{kerosene} (Equation (24)), γ_{TBP} (Equation (26)) and a_{TBP} (Equation (27)) were then evaluated.

4. Results and Discussion

There is not much known of the physical properties of TBP. Determination of the activity and the activity coefficient of TBP is, therefore, derived from the corresponding quantities for kerosene. The isothermal compressibility of kerosene at 298 K and 1 atm pressure is known to be 1.45×10^{-4} atm^{-1} [18]. Applying this value to Equation (23), the activity of kerosene is, therefore, being determined.

Figure 1 shows the integral molar volume of the binary kerosene-TBP solution as a function of the kerosene mole fraction; X_{kerosene}. As shown in this figure, the integral molar volume of the solution has a positive deviation from the ideal behavior (dashed line). This quantity is defined by Equation (4). **Figure 2** illustrates the relative molar volume of the binary solution $\left(\Delta \bar{V}^S\right)$ versus kerosene mole fraction. It is seen from this figure that, the molar volume of the formation of the binary solution has a positive deviation from ideality and has a maximum near $X_{\text{kerosene}} = 0.45$. The predicted value of the

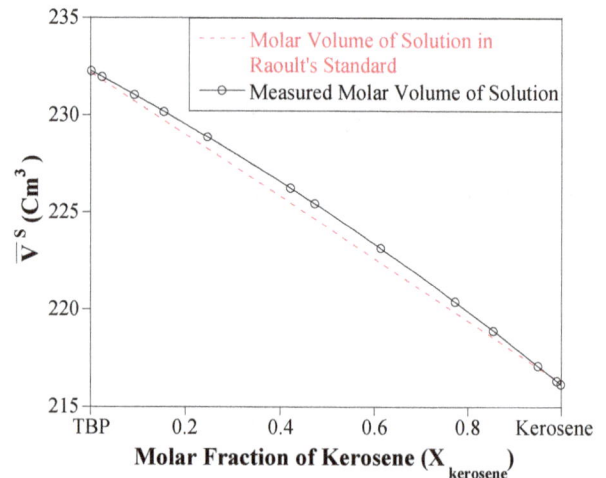

Figure 1. The integral molar volume of TBP-kerosene binary solution as a function of measured X_{kerosene} at room temperature.

volume change has a correlation with the experimental data:

$$\Delta \overline{V}^S$$
$$= 1.75 X_{\text{kerosene}} - 3.49 X_{\text{kerosene}}^{2.5} + 1.24 X_{\text{kerosene}}^3 + 0.46 X_{\text{kerosene}}^{0.5}$$
(29)

The molar volume change of the kerosene dissolved in the binary solution is determined from Equation (7) as a function of X_{kerosene}. **Figure 3** illustrated the partial molar volume change of kerosene $\left(\Delta \overline{V}_{\text{kerosene}}^S \right)$ as a function of X_{kerosene}. The partial molar volume changes of the kerosene in the pure TBP $\left(X_{\text{kerosene}} = 0 \right)$ and in the pure kerosene $\left(X_{\text{kerosene}} = 1 \right)$ are equal to infinity and zero, respectively.

Based on Equation (14), the activity of kerosene can be determined by using a graphical method. **Figure 4**

illustrates that the change of $-1/RT\beta_{\text{kerosene}}$ vs $\Delta \overline{V}_{\text{kerosene}}^M$ at the constant temperature 298°K. We have assumed that the isothermal compressibility constant of the kerosene at every composition of binary solution is constant and that the compositional change has no significant effect on its value. So the activity of the kerosene is determined from the area under the curve plotted in **Figure 4**, as was stated by Equation (23). The results are given in **Figure 5** (solid line). As shown in this figure, the activity of the kerosene indicates a very positive deviation from ideality, especially at low concentrations. A curve fitting method can be used to determine the activity of the kerosene at

$$X_{\text{kerosene}} < 0.03$$
$$a_{\text{kerosene}} = 213.045 - 213.045 X_{\text{TBP}} = 213.045 X_{\text{kerosene}}$$
(30)

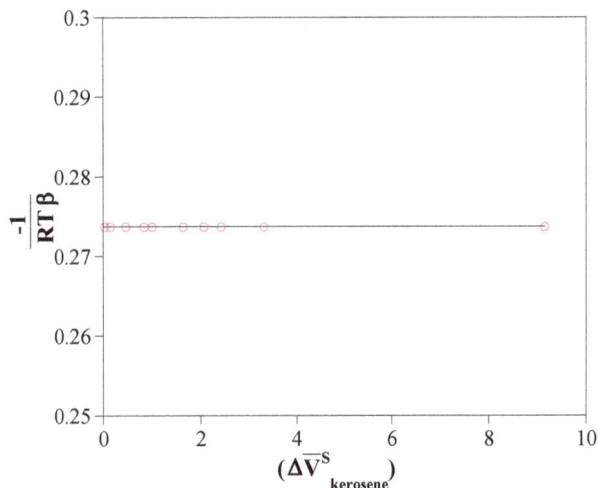

Figure 2. The relative integral molar volume of the kerosene-TBP solution versus the solution composition.

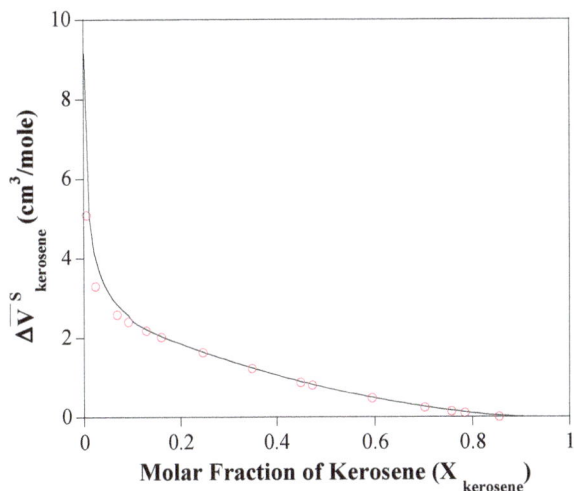

Figure 4. The value of $-1/RT\beta_{\text{kerosene}}$ vs molar volume change of kerosene in binary kerosene-TBP solution.

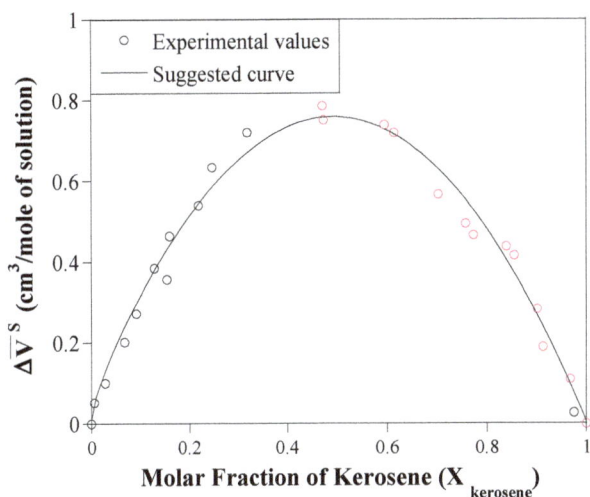

Figure 3. The relative partial molar volume of kerosene as a function of composition.

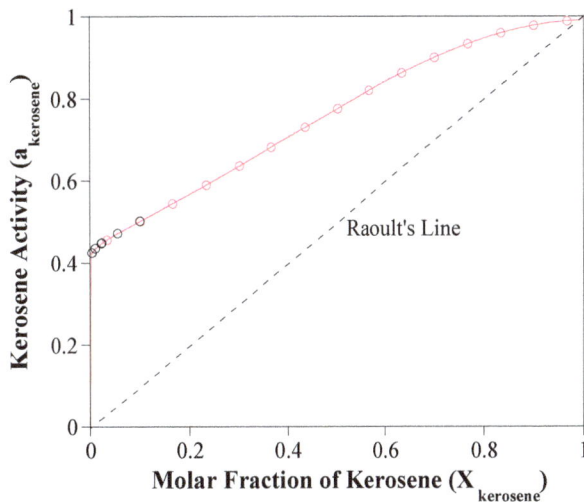

Figure 5. Activity of kerosene as a function of composition of the solution.

and at the higher kerosene concentration, the activity of kerosene is determined by:

$$a_{kerosene} = \frac{0.993 - 1.210 X_{TBP} + 0.23087 X_{TBP}^2 - 0.002 X_{TBP}^3}{1 - 1.169 X_{TBP} + 1.068 X_{TBP}^2 - 0.870 X_{TBP}^3} \quad (31)$$

The activity coefficient of kerosene is determined from Equation (24). **Figure 6** shows the change of $\gamma_{kerosene}$ as a function of $X_{kerosene}$. With the method of curve fitting, the activity coefficient of the kerosene at $0.03 < X_{kerosene} < 0.2$ is determined by:

$$\ln \gamma_{kerosene} = \frac{0.2796 - 0.5576 X_{TBP} + 0.2784 X_{TBP}^2}{1 - 2.9145 X_{TBP} + 2.8292 X_{TBP}^2 - 0.9146 X_{TBP}^2} \quad (32)$$

at the higher kerosene concentrations $\left(0.2 < X_{kerosene} < 0.99\right)$, the activity coefficient of kerosene is determined by:

$$\ln \gamma_{kerosene} = \frac{-0.0006 - 0.1858 X_{TBP}}{1 - 1.8296 X_{TBP} + 1.1551 X_{TBP}^2} \quad (33)$$

The Henry's constant for the kerosene $\left(\gamma_{kerosene}^\circ\right)$ is determined by:

$$\gamma_{kerosene}^\circ = \lim_{X_{kerosene} \to 0} \gamma_{kerosene} = \lim_{X_{kerosene} \to 0} \frac{a_{kerosene}}{X_{kerosene}} = 214.35$$

$$(34)$$

The Gibbs-Duhem Equation and its results (Equations (23) and (26)) help to determine the activity coefficient of the TBP with a graphical method. **Figure 7** shows the value of $\frac{X_{kerosene}}{X_{TBP}}$ vs $\ln \gamma_{kerosene}$. The results of the calculations are shown in **Figure 8**. With the curve fitting method, the activity coefficient of the TBP at

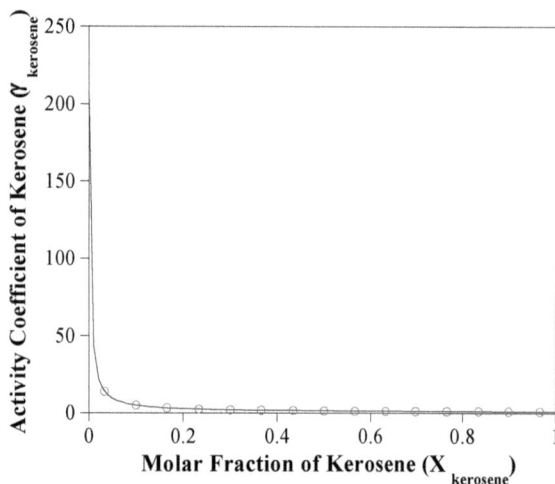

Figure 6. Activity coefficient of kerosene as a function of concentration.

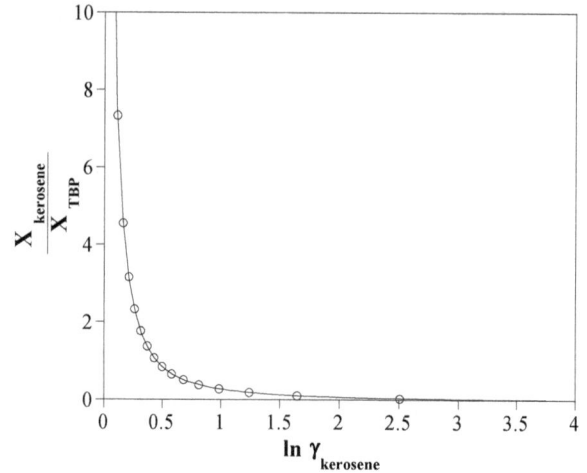

Figure 7. The value of $\frac{X_{kerosene}}{X_{TBP}}$ vs $\ln \gamma_{kerosene}$.

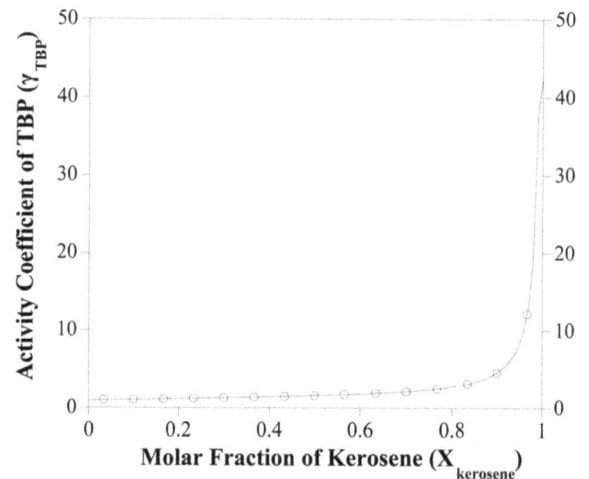

Figure 8. The activity coefficient of TBP as a function of the molar fraction of the kerosene.

$X_{TBP} < 0.01$ is determined by:

$$\gamma_{TBP} = 42.530, \quad (35)$$

at $0.01 < X_{TBP} < 0.2$ is given by:

$$\ln \gamma_{TBP} = \frac{0.2913 - 0.2843 X_{kerosene}}{1 - 1.8674 X_{kerosene} + 0.8687 X_{kerosene}^2} \quad (36)$$

and at the higher TBP concentrations $\left(0.2 < X_{TBP} < 0.97\right)$, the activity coefficient of TBP is evaluated by:

$$\ln \gamma_{TBP} = \frac{-0.0146 + 1.2826 X_{kerosene}}{1 + 1.5595 X_{kerosene} - 1.9594 X_{kerosene}^2} \quad (37)$$

So the Henry's constant for the TBP is equal to 42.530. The values of the activity of the TBP evaluated from an equation similar to Equation (24) are shown in **Figure 9**. The activity of the TBP has a very positive deviation

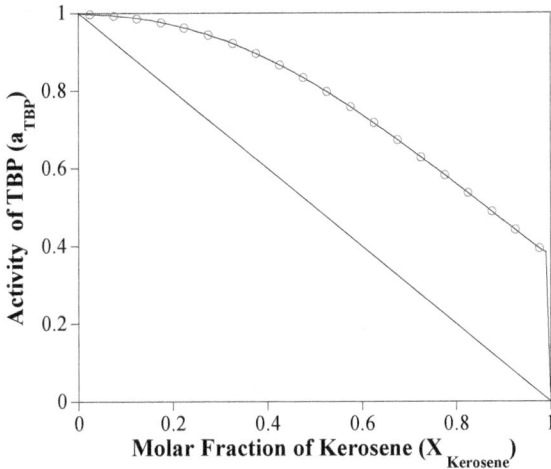

Figure 9. The activity of TBP vs $X_{kerosene}$.

from ideality. With the curve fitting method, the activity of the TBP at $X_{TBP} < 0.01$ is determined by:

$$a_{TBP} = 42.35 - 42.35 X_{kerosene} = 42.35 X_{TBP}, \qquad (38)$$

at $0.01 < X_{TBP} < 0.2$. It is evaluated by:

$$\ln a_{TBP} = \frac{-0.1013 - 0.0134 X_{kerosene}}{1 - 1.5101 X_{kerosene} + 0.6266 X_{kerosene}^2} \qquad (39)$$

at the higher TBP concentration $\left(0.2 < X_{TBP} < 0.97\right)$, the activity of TBP is determined by:

$$\ln a_{TBP} = \frac{0.0031 - 0.1264 X_{kerosene}}{1 - 2.0240 X_{kerosene} + 1.2479 X_{kerosene}^2} \qquad (40)$$

and at Raoultian range $\left(0.97 \le X_{TBP}\right)$: $a_{TBP} = X_{TBP}$.

5. Summary

An analytical method is presented in this paper for determination of the activity of the TBP dissolved in the kerosene through a simple physical property measurement. The results show that the binary solution of the kerosene with the TBP has a positive deviation from the Raoult's law behavior. The Henry's constant of very dilute TBP in kerosene is equal to 42.350. This constant for very dilute kerosene in TBP is equal to 214.35. The activity coefficient of the TBP at $X_{TBP} < 0.01$ is determined by: $\gamma_{TBP} = 42.530$, at $0.01 < X_{TBP} < 0.2$ is given by:

$$\ln \gamma_{TBP} = \frac{0.2913 - 0.2843 X_{kerosene}}{1 - 1.8674 X_{kerosene} + 0.8687 X_{kerosene}^2}$$

and at the higher *TBP* concentrations the activity coefficient of the TBP is determined by:

$$\ln \gamma_{TBP} = \frac{-0.0146 + 1.2826 X_{kerosene}}{1 + 1.5595 X_{kerosene} - 1.9594 X_{kerosene}^2}.$$

Also the activity of TBP at $X_{TBP} < 0.05$ is deter-

mined by:

$$a_{TBP} = 42.35 - 42.35 X_{kerosene} = 42.35 X_{TBP}$$

at $0.01 < X_{TBP} < 0.2$ is given by:

$$\ln a_{TBP} = \frac{-0.1013 - 0.0134 X_{kerosene}}{1 - 1.5101 X_{kerosene} + 0.6266 X_{kerosene}^2}$$

and at the higher TBP concentrations the activity of TBP is determined by:

$$\ln a_{TBP} = \frac{0.0031 - 0.1264 X_{kerosene}}{1 - 2.0240 X_{kerosene} + 1.2479 X_{kerosene}^2}.$$

REFERENCES

[1] E. Keshavarz Alamdari, D. Darvishi, D. F. Haghshenas, N. Yousefi and S. K. Sadrnezhaad, "Separation of Re and Mo from Roasting-Dust Leach-Liquor Using Solvent Extraction Technique by TBP," *Separation and Purification Technology*, Vol. 86, 2012, pp. 143-148. doi:10.1016/j.seppur.2011.10.038

[2] D. Darvishi, D. F. Haghshenas, E. Keshavarz Alamdari and S. K. Sadrnezhaad, "Extraction of ZN, MN and CO from ZN-MN-CO-CD-NI Containing Solution Using D2EHPA, Cyanex® 272 and Cyanex® 302," *International Journal of Engineering, Transactions B: Applications*, Vol. 24, No. 2, 2011, pp. 183-192.

[3] D. F. Haghshenas, D. Darvishi, S. Etemadi, A. R. Eivazi Hollagh, E. Keshavarz Alamdari and A. A. Salardini, "Interaction between TBP and D2EHPA during Zn, Cd, Mn, Cu, Co and Ni Solvent Extraction: A Thermodynamic and Empirical Approach," *Hydrometallurgy*, Vol. 98, No. 1-2, 2009, pp. 143-147.

[4] D. F. Haghshenas, D. Darvishi, H. Rafieipour, E. Keshavarz Alamdari and A. A. Salardini, "A Comparison between TEHA and Cyanex 923 on the Separation and the Recovery of Sulfuric Acid from Aqueous Solutions," *Hydrometallurgy*, Vol. 97, No. 3-4, 2009, pp. 173-179. doi:10.1016/j.hydromet.2009.02.006

[5] D. Darvishi, D. F. Haghshenas, S. Etemadi, E. Keshavarz Alamdari and S. K. Sadrnezhaad, "Water Adsorption in the Organic Phase for the D2EHPA-Kerosene/Water and Aqueous Zn²⁺, Co²⁺, Ni²⁺ Sulphate Systems," *Hydrometallurgy*, Vol. 88, No. 1-4, 2007, pp. 92-97. doi:10.1016/j.hydromet.2007.02.010

[6] D. Darvishi, D. F. Haghshenas, E. Keshavarz Alamdari, S. K. Sadrnezhaad and M. Halali, "Synergistic Effect of Cyanex 272 and Cyanex 302 on Separation of Cobalt and Nickel by D2EHPA," *Hydrometallurgy*, Vol. 77, No. 3-4, 2005, pp. 227-238. doi:10.1016/j.hydromet.2005.02.002

[7] E. Keshavarz Alamdari, D. Moradkhani, D. Darvishi, M. Askari and D. Behnian, "Synergistic Effect of MEHPA on Co-Extraction of Zinc and Cadmium with DEHPA," *Minerals Engineering*, Vol. 17, No. 1, 2004, pp. 89-92. doi:10.1016/j.mineng.2003.10.003

[8] S. K. Sadrnezhaad and E. Keshavarz Alamdari, "Thermodynamics of Extraction of Zn²⁺ from Sulfuric Acid Media with a Mixture of DEHPA and MEHPA," *Interna-*

tional Journal of Engineering, Transactions B: Applications, Vol. 17, No. 2, 2004, pp. 191-200.

[9]   R. E. Blanco, C. A. Blake Jr., W. Davis Jr. and R. H. Rainey, "Survey of Recent Developments in Solvent Extraction with Tri-Butyl-Phosphate," Oak Ridge National Laboratory (ORNL), 1963. www.ornl.gov/info/reports/1963/3445605494266.pdf

[10] W. Davis Jr., "Thermodynamics of Extraction of Nitric Acid by Tri-N-Butyl Phosphate—Hydrocarbon Diluent Solutions I. Distribution Studies with Tbp in Amsco 125-82 at Intermediate and Low Acidities," Oak Ridge National Laboratory (ORNL), 1961. www.ornl.gov/info/reports/1963/3445605700033.pdf

[11] X. Liu, D. Fang, J. Li, J. Yang and S. Zang, "Thermodynamics of Solvent Extraction of Thallium(I)," *Journal of Phase Equilibria and Diffusion, Section I: Basic and Applied Research*, Vol. 26, 2005, pp. 342-346. doi:10.1361/154770305X56791

[12] D. R. Gaskell, "Introduction to the Thermodynamics of Materials," 5th Edition, Taylor & Francis Publisher, New York, 2008.

[13] D. V. Ragon, "Thermodynamics of Materials," John Wiley & Sons Inc., New York, 1995.

[14] R. T. Dehoff, "Thermodynamics in Materials Science," 2nd Edition, Mc Graw-Hill, New York, 1993.

[15] J. B. Hudson, "Thermodynamics of Materials," John Wiley & Sons Inc., New York, 1996.

[16] F. Daniels, J. W. Williams, P. Bender, R. A. Alberty and C. D. Cornwell, "Experimental Physical Chemistry," 7th Edition, McGraw Hill, New York, 1970.

[17] G. W. Castellan, "Physical Chemistry," 3rd Edition, Addison-Wesley, Menlo Park, 2004.

[18] R. H. Perry, "Perry's Chemical Engineers' Handbook," 7th Edition, McGraw Hill, New York, 1997.

9

Argentinean Copper Concentrates: Structural Aspects and Thermal Behaviour

Vanesa Bazan[1*], Elena Brandaleze[2], Leandro Santini[2], Pedro Sarquis[3]

[1]CONICET—Instituto de Investigaciones Mineras, Universidad Nacional de San Juan, San Juan, Argentina
[2]Metallurgical Department and Technology and Materials Develop Center, DEYTEMA-Universidad Tecnológica Nacional, Facultad Regional de San Nicolás, Colón, Argentina
[3]Instituto de Investigaciones Mineras, Universidad Nacional de San Juan, San Juan, Argentina
Email: *bazan@unsj.edu.ar, ebrandaleze@frsn.utn.edu.ar, psarquis@unsj.edu.ar

ABSTRACT

In Argentina, there are many sources of copper concentrates. Some of them are currently in operation, while others are in the exploration stage. All copper concentrates produced are exported to other countries for copper refinement and to create various finished products. It is desirable that in the near future, these copper concentrates will be processed in an Argentinean industrial plant. The aim of this paper was to present the results of a characterisation study carried out on five different copper concentrate samples. The thermal decomposition of the copper concentrates was determined by differential thermal analysis and thermogravimetry (DTA TG). The information was correlated with the chemical composition and the mineralogical phases of the samples identified by X-ray diffraction. A melting test at temperatures of up to 1300°C was performed to complete the study of the concentrate's behaviour during heating. After the test, all of the samples were observed by light and electronic scanning microscopy to identify the different phases generated under high-temperature conditions.

Keywords: Copper Concentrates; Thermal Analysis; Pyrometallurgy; Mineral Phases

1. Introduction

The copper market is undoubtedly one of the most important metallic markets in the world. This statement is based on the significantly increased global demand for this metal and its alloys in recent years.

Argentina has large reserves of copper ores and constitutes a strong supplier of concentrates in our region [1-4]. The ores selected for this study contain between 0.5% - 0.8% copper and are free of impurities such as Sb, As, Te and Se. Currently, concentrates obtained by flotation operations are exported to other countries to produce final copper products such as wires, pipes and sheets [5]. It is important to promote the installation of a copper pyrometallurgy plant in our country to allow for the ability to process concentrates and create final products for Argentina's domestic market. Copper industrial processes include melting, conversion and refining operations [6]. To design the pyrometallurgy process, it is necessary to completely characterise the concentrates. Thus, it is necessary to obtain information about the current phases that exist in concentrate particles and the type of transformations and reactions that occur under processing conditions [7].

Winkel *et al.* [8] described the importance of enhancing the knowledge about the high-temperature behaviour of copper iron sulphides, as well as the volatile impurities they contain. This information is essential to develop new processes of extractive metallurgy.

Thermal analysis techniques, such as differential thermal analysis (DTA) and thermogravimetry (TG), represent important tools to determine the concentrates' behaviour during the decomposition of copper iron sulphides and sulphur volatilisation rates [8-12].

In this paper, information regarding the thermal behaviour of five samples of Argentinean copper concentrates is presented. The results obtained by DTA and TG were correlated with information determined by X-ray diffraction (XRD) and a microscopy study. Structural analyses of the concentrate samples were performed after the melting tests. The test products were examined by electron microscopy (SEM), and the phases were identified by XRD. The authors describe the mechanisms, such

---

*Corresponding author.

as mass transport, which allow for the separation of metallic copper.

2. Materials and Methods

2.1. Samples

Five samples of copper concentrates obtained by Rougher flotation were selected for this study. The copper ores were obtained from sources located in northern Argentine.

2.2. Methods

The concentrate characterisation was carried out. The chemical composition of each sample was determined by acid attack and atomic absorption spectrometry using a Perkin Elmer AA110 instrument. The crystal phases were identified by X-ray diffraction (XRD) at room temperature using a Philips X'Pert diffractometer. The sulphide quantification was conducted by applying the ASTM standard C25 [13].

A microscopy study performed using an Olympus GX51 light microscope and applying a LECO IA32 analysis system, which allowed for the observation of the types of particles and their morphological characteristics.

Another objective of this paper was to evaluate the behaviour of the concentrates at high temperatures. To obtain information about the reactions that occur during-heating, melting tests were carried out at temperatures of up to 1300°C in air. Samples were melted in a porcelain

crucible using an electric furnace. All of the melted samples were prepared for microscopy observation. Their structure was studied using light and scanning electron microscopy (SEM). Finally, the results were correlated with the DTA-TG results.

3. Results

3.1. Chemical Composition

The chemical composition of the five concentrates samples is presented in **Table 1**.

3.2. Mineral Phases

The X-ray diffraction results provide information regarding the crystalline phases present in the five samples studied. The major minerals present in the concentrates were chalcopyrite ($CuFeS_2$), pyrite (Cu_2S) and iron sulphide (FeS). Nevertheless, other minerals such as iron bisulphide (FeS_2), geerite (Cu_8S_5), enargite (Cu_3AsS_4) and dicopper zinc silicon tetrasulphide ($Cu_2Zn\ SiS_4$), among others, were identified. However, sample E was observed to contain a lower chalcopyrite content and higher pyrite content. In sample B, geerite (Cu_8S_5), enargite (Cu_3AsS_4) and dicopper zinc silicon tetrasulphide ($Cu_2Zn\ SiS_4$) were identified, and in sample C, PbS was identified. Traces of cassiterite (SnO_2) were identified in samples A and D. **Figures 1-5** show the diffractograms of all of the samples, specifically the principal peaks corresponding to the crystal phases observed.

Table 1. Chemical composition of the copper concentrates samples.

(a)

Sample	Cu%	Fe%	Ins%	SiO$_2$%	R$_2$O$_3$%	CaO%	MgO%	S%	Al$_2$O$_3$%
A	16.9	32.1	2.76	1.64	53.12	0.15	0.08	37.6	8.78
B	22.3	31.4	1.70	0.88	51.58	0.18	0.08	35.2	8.25
C	18.4	31.5	1.64	1.04	52.96	0.14	0.05	37.8	9.43
D	14.6	33.1	2.52	1.28	55.06	0.11	0.07	38.9	9.35
E	18.1	35.3	2.04	0.52	59.36	0.13	0.05	33.3	10.59

(b)

Sample	Sb ppm	As ppm	Se ppm	Te ppm
A	<50	52	<50	<30
B	<50	68	<50	<30
C	<50	48	<50	<30
D	<50	35	<50	<30
E	<50	49	<50	<30

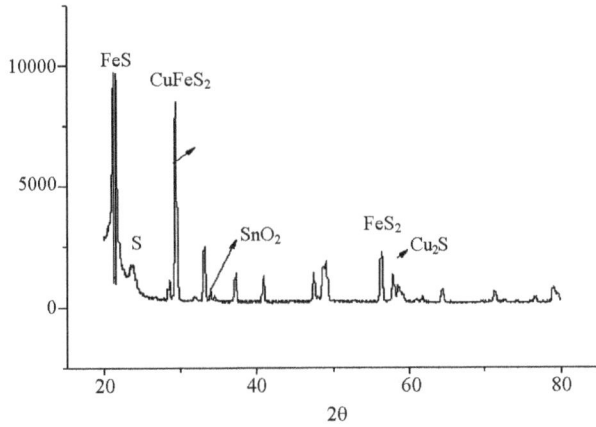

Figure 1. Sample A diffractogram.

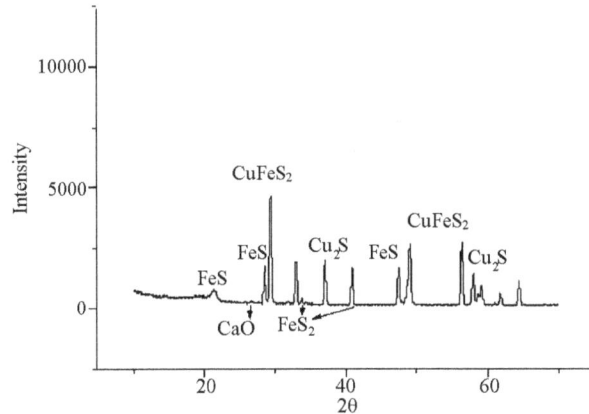

Figure 4. Sample D diffractogram.

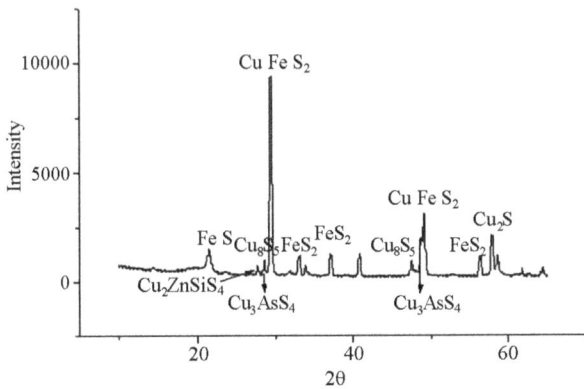

Figure 2. Sample B diffractogram.

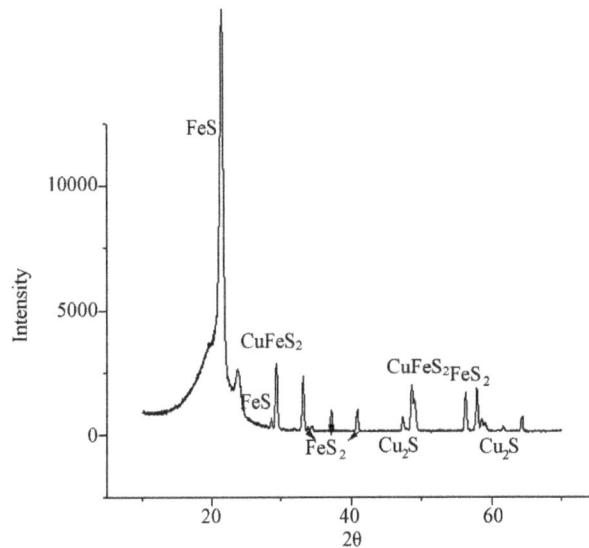

Figure 3. Sample C diffractogram.

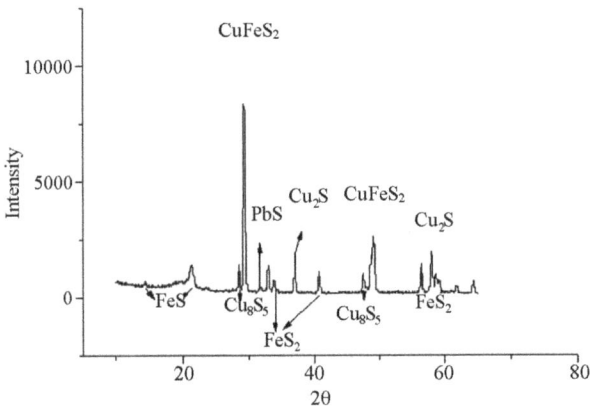

Figure 5. Sample E diffractogram.

3.3. Thermal Analysis Tests

DTA and TG tests were carried out at a heating rate of 10°C/s using a JENCK instrument. In this manner, the mechanisms of oxidation during heating up to 1000°C were characterised for all of the copper concentrate samples. For each sample, the first exothermic peak temperature, T_{in}, was determined. This temperature establishes the beginning of the oxidisation reaction. The exothermic heat, Q_{ex}, was determined by the area below the DTA curve. These values represent the total heat that developed during all of the reactions. Both values are very useful for understanding the flash-melting process [11,12,14,15].

Figure 6 shows the DTA curves of the five samples. The T_{in} and Q_{ex} values calculated by numerical integration are presented in **Table 2**. Notably, both values are related to the oxygen content of the copper, which in this case, are constant.

The thermogravimetric curves (TG) (**Figure 7**) allow for the observation of mass changes and losses in the concentrate samples during heating.

3.4. Melting Test

Samples of each copper concentrate were melted in porcelain crucibles at 1200°C at a heating rate of 5°C/min in air. The structure of the solidified product of each sample was studied by light and scanning electron microscopy

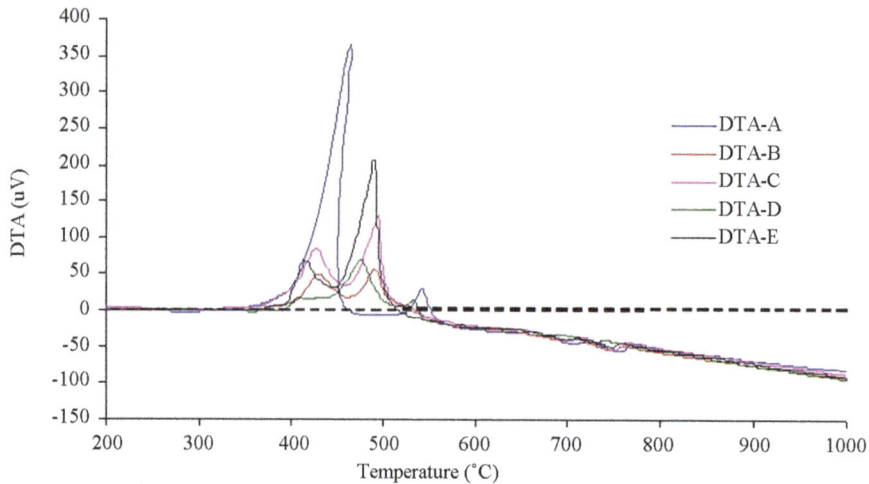

Figure 6. DTA curves of the five copper concentrates samples.

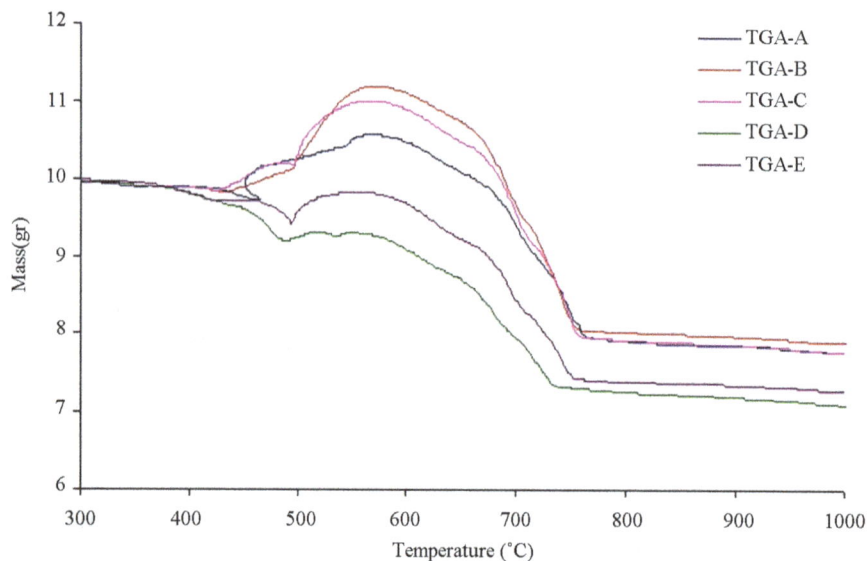

Figure 7. Thermogravimetric curves of the copper concentrates.

Table 2. Values of the initial oxidation temperature T_{in}, exothermic heat Q_{ex} and the copper law of the concentrates.

Sample	Copper law (%)	T_{in} (°C)	Q_{ex} (KJ/Kg)
A	16.86	309 ± 10	13880 ± 300
B	22.28	344	5185
C	18.38	340	7328
D	14.65	390	3254
E	18.03	415	12225

(SEM). In all cases, three different layers were recognised: a greyish layer with a fine dispersion of white particles (located in the bottom of the crucible), a white (intermediate) layer that contained a large number of dendritic crystals and extensive porosity and finally a top grey layer composed of spherical and irregular white

particles of a larger size with respect to the bottom-layer particles. **Figure 8** shows a vertical cross section of the solidified layers in sample A.

Notably, in the intermediate white layer, native copper bands (**Figure 9**) and white dendrites were observed. A higher proportion of native copper bands and globular particles was observed in sample B. In the grey top layer, white crystals with different morphologies, irregular, needle-shaped and dendritic, were observed.

In samples B, C, D and E, the bottom and top grey layers were observed to be thicker than those in sample A and to contain a low proportion of white particles. However, in all of the samples, the intermediate white layer was observed to contain metallic copper bands.

The phases were examined by SEM in the three layers. Elemental mapping revealed the elemental distribution in each layer. **Table 3** shows the elemental distribution ob-

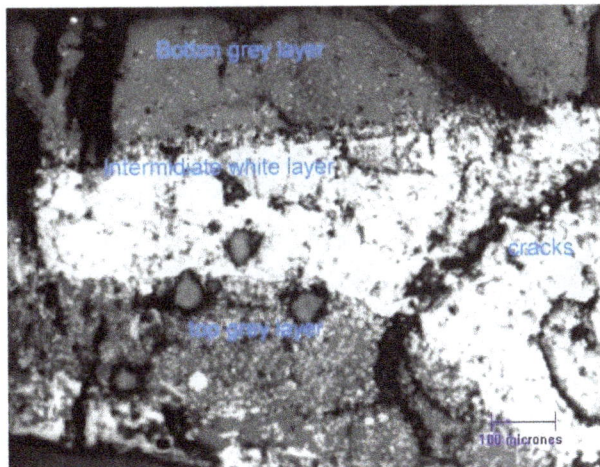

Figure 8. Three layers present in the structure of solidified product A.

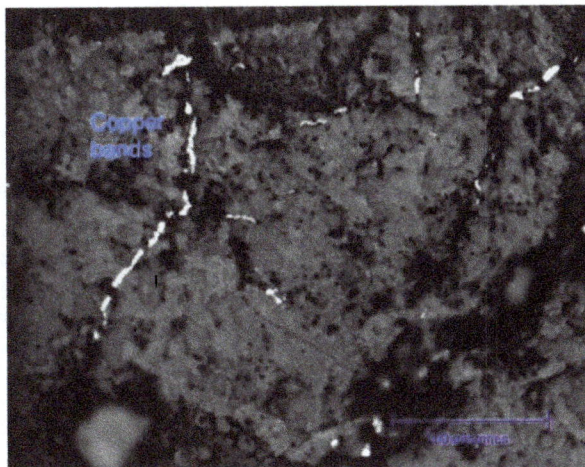

Figure 9. Copper band observed in the intermediate.

served in sample A.

The bottom grey layer of the samples contains white particles of silicoaluminate. A few particles of CuS and phases with higher iron contents were observed near the interface in contact with the intermediate white layer. Also, traces of other elements such as Mg and Cr in the silicoaluminate matrix were detected.

The intermediate white layer contains metallic copper bands, white dendrites or irregular crystals of (Cu, Fe)S, and the matrix has a high iron sulphide (FeS) content.

The white and the grey top layers were observed to be porous. Sample D showed the highest porosity among the samples studied.

The top grey layer matrix consists of calcium silicoaluminate and copper sulphide particles (with dendritic and globular morphologies). Traces of K, Ti and Cr were identified. Irregular copper particles isolated in the layer and in the microcracks were also observed.

If we consider copper, iron and sulphur to be the main elements and approximate their content to be 100%, it is possible to plot on a ternary diagram the sulphide composition identified in the intermediate white layer and in the grey top layer. For simplicity, the plotted compositions correspond to samples A and D (**Figure 10**).

These two samples were selected because of the phases differences observed. Samples B, C and E contain phases similar to those observed in sample A but in different proportions. Notably, the sulphide phases developed under the test conditions (1200°C) show copper contents up to 52%.

4. Discussion

The chemical composition of the concentrates show that sample B possessed a higher Cu content (22.28%) and a lower content of Fe (31.40%). Importantly, sample B also exhibited the lowest S/Cu ratio (1, 58).

Sample D contained the lowest concentration of Cu (14.65%) and the highest S and Fe concentrations. Moreover, the S/Cu ratio of sample D was 2.65, which is highest value observed in the concentrate samples. The XRD results are consistent with the chemical composition results.

Based on the DTA results, the correlation between the T_{in} and Q_{ex} values with respect to the law of copper concentration ($Cu(X)$) can be modelled by third-degree polynomial equations, (Equations (1) and (2)) for which $R = 0.9946$.

$$T_{in}\left(°C\right) = 1.425\left(X\right)^3 - 78.488\left(X\right)^2 \\ + 1377.8\left(X\right) - 7490.8 \tag{1}$$

$$Q_{ex}\left(KJ/Kg\right) = -215.6\left(X\right)^3 + 14335\left(X\right)^2 \\ - 268157\left(X\right) + 2 \cdot E^{+6} \tag{2}$$

The T_{in} values vary with mineralogical composition and respect the following order of the samples: A < C < B < D < E. As observed experimentally, sample B has an intermediate initial oxidation temperature. A higher content of pyrite indicates an increase in T_{in} because of the high melting temperature of the compound. Nevertheless, the other phases in the concentrates also affect T_{in}.

The exothermic heat Q_{ex} values were obtained by the numerical integration of the exothermic peaks on the DTA curves (**Figure 6**). The exothermic heat is generally associated with the oxygen content present during the test. However, in this case, the oxygen concentration is constant. Thus, it is possible that the higher value is due to the chalcopyrite ($CuFeS_2$) and coveline (CuS) contents.

The DTA and TG curves show that the sample-B reactions develop with low Q_{ex}, which is associated with a significant loss of mass during heating between room temperature and 1000°C. A gradual weight loss in the

Table 3. Element distribution in the three layers of the post melted sample A.

Botton grey layer	Intermidiate white layer	Top grey layer

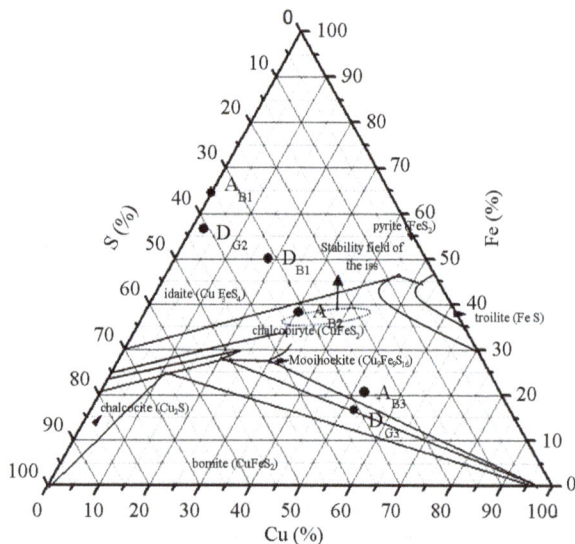

Al	Ca	Al	Ca	Al	Ca
Fe	Si	Fe	Si	Fe	Si
S	Cu	S	Cu	S	Cu
Fe-Cu-S	Fe-Si		S-Fe-Cu	Fe-Cu-Si	Fe-S-Cu

Figure 10. Sulphide composition determined in the white ($_B$) intermediate layer and in the grey ($_G$) top layer of samples A and D.

range of 300°C to 450°C was observed for all of the concentrates. This change is attributed to humidity loss. In particular, sample E exhibited a drastic loss of mass at 500°C; this is because the range 450°C - 550°C favours the direct oxidation of pyre to loadstone, while at temperatures above 650°C the complete elimination of any formed ferrous sulphate is favoured; this is manifested as weight loss [16].

A gradual increase in mass up to 600°C was observed for all of the concentrate samples. The transformation of the sulphides into oxides coincides with these mass-change exothermic peaks (in DTA curves). A new and pronounced loss of mass up to 750°C was observed for the five concentrates. At these temperatures, the trans-

formations are probably associated with the sulphide and SO_2 dissociation.

The proposed reaction mechanisms, based on the behaviour indicated by the DTA and TG curves and the equilibrium information obtained by calculation using the software program HSC (Outokumpu), are detailed in **Table 4**.

The DTA results are consistent with proposed reaction 5, which implies a loss of mass because of the sulphide-to-oxide transformation. In addition, it is important to consider the chalcopyrite ($CuFeS_2$) transformation (reactions 3 and 4).

The initial mass loss revealed by TG is based on the decomposition of a pure Cu_2S sample and the oxidation of pyrite, as observed in reactions 5 and 6.

The gradual increase in mass determined by the TG test at 600°C was also reported by Perez-Tello *et al.* [17]. They believed that the concentrates quickly gain mass through the copper sulphate formation. However, **Figure 11** shows that reactions 11 and 10 contribute to the increase in mass through the formation of iron sulphide. Although this species is later dissociated, as indicated by reaction 14, the reaction speed is slow due to that the formation of sulphates, which closes the pores of the particle and prevents the reaction from occurring, and thus creates a gain in mass, as shown in **Figure 7** [16].

The gradual final loss of mass is explained by the decomposition reactions of the sulphate into oxides.

Samples B and C showed the same behaviour during heating because of their similar mineralogical composition: high content of chalcopyrite ($CuFeS_2$) and low copper and iron sulphides in the concentrate. These samples developed intermediate T_{in} and Q_{ex} during heating.

Sample A presented a lower T_{in} and higher Q_{ex} because of the higher contents of chalcopyrite ($CuFeS_2$)

Table 4. Reaction mechanisms.

Reaction	$-\Delta G^\circ$ (kcal)$_{reacc}$	Mass changes %	Reaction number
$5CuFeS_2 + 9O_2 (g) = Cu_5FeS_4 + 2Fe_2O_3 + 6SO_2 (g)$	$653 - 621$	-10.50	1
$2Cu_5FeS_4 + 14.5O_2 (g) = 10CuO + Fe_2O_3 + 8SO_2 (g)$	$876 - 788$	-4.85	2
$2CuFeS_2 + 6.5O_2 (g) = 2CuO + Fe_2O_3 + 4SO_2 (g)$	$436 - 406$	-8.59	3
$2CuFeS_2 + 6O_2 (g) = Cu_2O + Fe_2O_3 + 4SO_2 (g)$	$410 - 387$	-17.50	4
$2FeS_2 + 5.5O_2 (g) = Fe_2O_3 + 4SO_2 (g)$	$388 - 377$	-33.44	5
$2FeS + 3.5O_2 (g) = Fe_2O_3 + 2SO_2 (g)$	$273 - 253$	-9.17	6
$Cu_2S + 2O_2 (g) = 2CuO + SO_2 (g)$	$113 - 98$	-0.04	7
$Cu_2S + 1.5O_2 (g) = Cu_2O + SO_2 (g)$	$86 - 79$	-10.09	8
$2Cu_5FeS_4 + 2SO_2 (g) + 19.5O_2 (g) = 10CuSO_4 + Fe_2O_3$	$1066 - 781$	74.91	9
$2CuFeS_2 + 9O_2 (g) + SO_2 (g) = 2CuSO_4 + Fe_2(SO_4)_3$	$532 - 403$	95.92	10
$FeS_2 + 3O_2 (g) = FeSO_4 + SO_2 (g)$	$203 - 176$	233.33	11
$Cu_2S + 2.5O_2 (g) = CuO * CuSO4$	$125 - 90.61$	50.26	12
$2Cu_2O + O_2 (g) = 4CuO$	$14 - 5$	11.17	13
$2FeSO_4 + 1.5O_2 (g) = Fe_2O_3 + 2SO_2 (g) + 2O_2 (g)$	$6 - 26$	-47.44	14
$Fe_2(SO4)_3 = Fe_2O_3 + 3SO_2 (g) + 1.5O_2 (g)$	$3 - 41$	-60.06	15
$2CuSO_4 + O_2 (g) = 2CuO + 2SO_2 (g) + 2O_2 (g)$	$+10 - 14$	-50.16	16
$2CuO * CuSO_4 = 4CuO + 2SO_2 (g) + O_2 (g)$	$+13 - 10$	-33.48	17

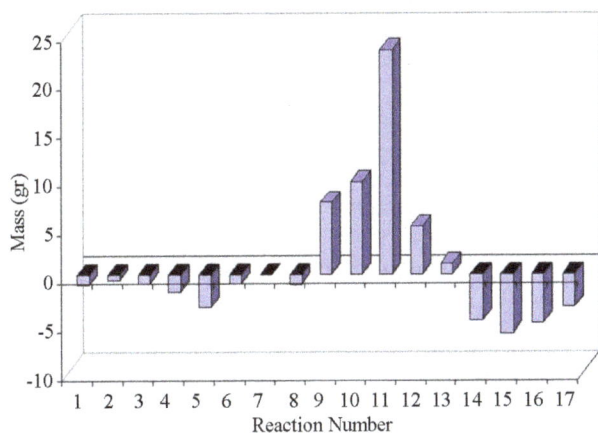

Figure 11. Mass evolution with respect to reaction mechanisms.

and iron sulphide (FeS), promoting an intermediate change in weight.

Sample E, which contained the highest percentage of FeS, showed a combination of higher Q_{ex} and T_{in}. The mass lost by this sample is lower than that by samples A, B and C.

The microscopy study of the five concentrates solidified samples showed that the contents of chalcopyrite and iron sulphide (*i.e.*, sample A) in the concentrates increase with the thickness of the white layer. In all samples, this white intermediate layer featured native copper bands or globular particles because of the decomposition of sulphides. In this case, the (Cu, Fe)S exhibited 31% Cu.

In the top grey layer matrix, white dendrites or irregular crystals of (Cu, Fe)S in a matrix with a higher iron sulphide (FeS) content were identified. These phases were observed in the greatest quantity in samples D and E. (Cu, Fe)S contained approximately 52% Cu.

Based on the S-Cu-Fe system, it was possible to conclude that different sulphide phases formed at the selected melting test conditions (1200°C, 1 h) in both layers (A_B or D_B in the white intermediate layer and A_G or D_G of the grey top layer) contain up to 52% copper. Chalcopyrite decomposes at 623 K to an intermediate solid solution (ISS) (see **Figure 10**) or to chalcopyrite with a slightly different composition ($CuFeS_{2ssb}$); moreover, an iron sulphide transformation occurred. The presence of native copper in all of the samples indicates that sulphur was only partly removed.

5. Conclusions

To summarise, the characterisation of the five concentrates studied in this work revealed that these materials contained about 14% to 22% copper and that sample B was the richest in copper. The major copper minerals were sulphides: chalcopyrite ($CuFeS_2$) and coveline (CuS). Iron was present in the concentrates in the form of FeS_2 or FeS. Samples A and E showed the highest iron sulphide contents

The thermal analysis techniques applied in this study provided information that allowed us to determine the initial oxidation temperature T_{in} and the exothermic heat Q_{ex} of the samples, as well as the mass changes during heating to 1000°C.

By correlating the DTA and TG results with the crystal phases identified by XRD and the information obtained though the thermodynamic equilibrium study performed using the software program HSC (Outokumpu), it was possible to identify the complex reaction mechanisms involved during concentrate heating. The reaction evolution features an initial transformation of the sulphides to oxides; then, the oxides react to form sulphates and are finally transformed to oxides. These results were confirmed by the phases identified by microscopy and the thermodynamic study on the Cu-Fe-S system.

REFERENCES

[1] "Peralta e. Actas de Encuentro Internacional de Minería," Secretaria de Minería de la Nación. Buenos Aires, Argentina, 1994, pp. 5-15.

[2] Segemar, Servicio Geológico Minero de Argentina. http://www.segemar.gov.ar/P_Oferta_Regiones/Oferta/Introducción.htm

[3] A. Beretta and V. Bazan, "Algunos Fundamentos de las Ventajas Estratégicas de la Instalación de una Fundición de Cobre en Argentina," SAM/CONAMET, San Nicolás, Buenos Aires, 2007, pp. 7-12.

[4] A. Guitierrez, D. Chong and R Espinoza, "Niveles de Exposición de Yacimientos del Distrito Minero de agua de Dionisio (YMAD)," *Revista de la Asociación Geológica Argentina, Catamarca*, Vol. 61, No. 2, 2006, pp. 269-277.

[5] V Bazan, P Sarquis and E. Brandaleze, "Caracterización de un Mineral de Cobre en Argentina para la Producción de Matte," Revista Dyna, No. 167, 2011, pp. 220-228.

[6] V. Bazán, P. Sarquis and E. Brandaleze, "Factibilidad de una Industria Pirometalurgia con Mineral Argentino 1ras," Jornadas de Investigación de la Minería del norte Argentino Editorial Científica Universitaria, Universidad Nacional de Catamarca, 2009.

[7] J. Dunn and C. Muzenda, "Thermal oxidation of covellite (CuS)," *Thermochimica Acta*, Vol. 369, No. 1-2, 2001, pp. 117-123. http://dx.doi.org/10.1016/S0040-6031(00)00748-6

[8] L. Winkel, I. Alxneit and M. Sturzenegger, "Thermal Decomposition of Copper Concentrates under Concentrated Radiation-Mechanistic Aspects of the Separation of Copper from Iron Sulfide Phases," *International Journal of Mineral Processing*, Vol. 88, No. 1-2, 2008, pp. 24-30. http://dx.doi.org/10.1016/S0040-6031(00)00748-6

[9] H. Tsukada, Z. Asaki, T. Tanabe and Y. Kondo, "Oxidation of Mixed Copper-Iron Sulfide," *Metallurgical Transactions B-Process Metallurgy*, Vol. 12, No. 3, 1981, pp. 603-609. http://dx.doi.org/10.1007/BF02654333

[10] M. Perez-Tello, H. Y. Sohn and J. Lottiger, "Determination of the Oxidation Characteristics of Solid Copper Matte Particles by Differential Scanning Calorimetry and Thermogravimetric Analysis," *Minerals & Metallurgical Processing*, Vol. 16, No. 2, 1999, pp. 1-7.

[11] J. Dunn and S. Jayaweera, "Applications of Thermoanalytical Methods to Studies of Flash Smelting Reactions," *Thermochimica Acta*, Vol. 85, 1985, pp. 115-118. http://dx.doi.org/10.1016/0040-6031(85)85543-X

[12] Z. Zivkovic, N. Strbac, D. Zivkovic, V. Velinovski and I. Mihajlovic, "Kinetic Study and Mechanism of Chalcocite and Covellite Oxidation Proce*ss,*" *Journal of Thermal Analysis and Calorimetry*, Vol. 79, No. 3, 2005, pp. 715-720. http://dx.doi.org/10.1007/s10973-005-0601-1

[13] ASTM C25-99, "Stándar Test Methods for Chemical Analysis of Limestone, Quicklime and Hydrated Lime," 1999.

[14] F. Jorgensen and P. Koh, "Combustion in Flash Smelting Furnaces," *JOM*, Vol. 53, No. 5, 2001, pp. 16-21. http://dx.doi.org/10.1007/s11837-001-0201-x

[15] S. Perez-Fontes, "Determinación de las Características de Oxidación de Minerales Sulfurosos a Altas Temperaturas," Tesis de Licenciatura en Ingeniería Química, Universidad de Sonora, Hermosillo Sonora, 2004.

[16] V. Arias, R. Coronado, L. Puente and D. Lovera, "Refractariedad de Concentrados Auriferous," *Revista del Instituto de Investigación de la Facultad de Ingeniería Geológica, Minera, Metalúrgica y Geográfica*, Vol. 8, No. 16, 2005, pp. 5-14.

[17] S. Perez-Fontes, M. Perez-Tello, L. Prieto, F. Brown and F. Castillon-Barraza, "Thermoanalytical Study on the Oxidation of Sulfide Minerals at High Temperatures," *Minerals & Metallurgical Processing*, Vol. 24, No. 4, 2007, pp. 275-283.

10

Kinetics Analyzing of Direction Reduction on Manganese Ore Pellets Containing Carbon

Bo Zhang*, Zheng-Liang Xue

Key Laboratory for Ferrous Metallurgy and Resources Utilization of Ministry of Education, Wuhan University of
Science and Technology, Wuhan, China
Email: *tale-2002@163.com

ABSTRACT

Using high temperature carbon tube furnace, reduction of manganese ore pellets containing carbon was investigated. The reaction was divided into two stages at five minutes after reaction, and the kinetics model of reduction process was established. The experimental results showed that, the reaction rate in the earlier stage was controlled by the chemical reactions between FeO, MnO and carbon reductant, and the activation energy was 28.85 KJ/mol. In the later stage, as the carbon reductant replaced by CO, the reaction rate was controlled by CO-diffusing in solid products, and the corresponding activation energy was 86.56 KJ/mol. Reaction rate of the later stage was less than the earlier one.

Keywords: Kinetics Model; Manganese Ore Pellets Containing Carbon; Self-Reduction

1. Introduction

The technology of hot metal pretreatment and less slag steelmaking are mature increasingly since 1990s. The basic theory and industrial application of direct alloying by manganese ore under the condition of less slag steelmaking were researched systematically and deeply, by the main steel enterprises and research institutes in Japan [1-4]. In China, however, due to the condition restriction of hot metal pretreatment and the less slag steelmaking, the yield of manganese fluctuates in 5% - 25% generally, so that the economic benefits is not obvious. Therefore, the new technology of direct alloying by manganese ore in converter needs to be explored. In this paper, based on the new process of direct alloying using manganese oxide composite pellets in converter which were proposed by Xue Zheng Liang et al. [5], the kinetics model was established according to the kinetics analyzing in reduction process, which provided a theoretical basis for direct alloying by manganese ore.

2. Experimental Research Method

The manganese ore grain were crushed to 10 mm size collected from the mine, and then heated to 1000°C and kept for 30 min for high temperature treatment under the protection of N$_2$, in which water of crystallization and

carbonate fully decomposed. The chemical composition of manganese ore after high temperature treatment is shown in **Table 1**.

The samples were pressed to $\varphi 10 \times 10$ mm cylindrical with the well mixed and wetted manganese ore, which were crushed to −0.074 mm after high temperature treatment and reductant by using sodium silicate. The samples were stoved for reservation. The C/O molar ratio is expressed by the molar proportion of C in carbon black and O in manganese oxide and ferric oxide, calculate the amount of carbon black according to the C/O molar ratio is 1.0.

Reduction experiments were carried out in 25 kW high temperature carbon tube furnace with rapid heating function, and the internal diameter of carbon tube furnace is 70 mm. Corundum crucible was built-in furnace in Ar protection and increased temperature within furnace. When the predetermined temperature reached, cylindrical samples were put in (sample weight is approximately 100 g). The reaction temperature is set as 1823 K 1873 K and 1923 K, and reaction time is set as 3 min, 4 min, 5 min, 10 min and 15 min. Corundum crucible and specimen

Table 1. Raw Material Components, %.

TFe	FeO	Fe$_2$O$_3$	Mn	SiO$_2$	CaO	Al$_2$O$_3$	MgO
13.3	3.27	15.37	28.25	34.05	0.19	5.92	0.29

*Corresponding author.

were hanged out from the high temperature zone quickly, and cooled under an Ar atmosphere. The reduction degree was calculated according to the weight change of the pellets and the detection result of chemical composition.

3. Reduction Kinetics Analysis

3.1. Reduction Degree Calculation

The oxygen loss ratio of the reduction is calculated through the following formula according to the chemical test results before and after the reaction:

$$\xi = \left[\left(w_{C \cdot 0} \cdot m_0 - w_{C \cdot T} \cdot m_T \right) 16/12 \right] / m_0 w_{O \cdot 0} \qquad (1)$$

where, m_0 and m_T are the weight before and after the reaction, respectively, $w_{C \cdot 0}$, $w_{C \cdot T}$ are carbon content in the sample before and after the reaction, respectively, and $w_{O \cdot 0}$ is oxygen content of the sample before the reaction.

3.2. Reduction Reaction

The oxide reduction rate ξ is calculated according to the above Equation (1), thus the relationship between reduction rate and reduction time under different temperature can be achieved as shown in **Figure 1**.

Figure 1 shows that the reduction rate of the carbon-manganese pellets had little changed with the reduction temperature increases. The change of reduction rate in the slope can be divided into two phases, the initial reduction rate is relatively fast, while after 5 minutes reduction rate decreases and eventually gentle. Therefore, it is possible to divide reduction process into two stages at 5 minutes, earlier stage and later stage. At the earlier stage, the reaction can be considered as a typical direct reduction process between reductant carbon and FeO, MnO due to the lack of CO; at the later stage, as the solid

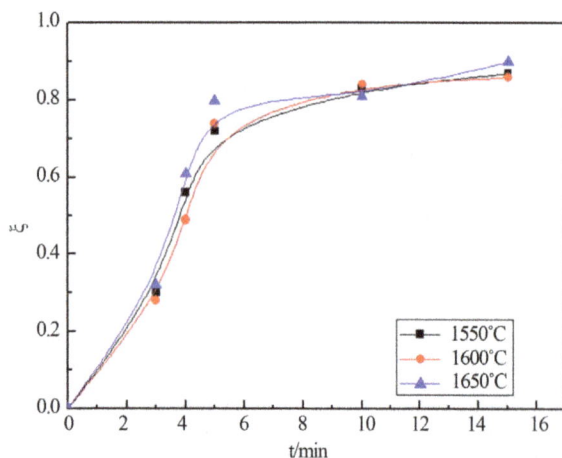

Figure 1. degree reduction-time curve at different temperatures.

products increases, reductant carbon cannot be contacted with the oxide, so CO substituted carbon to take part in reduction reaction with FeO and MnO.

4. Kinetics of Reduction Reaction

4.1. Kinetic Model of the Earlier Stage

In the earlier stage, the reduction rate depends on the contact area between the solid carbon and the metal oxide [6], which can be expressed as follows:

$$-\frac{dm}{dt} = K'S \qquad (3)$$

where, K' is a constant; t is reaction time; m is ore quality, $m = n(4/3)\pi r^3 \rho$; S is the surface area of slag, $S = n4\pi r^2$.

Substitution of m and s into Equation (3) can be obtained Equation (4) as follows:

$$-\frac{d\left(n(4/3)\pi r^3 \rho \right)}{dt} = K'\left(n4\pi r^2 \right) \qquad (4)$$

Hence:

$$-dr = \left(\frac{K'}{\rho} \right) dt \qquad (5)$$

By taking the integral for t of the Equation (5): $\int -dr = \int \left(\frac{K'}{\rho} \right) dt$,so a new equation can be obtained as follows:

$$r_0 - r = \left(\frac{K'}{\rho} \right) t \qquad (6)$$

And the reduction rate can be expressed as:

$$\xi = 1 - \frac{n(4/3)\pi r^3 \rho}{n(4/3)\pi r_0^3 \rho} = 1 - \left(\frac{r}{r_0} \right)^3 \qquad (7)$$

So:

$$r = r_0 \left(1 - \xi \right)^{1/3} \qquad (8)$$

Equation (9) can be obtained by substituting Equation (8) into Equation (6).

$$f(\xi) = 1 - (1 - \xi)^{1/3} = \left(\frac{K'}{r_0 \rho} \right) t = K_1 t = A e^{-E/RT} t \qquad (9)$$

where, $f(\xi)$ is only related with ξ; $K_1 = K'/(r_0 \rho)$, is rate constant; A is the frequency factor; E is the apparent activation energy; T is the absolute temperature. Both sides of the equation can be obtained for t differential:

$$f(\xi)\frac{d\xi}{dt} = A e^{-E/RT} \qquad (10)$$

$$\ln\left(f\left(\xi\right)\frac{d\xi}{dt}\right)=\ln A-\frac{E}{RT} \tag{11}$$

According to the test data, calculate the value of k in the respective temperature, and taking the average, the relation between $\ln k$ and T^{-1} is plotted in **Figure 2**. It is clearly that $\ln k$ and T^{-1} has a very good linear correlation, and we can calculate the frequency factor $A = 0.0038$ and the apparent activation energy $E = 28.85$ KJ/mol can be achieved. Therefore, in the restored earlier stage, the pellet reduction kinetics equation can be expressed as:

$$1-\left(1-\xi\right)^{1/3}=0.0038e^{-28850/RT}t \tag{12}$$

4.2. Kinetic Model of the Later Stage

In the later stage of reaction, the reaction rate mainly depends on CO diffusion through the metal and the solid product layer, so the diffusion step is restrictive step. The mainly practical equations on diffusion control are Jander Equation (13) and Ginstling-Brounshtein Equation [7,8] (14).

$$\left[1-\left(1-\xi\right)^{1/3}\right]^2=kt \tag{13}$$

$$1-\left(2/3\right)\xi-\left(1-\xi\right)^{2/3}=kt \tag{14}$$

where: ξ is the reduction rate of the pellets; k is the reaction rate constant; t is the reaction time. The relation between $\ln k$ and T^{-1} is plotted in **Figure 3**, it can be seen from the figure that the $\ln k$ does not have a good linear correlation with T^{-1}. Therefore, it is not suitable to calculate the kinetic parameters in the later stage of reduction by this method, and has some limitations to calculate the reaction kinetic through the above methods.

In recent years, Zhou Guozhi [9] deduced a new dynamic model for gas-solid phase reaction system (as shown in Equation (15)), which assumed the diffusion process as the reaction restrictive step, the new model introduced a new concept of characteristic time to analysis reaction rate, and expressed analytically the

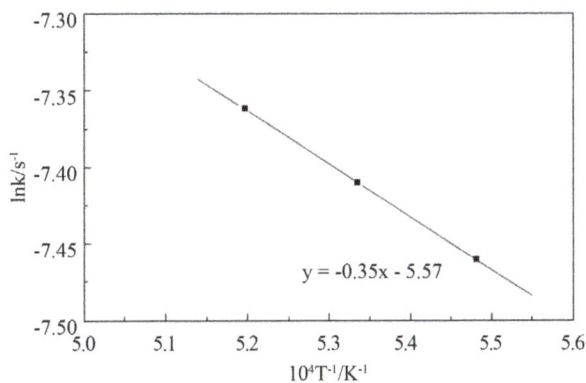

Figure 2. Restore the initial kinetic parameters illustrations.

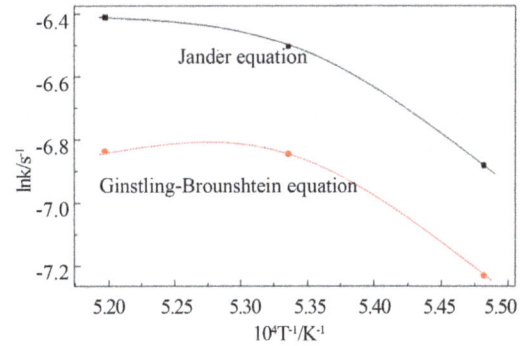

Figure 3. Kinetic parameters in the late period of reduction.

impact of factors on gas-solid phase reaction kinetics (temperature, pressure, and particle radius, etc.) through a simple explicit function. According to the research of Cheng GongWei [10] and Liu SongLi [6], it shown that the model had a characteristic of simple, accurate and practical, and had very good guidance on dynamics analysis that based on carbon pellet reduction process.

$$t_{\xi=\phi}\left(T_0\right)=-\frac{[1-\left(1-\phi\right)^{1/3}]^2}{\dfrac{2\left(C_H^{n\beta}-C_H^{m\beta}\right)D_H^0\exp\left(-\dfrac{E}{RT_0}\right)}{r_0^2v_m}} \tag{15}$$

where: r_0 is the radius of particles; D_H^0 is the diffusion constant of the diffusion materials; $C_H^{n\beta}$ is the diffusion concentration of total diffusion materials of the resultant phase in vapor balance; $C_H^{m\beta}$ is the diffusion concentration of diffusion materials of the resultant phase close to the unreacted materials interface; v_m is correlation coefficient that depends on the material reaction; $t_{\xi=\phi}\left(T_0\right)$ is the characteristic time of the reaction, and the physical meaning is the time that the reaction fraction reach to ϕ while $T = T_0$,it will decrease with the increasing of temperature and pressure and the decrease of particle diameter, it can be used to show the speed of the reaction.

$$\xi=1-\left\{1-\left[1-\left(1-\phi\right)^{1/3}\right]^2\sqrt{\frac{t}{t_{\xi=\phi}\left(T_0\right)}}\right.$$
$$\left.\times\exp\left(-\frac{E}{2R}\left(\frac{1}{T}-\frac{1}{T_0}\right)\right)\right\}^3 \tag{16}$$

Where, E is the activation energy; R is the gas constant; T is the absolute temperature; rate; t is reaction time. The reaction time is constant, the equation of activation energy can be calculated from measuring T_1 and T_2, ξ_1 and ξ_2.

$$E=-\frac{2R\ln\left[\dfrac{1-\left(1-\xi_2\right)^{1/3}}{1-\left(1-\xi_1\right)^{1/3}}\right]^3}{\dfrac{1}{T_2}-\dfrac{1}{T_1}} \tag{17}$$

According to the test results, the activation energy E can be calculated under different temperature in the following three groups in **Table 2**.

When $t_{\xi=83}(1823\,\text{K})=10\,\text{min}$, it can be found that the reduction kinetics equation of pellets in the later stage of reaction can be expressed as fellow:

$$\xi = 1 - \left\{1 - 0.199\sqrt{\frac{t}{10}} \times \exp\left(-\frac{86560}{2\times 8.314}\left(\frac{1}{T}-\frac{1}{1823}\right)\right)\right\}^3$$

(18

5. Kinetics Model Result and Discussion

United Equation (12) and Equation (18), predicting the reduction degree changes of carbon-bearing pellet respectively in temperature $T = 1823$ K, $T = 1873$ and $T = 1923$ K. The forecast result and the actual experimental data are shown in the follows.

Figure 4 shows that calculated value and actual value agrees well under the 1823 K, 1873 K and 1923 K. In the earlier stage of reaction, calculated value was higher than the actual value slightly. This may be related with the experimental conditions and operation. The samples were putted into the furnace when the temperature rise to a predetermined, so there are need some preheating time before, and carbon and metal oxide also need certain incubation period to begin to reaction. But model calculation is in a kind of ideal state. So calculated value is more than test value at the beginning. The later reaction activation energy is far less than the reaction activation energy of earlier, so reaction rate of the later stage is less than the earlier one.

6. Conclusion

According to the reduction results of carbon manganese ore pellets, reduction process can be divided into two stages. The reaction of direct contact of carbon reductant and FeO, MnO are the main reaction during the earlier stage of reduction process. Reaction rate is controlled by chemical reaction and its kinetics equation is: $1-(1-\xi)^{1/3} = 0.0038e^{-28850/RT}t$. CO replace carbon as the reducing agent at the later of reaction. Reaction rate at the later stage of the reaction is controlled by CO-diffusion reaction in solid products and its kinetics equation is:

Table 2. activation energy in different temperature ranges.

temperature range/°C	1550 - 1600	1600 - 1650	1550 - 1650
activation energy /KJ·mol⁻¹	84.73	89.81	85.15
average activation energy /KJ·mol⁻¹		86.56	

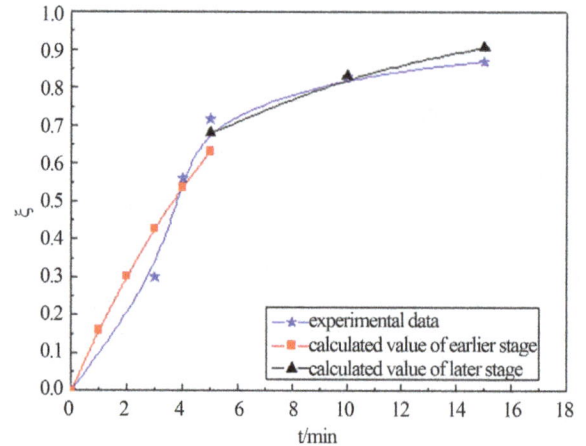

(a)

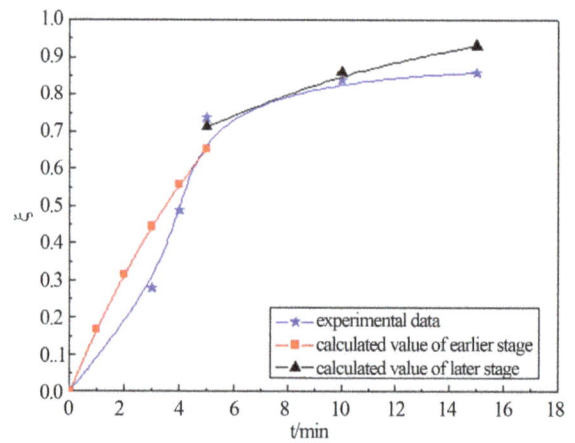

(b)

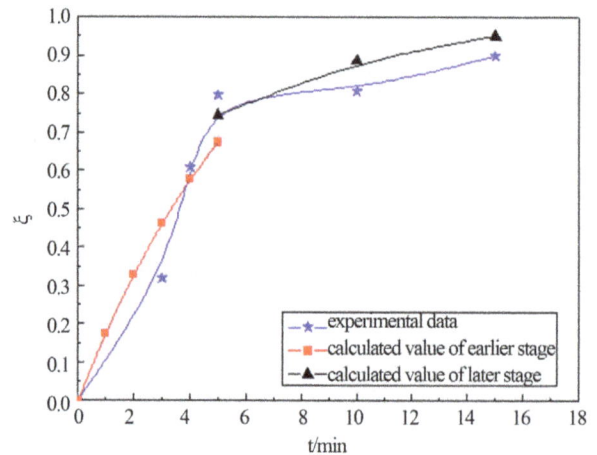

(c)

Figure 4. calculated values comparison of actual value for degree reduction-time curve at different temperatures; (a) 1823 K; (b) 1873 K; (c) 1923 K.

$$\xi = 1 - \left\{1 - 0.199\sqrt{\frac{t}{10}} \times \exp\left(-\frac{86560}{2\times 8.314}\left(\frac{1}{T}-\frac{1}{1823}\right)\right)\right\}^3 .$$

REFERENCES

[1] L. Z. Shan, *et al.*, "Industrialization of a New Steelmaking Process Utilizing Hot Metal Pretreatment and Smelting Reduction," *Iron & Steel*, Vol. 2, 1988, pp. 64-71.

[2] A. M. H. Jin, "Improvement of Mn Yield in Less Slag Blowing at BOF by Use of Sintered Ore," *Iron & Steel*, Vol. 8, 1993, pp. 45-51.

[3] R. S. Li, Y. M. Feng and G. Z. Wei, "Application of Poor Manganese Ore in the Converter Steelmaking Process," *Steelmaking*, Vol. 20, No. 1, 2004, pp. 13-15.

[4] Z. L. Meng and X. U. Yang, "Discussion on Manganese Alloyed in Top-Blown Converter," *Laiwu Steel Technology*, Vol. 4, No. 8, 2002, pp. 33-35.

[5] Z. L. Xue, Y. Yu, W. Wang, *et al.*, "Steelmaking Process with Manganese Oxide Composite Mass of Direct Alloying by the Combined Blowing Converter," Invention Patent: ZL201010245102.1.

[6] S. L. Liu and C. G. Bai, "Kinetics of Vanadium-Titanium Iron Concentrate Pellets Containing Carbon Direct Reduction," *Journal of iron and Steel Research*, Vol. 4, No. 23, 2011, pp. 5-8.

[7] W. Jander, "Reactions in the Solid State at High Temperatures," *Zeitschrift für Anorganische und Allgemeine Chemie*, Vol. 163, No. 1-2, 1927, pp. 1-30. doi:10.1002/zaac.19271630102

[8] A. M. Ginstling and V. I. Brounshtein, "Concerning the Diffusion Kinetics of Reactions in Spherical Particles," *J. Appl. ChcilL USSR*, Vol. 23, 1950, pp. 1327-1338.

[9] K. C. Chou, Q. Li, Q. Lin, *et al.*, "Kinetics of Absorption and Desorption of Hydrogen in Alloys Powder," *International Journal of Hydrogen Energy*, Vol. 30, 2005, pp. 301-309. doi:10.1016/j.ijhydene.2004.04.006

[10] H. W. Cheng and X. G. Lu, "Reduction Kinetics Mechanism of Carbon Containing Pellets," *Chinese Journal of Rare Earths*, Vol. 8, No. 26, 2008, pp. 103-108.

11

Alumina/Iron Oxide Nano Composite for Cadmium Ions Removal from Aqueous Solutions

Mona Mahmoud Abd El-Latif[1*], Amal M. Ibrahim[2,3], Marwa S. Showman[1], Rania R. Abdel Hamide[1]

[1]Fabrication Technology Department, Advanced Technology and New Materials Research Institute (ATNMRI),
City of Scientific Research and Technological Applications (SRTA-City), Previously
"Mubarak City for Scientific Research and Technology Applications (MuCSAT)", Alexandria, Egypt.
[2]Dharyya College for Arts and Sciences, Qassim University, Al Qassim, Saudi Arabia
[3]Surface Chemistry and Catalysis Laboratory, Physical Chemistry Department,
National Research Center, Cairo, Egypt
Email: *amona1911@yahoo.com

ABSTRACT

Magnetic alumina nano composite (MANC) was prepared for combination of the adsorption features of nano activated alumina with the magnetic properties of iron oxides to produce a nano magnetic adsorbent, which can be separated from the medium by a simple magnetic process after adsorption. MANC was characterized using XRD, SEM, TEM, EDX and surface area (BET). Quantum design SQUID magnetometer was used to study the magnetic measurement. The present study was conducted to evaluate the feasibility of MANC for the removal of cadmium ions from aqueous solutions through batch adsorption technique. The effects of pH, adsorbent dose, temperature, contact time and initial Cd^{2+} concentration on cadmium ions adsorption were studied. Equilibrium data were fitted to Langmuir, Freundlich and Temkin isotherms. The equilibrium data were best represented by the Langmuir isotherm. The kinetic data were fitted to pseudo-first-order, pseudo-second-order, Elovich and intraparticle diffusion models, and it was found to follow closely the pseudo-second-order model. Thermodynamic parameters were calculated for the Cd^{2+} ion-MANC system and the positive value of $\Delta H°$ showed that the adsorption was endothermic in nature. Furthermore, a single-stage batch adsorber was designed for the removal of Cd^{2+} ions by MANC based on the equilibrium data obtained.

Keywords: Activated Alumina; Magnetic Properties; Cadmium Ions Removal; Equilibrium Isotherms; Kinetics

1. Introduction

Water is the most precious natural resource that exists on our planet although we as humans recognize this fact; we discharged it by polluting our rivers, lakes and underground water. The release of heavy metals into the environment is a potential threat to water and soil quality as well as to plant, animal and human health. Heavy metals can be bioaccumulated through food chain transfers and unlike organic toxicants are not amenable to biological degradation [1]. Over the last two decades there has been a sharp rise in the global use of Cd for batteries and a steady decline in its use for other applications, such as pigments, polyvinyl chloride stabilizers, and plating. This trend in the use of Cd products and compounds has inspired a number of international agreements to manage

and control the release of Cd to the environment and limit human and environmental exposure to Cd can cause kidney damage in mammals and humans [2,3]. Also, cadmium (Cd) is one toxic heavy metal of particular environmental concern, because it can be introduced into and accumulated in soils through agricultural application of sewage sludge, fertilizers, and/or through land disposal of Cd-contaminated municipal and industrial wastes. Cd is a known human carcinogen and may induce lung insufficiency, bone lesions and hypertension [4]. The high toxicity of cadmium has resulted in governments imposing ever tighter environmental legislation limiting wastewater discharge and the removal of heavy metals such as Cd from wastewater has been a major preoccupation of environmental professionals for many years. In particular, ever increasing world populations are likely to place increasing stress on a limited clean water

---

*Corresponding author.

resource placing a greater focus on clean up and reuse of contaminated wastewater streams [1]. Among the various methods proposed for this purpose adsorption proved to be of the most promising ones [5,6]. Several natural (e.g. natural zeolites, bentonites, metal oxides) and synthetic (e.g. synthetic zeolites, resins, metal phosphates and silicates, synthetic oxides/hydroxides/hydroxyoxides) materials have been investigated as sorbents for heavy metal removal from solutions achieving different levels of success [7-13]. Moreover, considerable research work has been done on various industrial waste materials in order to develop suitable sorbents for water treatment; so fly ash [14,15], blast fumace slug [16], biomass [17,18] and bagasse fly ash [19], among others have been tested as sorbents for heavy metal removal with various levels of success. Although, Adsorption processes are widely used for treatment of polluted surface and groundwaters and also play a significant role in advanced wastewater treatment. In adsorption, Adsorbents remove adsorbates by means of concentrating them on the large inner surface so that adsorbents with higher specific surface area possess superior adsorption capacity. A small particle size of the adsorbent can offer not only a greater specific surface area but also better mass transfer efficiency. However, a high pressure drop may be encountered with the compact adsorbent in a packed bed adsorption column if the particle diameter is smaller than 0.5 mm [5]. Because of the disadvantages caused by the small particle size, the batch stirred adsorption system with intensive mixing, which enhances the adsorption rate, is usually adopted for the use of powder adsorbents. However, the separation technology of ultra-fine particles, especially nano-scale particles, is still under development and presents difficulties to a certain extent. As a result of these disadvantages using such small particles as adsorbents, there is a great deal of interest in the preparation of nanosized magnetic particles and understanding of their properties, which are drastically different from those of the corresponding bulk materials [20]. The magnetic particle technology has a high potential to be applied in adsorption systems. In order to remove the target pollutants or compounds from the streams, the magnetic particles can be modified by the combination or modification of functional groups or inorganic compounds yielding magnetic adsorbents [21]. It is well known that porous ceramics are used in a wide range of applications including catalysts, effective absorbents, ionic conductors, filtering membranes, coatings and insulating aerogels [22-25]. Alumina is one of the most widely used ceramics due to its high specific surface area, very good thermal stability and amphoteric properties [26]. Due to these characteristics porous alumina is generally a very good candidate as a catalyst carrier as well as an adsorbent. Although alumina is used very fre-

quently as a ceramic matrix for metal-composite materials, only a few works have been published concerning its participation in magnetic composites.

This work aims to evaluate the prepared magnetic alumina nano-composite (MANC) in removal of cadmium ions from aqueous solutions. The effects of various operating parameters such as solution pH, adsorbent dose, temperature, initial Cd^{2+} concentration and contact time on cadmium ions adsorption were investigated. Also the effect of loading of alumina by iron oxide on cadmium ions adsorption was studied. Adsorption isotherms, kinetics and thermodynamics of the sorption process were studied. Further, a single-stage batch adsorber was designed for the removal of cadmium ions by MANC based on the equilibrium data obtained.

2. Materials and Methods

2.1. Materials

All chemicals used in this study were of analytical grade. Ferric chloride ($FeCl_3$) and ($FeSO_4 \cdot 7H_2O$) were supplied from Adwic Company, Cadmium chloride ($CdCl_2 \cdot 2.5H_2O$), was supplied from Aldrich Company and Aluminium oxide Al_2O_3, Activated Neutral, Brock Mann 1, STD Grade, CA, 150 Mesh.

2.2. Preparation of Magnetic Alumina Nano Composite (Adsorbent)

The nano composite (MANC) was prepared from a suspension of activated alumina in 400 mL solution of $FeCl_3$ (7.8 g, 28 mmol) and $FeSO_4 \cdot H_2O$ (3.9 g, 14 mmol) at 343 K. A solution of NaOH (100 mL, 5 mol·L^{-1}) was added drop wise to precipitate the iron oxides. The amount of activated alumina was adjusted in order to obtain the Activated Alumina/iron oxide with weight ratio of 3:1. The 3:1 weight ratio of activated alumina to Fe oxide was chosen to avoid a decreasing in adsorption capacity of the composites due the high content of iron oxide. The obtained materials were dried in a digital dryer of (Carbolite, Aston Lane, Hope Sheffield, 5302RP, England) at 373 K for 3 h.

2.3. Characterization of Magnetic Alumina

The adsorbent(MANC) was characterized by using X-ray diffraction analysis (XRD), Scanning electron microscopy (SEM), transmission electron microscope (TEM), Quantum design SQUID magnetometer and BET surface area. X-ray diffraction analysis was carried out using X-ray diffractometer (Schimadzu-7000, USA) to evaluate the phase composition, XRD spectra were obtained with a 30 kW rotating anode diffractometer fitted with a copper target. XRD spectra were obtained between 20° and 80° (2θ) in continuous scan with 4°/min using the standard θ - 2θ geometry.

The morphology of the synthesized powders was studied using the SEM/EDX analysis which was performed using scanning electron microscope Jeol JMS 6360 LA, and the sample was prepared by coating with gold. Also, Jeol transmission electron microscope (TEM) with Max. Mag. 600 k× and resolution 0.2 nm was used to study the morphology of the prepared MANC. The samples were prepared by sonication for 30 min. Also the textural characteristics of MANC including surface area, pore size analyzer (BET) were determined using standard N2-adsorption techniques (Beckman Coulter, SA3100, USA). A quantum design SQUID magnetometer was used to obtain hysteresis loops of products at 25°C and in fields up to 15 kOe nano composite.

2.4. Adsorbate

A stock solution of Cd^{2+} was prepared (1000 mg·L^{-1}) by dissolving required amount of, $CdCl_2$·2.5H$_2$O in distilled water. The stock solution was diluted with distilled water to obtain desired concentration ranging from 100 to 1000 mg/L. pH was adjusted using 0.1 N HCl or 0.1 N NaOH. The remaining concentration of Cd^{2+} in each sample before and after adsorption was determined by using prodigy prism high dispersion inductive coupled plasma-atomic emission spectroscopy (ICP-AES, USA).

2.5. Batch Mode Adsorption Studies

The effects of experimental parameters such as, pH (2 - 9) using pH meter (Denver Instrument Co., USA), adsorbent dosage (0.25 - 3.0 g·L^{-1}), temperature (22°C - 55°C), initial Cd^{2+} ions concentration (100 - 1000 mg·L^{-1}) and contact time (0 - 300 min) on the adsorptive removal of Cd^{2+} ions were studied in a batch mode operation. For kinetic studies, 250 mL of Cd^{2+} solution of known different initial concentrations and pH = 6 was taken in a 250 mL screw-cap conical flask with a fixed adsorbent dosage (1 g·L^{-1}) and was agitated in a orbital shaker (yellow line OS 10 Control, Germany) for a contact time varied in the range 0 - 300 min at a speed of 250 rpm at 295 K. At various time intervals, the adsorbent was separated and the concentration of Cd^{2+} was determined. For adsorption isotherms, 250 ml of different initial Cd^{2+} ion concentrations (100 - 1000 mg·L^{-1}) were agitated with 1 g·L^{-1} adsorbent dosage in an orbital shaker at 250 rpm for 240 min. The adsorbent was separated and the metal under consideration was determined as mentioned previously. The concentration retained in the adsorbent phase (q_e, mg·g^{-1}) was calculated by using the following equation

$$q_e = \left(C_o - C_e\right)V\big/W \tag{1}$$

where C_o is the initial Cd^{2+} concentration and C_e is the Cd^{2+} concentration (mg·L^{-1}) at equilibrium, V is the volume of solution (L) and W is the mass of the adsorbent (g).

3. Results and Discussion

3.1. Characterization of Magnetic Alumina Nano Composite

3.1.1. X-Ray Analysis and Microstructure

The X-ray diffractograph in **Figure 1(a)** represent the pattern of pure alumina used in the experimental work. **Figure 1(b)** represent the XRD pattern of iron oxide indicating the cubic iron oxide phase (d = 2.5, 2.91, 2.07, 1.60) which can be related to maghemite or magnetite. In **Figure 1(c)** the XRD pattern indicates the crystal structure of MANC particles, where the peak at $2\theta = 35.55°$ is corresponding to iron oxide phase JCPDS card (19-0629) and that at $2\theta = 43.22°$, 62.5° and 66.1° are corresponding to alumina JCPDS card (10-173). It is obvious that these peaks appear broader indicating smaller crystallite size of the produced MANC particles than that of the starting alumina particles and this may be devoted to the method of preparation of the composite which cause dispersion of alumina particles. The crystallite size of both phases of produced composite can be calculated from peaks at $2\theta = 35.55°$ corresponding to iron oxide phase and peak at $2\theta = 43.22°$ corresponding to Al$_2$O$_3$ phase using Sherrer Equation (2).

$$L = 0.89\lambda\big/\beta\cos\theta \tag{2}$$

where β is the FWHM of diffraction peak, λ is the wave length of X-ray (0.154 nm), L is the crystallite size, and θ is the Bragg peak position.

The crystallite size was found to be 13.6 nm and 17.5 nm for both Al$_2$O$_3$ and Fe$_3$O$_4$ respectively.

3.1.2. Magnetic Properties

The magnetization curve obtained from the SQUID magnetometer is shown in **Figure 2**. The saturation magnetization Ms value is 12.12 emu/g, the remanance

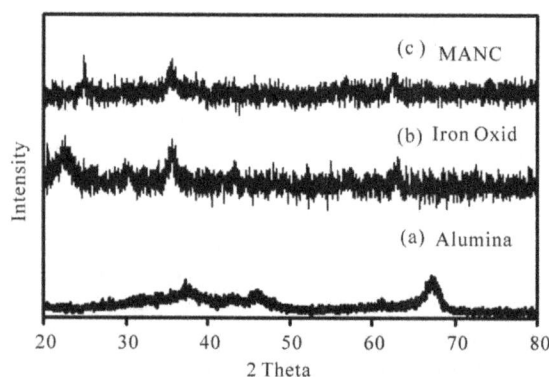

Figure 1. XRD diffractograph. (a) Alumina, (b) Iron oxide and (c) MANC.

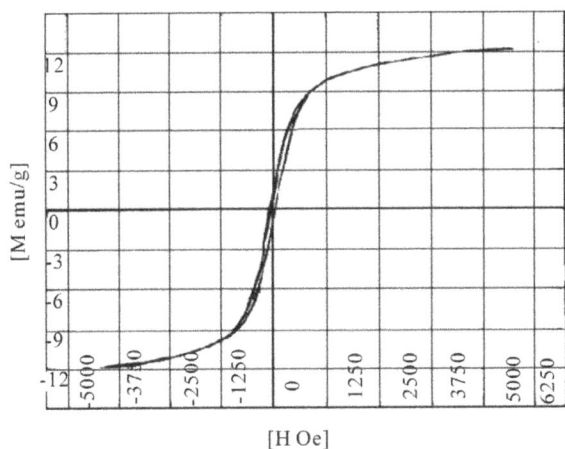

Figure 2. Hysteresis loop of MANC taken at 25°C.

Mr is 2.679 emu/g and the coericivity Hc is 44.89 Oe. This behaviour indicates that magnetic loaded alumina showed paramagnetic properties and its separation from treatment aqueous media can be achieved by applying magnetic field [27]. The magnetite content can be calculated on the basis of being the saturation magnetization of bulk magnetite is 94 emu/g. Because the magnetization of the produced composite is only due to Fe_3O_4, the mass percentage of Fe_3O_4 in MANC can be estimated from the value of Ms considering the saturation magnetization of Fe_3O_4 as reference saturation magnetization. So the content of Fe_3O_4 in MANC is 12.12 weight percent.

3.1.3. Morphology and Microstructure

The SEM image of MANC was shown in **Figure 3(a)**. The particles are nearly uniformly spherical and the particle size can be measured in the range of 25 - 29 nm. In addition the energy dispersive X-ray spectroscopy (EDX) is shown in **Figure 3(c)**, (EDX) shows the atomic percent of Al and Fe are 29.89 and 36.39 respectively. The EDX spectra reveal that the Fe_3O_4 phase is abundant on the surface of MANC deduced from the high atomic ratio of iron.

Transmission electron microscope (TEM) was used to investigate the nature of magnetic nanoparticles in the produced composite. The produced composite showed uniform spherical nanoparticles with diameter range between 7 and 18 nm, the magnetite nanoparticles are finally divided and well distributed within the composite as shown in **Figure 3(b)**. It was not possible to distinguish between the iron oxide and aluminium oxide phase's structure.

3.1.4. Textural Analysis

The surface area and the pore structure of MANC sample were determined from nitrogen isotherm analysis, as shown in **Figure 4**. MANC displayed a type-IV iso-

(a)

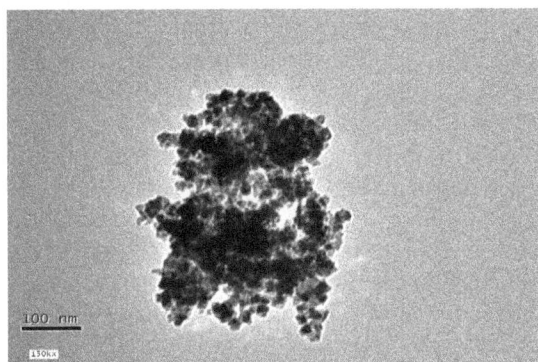

(b)

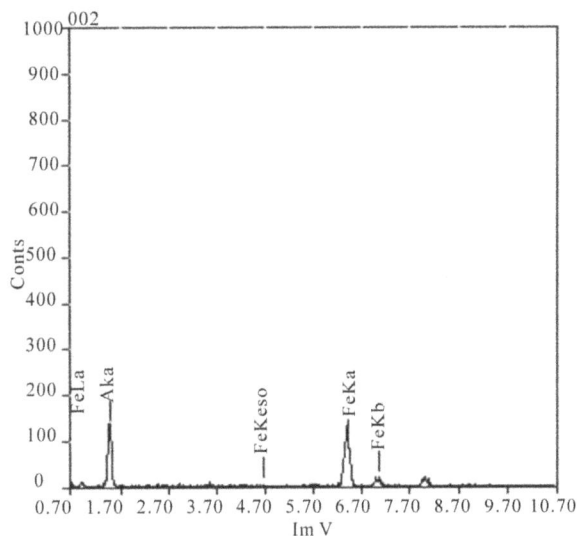

(c)

Figure 3. Morphology and microstructure of MANC. (a) SEM image; (b) TEM image and (c) EDX spectra.

therm, characteristic for mesoporous materials. Using these data, the specific surface area SBET was calculated to be 298 $m^2 \cdot g^{-1}$, the pore size 25 Å, and the total pore volume 0.29 $mL \cdot g^{-1}$.

3.2. The Effects of Experimental Parameters

3.2.1. Effect of pH

Metal-ion adsorption is known to be dependent on the pH

of solution. The effect of pH on the adsorption of Cd^{2+} by MANC was studied by varying the initial pH of the solution over the range of 2 - 9. The calculation from the solubility product equilibrium constant (Ksp) demonstrated that the best pH range of 2 - 9 for Cd^{2+} for adsorption [28]. **Figure 5** illustrated that removal efficiency increased with increase pH. The uptake of Cd^{2+} by MANC increased as the pH increased from 2 to 9. Although a maximum uptake was noted at a pH of 9, as the pH of the solution increased to >7, Cd^{2+} started to precipitate out from the solution. Therefore, experiments were not conducted over pH 7 to avoid precipitation [29]. The increased capacity of adsorption at pH > 7 may be a combination of both adsorption and precipitation on the surface of the adsorbent. It is considered that MANC had a maximum adsorption capacity at a pH = 6, if the precipitated amount is not considered in the calculation. Therefore, the optimum pH for Cd^{2+} was determined to be 6 and used in all experiments. The same trend has also been reported in the removal of Cd^{2+} ions by other adsorbent materials such as Bamboo charcoal [30] and crosslinked carboxymethyl konjac glucomannan [31].

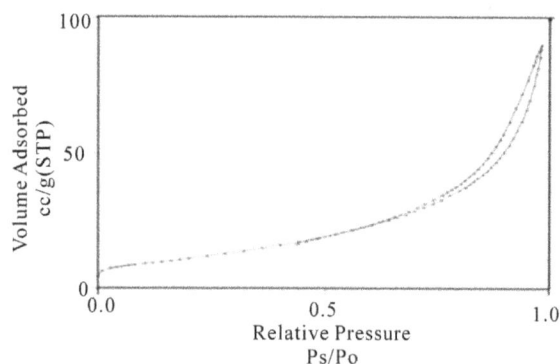

Figure 4. Nitrogen isotherm analysis of MANC.

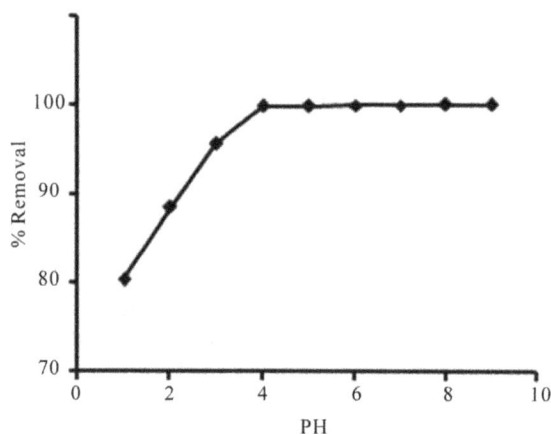

Figure 5. Effect of different pH on the removal of Cd^{2+} onto MANC (Initial Cd^{2+} ions concentration = 250 mg·L^{-1}, adsorbent dose = 1 g/l, solution temp. = 22°C ± 2°C, contact time = 60 min and agitation speed = 250 rpm.

The metal ions in the aqueous solution may undergo salvation and hydrolysis. The process involved for metal adsorption is as follows [32]:

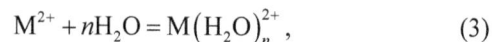

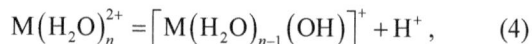

$$M^{2+} + nH_2O = M(H_2O)_n^{2+}, \qquad (3)$$

$$M(H_2O)_n^{2+} = \left[M(H_2O)_{n-1}(OH)\right]^+ + H^+, \qquad (4)$$

$$M^{2+} + nH_2O = K_a\left[M(H_2O)_{n-1}(OH)\right]^+ + H^+. \qquad (5)$$

The pK_a value for Cd^{2+} is 10.1. Perusal of the literature on metal speciation shows that the dominant species is $M(OH)_2$ at pH > 6.0 and M^{2+} and $M(OH)^+$ at pH < 6.0. Maximum removal of metal was observed at pH 6 for adsorption. On further increase of pH adsorption decreases probably due to the formation of hydroxide of cadmium because of chemical precipitation [33-35].

3.2.2. Effect of Adsorbent Dosage

Adsorbent dosage is an important parameter because it determines the capacity of an adsorbent for a given initial concentration of the adsorbate. The effect of adsorbent dosage was studied on Cd^{2+} ions removal from aqueous solutions by varying the amount of MANC from 0.25 to 3.0 g·L^{-1}, while keeping other parameters (pH, agitation speed, temperature, initial Cd^{2+} ion concentration and contact time) constant. **Figure 6** showed that the percent removal of Cd^{2+} ion increased from 96.06% to 99.99% for MANC as the adsorption dosage was increased from 0.25 to 3 g·L^{-1}. On the other hand, the amount adsorbed per unit mass of the adsorbent decreased considerably. The decrease in unit adsorption with increase in the dosage of adsorbent was due to adsorption sites remaining unsaturated during the adsorption process [36]. For the rest of the study 1 g·L^{-1} adsorbent dosage is considered as optimum dosage for cadmium removal using MANC.

3.2.3. Effect of Contact Time and Initial Cd^{2+} Ions Concentration

Figure 7 shows the effect of contact time on the

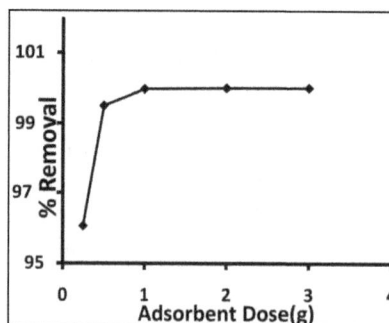

Figure 6. Effect of adsorbent dose on the removal of Cd^{2+} onto MANC (Initial Cd^{2+} ions concentration = 250 mg·L^{-1}, pH = 6, solution temp. = 22°C ± 2°C, contact time = 60 min and agitation speed = 250 rpm).

adsorbed amount of Cd^{2+} by MANC from solutions with different initial concentrations of Cd^{2+} (100 - 1000 mg/ L) at 22°C. The adsorption increased sharply with contact time in the first 40 min and attained equilibrium within 240 min. It is also clear from **Figure 7(a)** that by increasing the initial Cd^{2+} ions concentration the percentage of Cd^{2+} removal decreased, although the actual amount of Cd^{2+} adsorbed per unit mass of MANC increased as shown in **Figure 7(b)**. At low initial solution concentration, the surface area and the availability of adsorption sites were relatively high, and the Cd^{2+} ions were easily adsorbed. At higher initial solution concentration, the total available adsorption sites are limited, thus resulting in a decrease in percentage removal of Cd^{2+} ions. The increased in the equilibrium adsorption amount of Cd^{2+} at higher initial concentration can be-attributed to enhance the driving force. The equilibrium adsorption amount of Cd^{2+} was found to increase from 100 to 501 mg·g^{-1} as the initial concentration increased from 100 to 1000 mg·L^{-1}.

The same trend has also been reported in the removal of Cd^{2+} ions by Bamboo charcoal [30]. The higher adsorption capacity and rate of Cd^{2+} on MANC indicated its suitability to treat wastewater polluted with Cd^{2+}.

3.2.4. Effect of Temperature

Temperature plays key roles on the adsorption process [37]. First, increasing the temperature decreases the viscosity of the solution which, in turn, enhances the rate of diffusion of the adsorbate molecules across the external boundary layer of the adsorbent and resulted in higher adsorption. Second, changing the temperature may affect the equilibrium adsorption capacity of the adsorbent. For instance, the adsorption capacity will decrease upon increasing the temperature for an exothermic reaction; while it will increase for an endothermic one. Hence, a study of the temperature-dependent adsorption processes provides valuable information about the standard Gibbs free energy, enthalpy and entropy changes accompanying adsorption. In this study, a series of experiments were conducted at 22°C, 35°C, 45°C and 55°C to investigate the effect of temperature on Cd^{2+} adsorption and determine thermodynamic parameters. **Figure 8** shows the amount of Cd^{2+} adsorbed onto MANC nano adsorbents at different temperatures. As seen in the figure, the amount of Cd^{2+} adsorbed increases as the temperature increased. This increase suggests that the adsorption process is an endothermic one. The increase in the Cd^{2+} adsorption with temperature may be due to the increase in ions mobility, which in turn increases the number of ions that interacted with active sites at the adsorbent surfaces. Similar trends are also observed by other researchers for aqueous phase adsorption [38,39].

3.2.5. Effect of Loading of Alumina by Iron Oxide

The adsorption tests showed that the magnetic alumina nano-composite possesses the same adsorption capacity as pure activated alumina, suggesting that the presence of Fe oxide in the composite is not inhibiting the adsorption of metals. **Figure 9** compares the adsorption capacity of magnetic alumina nanocomposite and activated alumina. From this figure, the composites showed high adsorption capacities for the Cd^{2+} in aqueous solution and, more

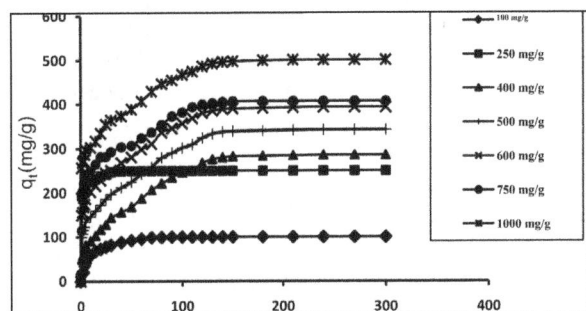

(a)

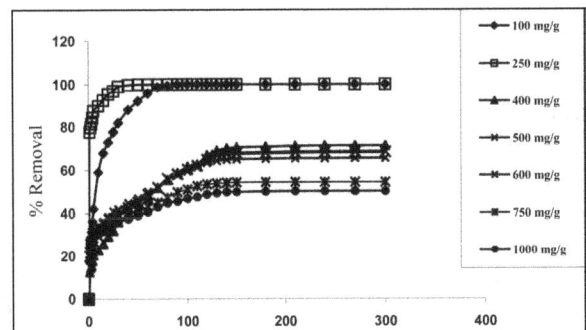

(b)

Figure 7. Effect of contact time and initial Cd^{2+} ions concentration on: (a) the removal of Cd^{2+} and (b) adsorption capacity onto MANC (adsorbent dose = 1 g/L, pH = 6, solution temp. = 22°C ± 2°C, contact time = 60 min and agitation speed = 250 rpm).

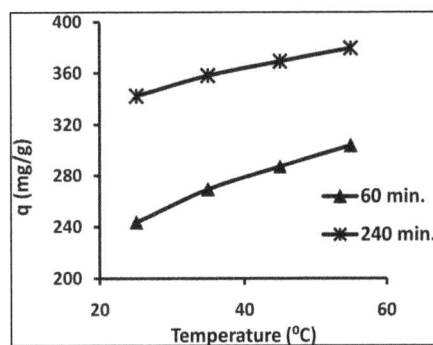

Figure 8. Effect of temperature on adsorption capacity onto MANC (Initial Cd^{2+} ions concentration 500 mg·L^{-1} adsorbent dose = 1g/L, pH = 6, agitation speed = 250 rpm and contact time = 60 min, 240 min).

important, no reduction of the adsorption was produced by the formation of the composite.

3.3. Adsorption Isotherms

Adsorption isotherms describe qualitative information on the nature of the solute-surface interaction as well as the specific relation between the concentration of adsorbate and its degree of accumulation onto adsorbent surface at constant temperature. Adsorption isotherms are critical in optimizing the use of adsorbents, and the analysis of the isotherm data by fitting them to different isotherm models is an important step to find the suitable model that can be used for design purposes [40]. There are several isotherm equations available for analyzing experimental sorption equilibrium data, the most famous adsorption models for single-solute systems are the Langmuir and Freundlich models. The experimental data obtained in the present work was tested with the Langmuir, Freundlich and Temkin, isotherm models. Linear regression is frequently used to determine the best-fitting isotherm, and the applicability of isotherm equations is compared by judging the correlation coefficients.

3.3.1. Langmuir Isotherm

The theoretical Langmuir sorption isotherm [41] is valid for adsorption of a solute from a liquid solution as monolayer adsorption on a surface containing a finite number of identical sites. The model is based on several basic assumptions: 1) the sorption takes place at specific homogenous sites within the adsorbent; 2) once a Cd^{2+} occupies a site; 3) the adsorbent has a finite capacity for the adsorbate (at equilibrium); 4) all sites are identical and energetically equivalent. Langmuir isotherm model assumes uniform energies of adsorption onto the surface without transmigration of adsorbate in the plane of the surface. Therefore, the Langmuir isotherm model was chosen for estimation of the maximum adsorption capacity corresponding to complete mono-layer coverage on the sorbent surface. The Langmuir isotherm model is

represented in the linear form as follows [41]

$$C_e/q_e = 1/(k_L Q_m) + C_e/Q_m \qquad (6)$$

where K_L is the Langmuir adsorption constant ($L \cdot mg^{-1}$) and Q_m is the theoretical maximum adsorption capacity ($mg \cdot g^{-1}$). **Figure 10** shows the Langmuir (C_e/q_e vs. C_e) plot for adsorption of Cd^{2+} ions. The value of Q_m and K_L constants and the correlation coefficient for Langmuir isotherm are presented in **Table 1**. The correlation coefficient was high ($R^2 = 0.976$) as shown in **Table 1**. This result indicates that the experimental data fitted well to Langmuir sorption isotherm with maximum adsorption capacity 625 mg/g.

The essential characteristics of Langmuir dimensionless constant separation factor or equilibrium parameter, R_L, which is defined by the following Equation [42]:

$$R_L = 1/1 + (K_L C_o) \qquad (7)$$

where C_o is the initial Cd^{2+} ions concentration, mg/L. The R_L parameter is considered as more reliable indicator of the adsorption. There are four probabilities for the R_L value: 1) for favorable adsorption $0 < R_L < 1$, 2) for unfavorable adsorption $R_L > 1$, 3) for linear adsorption $R_L = 1$ and 4) for irreversible adsorption $R_L = 0$. In the present study, the values of R_L (**Table 2**) are observed to be in the range 0 - 1, indicating that the adsorption process is favorable for MANC.

3.3.2. The Freundlich Isotherm

The Freundlich isotherm model is the earliest known relationship describing the sorption process [43]. The model applies to adsorption on heterogeneous surfaces with interaction between adsorbed molecules and the application of the Freundlich equation also suggests that sorption energy exponentially decreases on completion of the sorption centers of an adsorbent. This isotherm is an empirical equation which can be employed to describe heterogeneous systems and is expressed in the linear form as follows

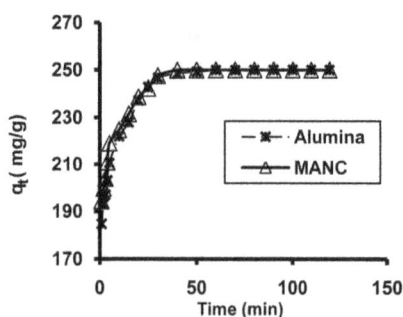

Figure 9. Effect of iron oxide loading on the adsorption of Cd^{2+} (Initial Cd^{2+} ions concentration = 250 mg·L^{-1} adsorbent dose = 1 g/L, pH = 6, solution temp. = 22°C ± 2°C, contact time = 60 min and agitation speed = 250 rpm).

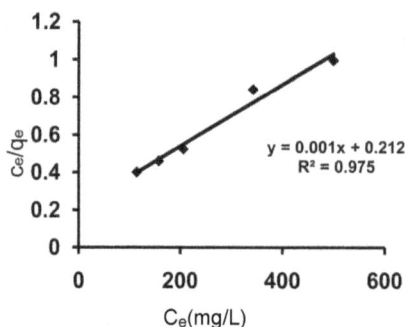

Figure 10. Langmuir isotherm plot for adsorption of Cd^{2+} onto MANC.

Table 1. Langmuir, Freundlich and Tempkin isotherm parameters and correlation coefficients for the adsorption of Cd^{2+} from aqueous solutions onto magnetic alumina nano-composite.

Langmuir			Freundlich			Temkin			
Q_m (mg·g^{-1})	K_L (L·mg^{-1})	R^2	K_F (L·mg^{-1})	$1/n$	R^2	K_T (L·mg^{-1})	B_T	b_T (J·mol^{-1})	R^2
625	0.0075	0.976	61.72	0.335	0.921	0.083	130.79	18.75	0.931

$$\log q_e = \log K_F + 1/n \log C_e \qquad (8)$$

where K_F is the Freundlich constant (L·g^{-1}) related to the bonding energy. K_F can be defined as the adsorption or distribution coefficient and represents the quantity of Cd^{2+} adsorbed onto adsorbent for unit equilibrium concentration. $1/n$ is the heterogeneity factor and n is a measure of the deviation from linearity of adsorption. Its value indicates the degree of non-linearity between solution concentration and adsorption as follows: if the value of n is equal to unity, the adsorption is linear; if the value is lower than unity, this implies that adsorption process is chemical; if the value is higher unity adsorption is a favorable physical process [44].

Figure 11 shows the plot of $\log(q_e)$ versus $\log(C_e)$ to generate the intercept value of K_F and the slope of $1/n$ (**Table 1**). The value of n is higher than unity, indicating that adsorption of Cd^{2+} onto MANC is a favorable physical process [44]. The correlation coefficients, $R^2 = 0.921$. This result indicates that the experimental data did not fit well to Freundlich model.

3.3.3. The Temkin Isotherm

Temkin isotherm model contains a factor that explicitly takes into account adsorbing species-adsorbate interactions [45]. This model assumes the following: 1) the heat of adsorption of all the molecules in the layer decreases linearly with coverage due to adsorbate-adsorbent interactions, and 2) adsorption is characterized by a uniform distribution of binding energies, up to some maximum binding energy. The derivation of the Temkin isotherm assumes that the fall in the heat of sorption is linear rather than logarithmic, as implied in the Freundlich equation. The Temkin isotherm has commonly been applied in the following form (Equation (9))

$$q_e = RT/b_T \ln \left(K_T C_e \right) \qquad (9)$$

Equation (9) can be linearized as

$$q_e = B_T \ln K_T + B_T \ln C_e \qquad (10)$$

where $B_T = RT/b_T$, T is the absolute temperature in degree K, R the universal gas constant, 8.314 J·mol^{-1}·K^{-1}, K_T is the equilibrium binding constant (Lmg^{-1}), B_T is related to the heat of adsorption and b_T is Temkin constant (J·mol^{-1}). A plot of q_e vs $\ln C_e$ at studied temperature is given in **Figure 12**. The constants obtained for Temkin isotherm are illustrated in **Table 1**. Examination of the data shows that the Temkin isotherm is not applicable to

Table 2. R_L values for different Cd^{2+} ions concentrations.

Initial Cd^{2+} concentrations (mg/L)	100	250	400	500	600	750	1000
R_L	0.571	0.348	0.25	0.267	0.182	0.151	0.118

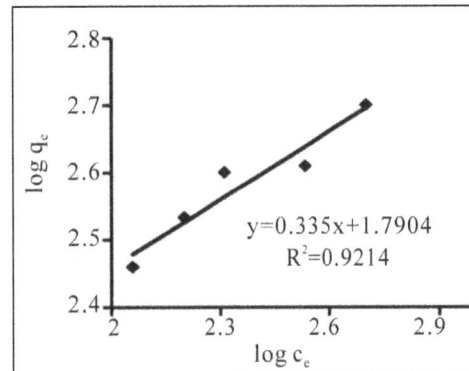

Figure 11. Freundlich isotherm plot for adsorption of Cd^{2+} onto MANC.

the Cd^{2+} adsorption onto MANC because of the small correlation coefficient.

Table 1 summarizes all the constants and correlation coefficients, R^2 of the three isotherm models. The Langmuir model yielded the best fit with R^2 which were higher than 0.97. Confirmation of the experimental data into the Langmuir isotherm equation indicated the homogeneous nature of magnetic alumina nano-composite surface, i.e., each Cd^{2+} ion/magnetic alumina nano composite (MANC) adsorption had equal adsorption activation energy. The results also demonstrated the formation of monolayer coverage of cadmium ions at the outer surface of MANC. The good correlation coefficients showed that Langmuir model is more suitable than Freundlich and Temkin for adsorption equilibrium of cadmium ions onto MANC. Values of the adsorption capacity of other adsorbents from the literature are given in **Table 3** for comparison [46-50]. It is clear from this table that the adsorption capacity of MANC for Cd^{2+} is higher than the other adsorbents.

3.4. Adsorption Kinetics

Kinetic models can be helpful to understand the mechanisms of metal adsorption and evaluate performance of the adsorbents for metal removal. The kinetics of Cd^{2+} adsorption onto MANC is required for selecting optimum operating conditions for the full-scale batch process. The kinetic parameters, which are helpful for the prediction of adsorption rate, give important information for designing and modeling the adsorption processes. Thus, Lagergren pseudo-first-order [51], pseudo-second-order [52], Elovich [53-55] and intraparticle diffusion [56,57] kinetic models were used for the adsorption of Cd^{2+} onto MANC. The conformity between experimental data and the model-predicted values was expressed by the correlation coefficients (R^2, values close or equal to 1, the relatively higher value is the more applicable model).

3.4.1. The Pseudo First-Order Equation

The Lagergren pseudo-first-order model [51] is the earliest known equation describing the adsorption rate based on the adsorption capacity, which can be expressed in a linear form as

$$\ln\left(q_e - q_t\right) = \ln\left(q_e\right) - k_1\left(t\right) \qquad (11)$$

where q_e and q_t are the amount of cadmium ions adsorbed ($mg \cdot g^{-1}$) on the MANC at the equilibrium and at time t, respectively, and k_1 is the rate of constant adsorp-

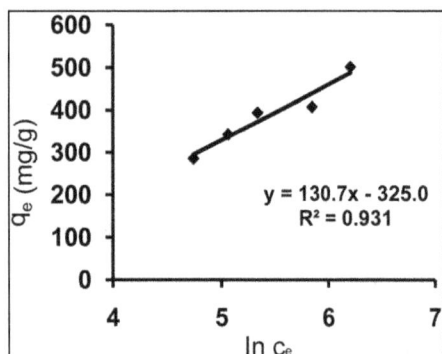

of $\ln(q_e - q_t)$ versus t for different concentrations of the tion (min^{-1}). Values of k_1 were calculated from the plots Cd^{2+} ion as shown in **Figure 13**. The values k_1 and q_e are given in **Table 4**.

The pseudo-first order equation was used to correlate the experimental data, based on the following mechanistic scheme:

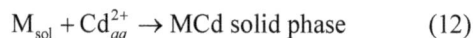

$$M_{sol} + Cd^{2+}_{aq} \rightarrow MCd \text{ solid phase} \qquad (12)$$

One cadmium ion was assumed to sorb onto one adsorption site of the MANC.

As noticed from **Table 4** pseudo-first order kinetic predicts a lower value of the equilibrium adsorption capacity than the experimental value. Hence, this equation cannot provide an accurate fit of the experimental data.

3.4.2. The Pseudo Second-Order Equation

A linear form of pseudo second-order model is shown in Equation (13)

$$t/q_t = 1/k_2 q_e^2 + 1/q_e\left(t\right) \qquad (13)$$

where k_2 is the rate constant of pseudo second-order adsorption ($g \cdot mg^{-1} \cdot min^{-1}$). The constants can be obtained

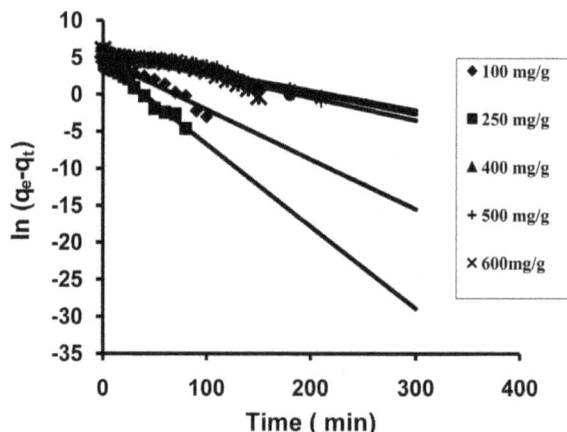

Figure 12. Temkin isotherm plot for adsorption of Cd^{2+} on to MANC.

Table 3. Materials used as adsorbent for Cd^{2+} ions.

Adsorbent	Capacity ($mg \cdot g^{-1}$)	Reference
Present study	625	-
Bamboo charcoal	12.08	[30]
Crosslinked carboxymethyl konjac glucomannan	23.6	[31]
Natural corncobs	5.09	[46]
Oxidized corncobs	55.2	[46]
Alcaligenes eutrophus	122	[47]
Leaves, platanus	110	[48]
Soil, haldimand	99.9	[49]
Yeast, baker's	91.74	[50]

Figure 13. First-order plots of Cd^{2+} adsorption onto MANC.

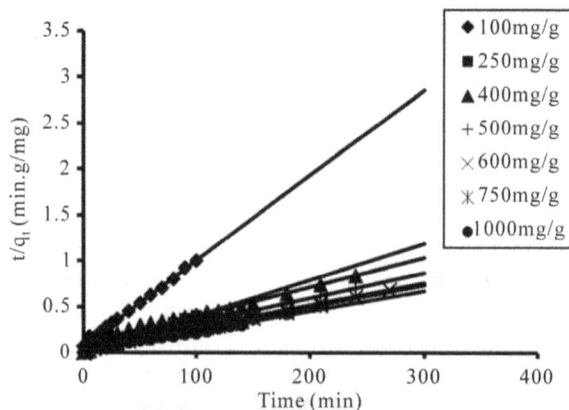

Figure 14. Second-order plots Cd^{2+} adsorption onto MANC.

Table 4. The pseudo first-order, second-order and Elovich kinetic parameters for Cd^{2+} at different initial cadmium ions concentration by MANC.

C_o (mg/L)	$q_{e\,exp.}$ (mg·g^{-1})	1st-order kinetic model			2nd-order kinetic model				Elovich kinetics model		
		K_1 (min^{-1})	$q_{e\,calc.}$ (mg·g^{-1})	R^2	K_2 (g·mg^{-1}·min^{-1})	$q_{e\,calc}$ (mg·g^{-1})	h	R^2	α (mg·g^{-1}·min^{-1})	β (g·mg^{-1})	R^2
100	100	0.067	104.64	0.958	1.17×10^{-3}	107.5	13.53	0.994	46.83	0.055	0.979
250	250	0.1111	80.7	0.977	5.71×10^{-3}	250	356.9	1	208.8×10^6	0.084	0.943
400	285.6	0.0263	314.4	0.952	1.37×10^{-4}	312.5	13.35	0.974	66.2	0.022	0.876
500	342.25	0.0277	336.84	0.946	1.79×10^{-4}	357.1	22.88	0.985	201.8	0.022	0.884
600	393.6	0.0279	327.9	0.959	2.04×10^{-4}	416.7	35.34	0.993	593.1	0.021	0.893
750	407.25	0.0305	282.9	0.901	2.98×10^{-4}	416.7	51.74	0.993	1589.	0.022	0.896
1000	501	0.0281	331.66	0.943	6.21×10^{-4}	454.5	128.2	0.993	5632.6	0.021	0.892

from plotting (t/q_t) versus t **Figure 14**. The values k_2, q_e and the initial adsorption rate h $\left(k_2 q_e^2\right)$ are given in **Table 4**.

The pseudo-second order model assumes that cadmium is sorbed onto two active sites:

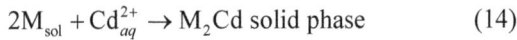

$$2M_{sol} + Cd_{aq}^{2+} \rightarrow M_2Cd \text{ solid phase} \qquad (14)$$

The equilibrium adsorption capacities, q_e, obtained with this model are slightly more reasonable than those of the pseudo-first order when comparing predicted results with experimental data. Moreover, the values of R^2 also indicated that this equation produced better results (**Table 4**) at all concentrations and adsorbent doses, R^2 values for pseudo-second-order kinetic model were found to be between 0.974 and 1. This indicates that the Cd^{2+} MANC adsorption system obeys the pseudo-second order kinetic model for the entire sorption period.

3.4.3. Elovich Kinetic Equation

The Elovich equation is of general application to chem. isorptions kinetics. The equation has been applied satisfactorily to some chemisorption processes and has been found to cover a wide range of slow adsorption rates. The same equation is often valid for systems in which the adsorbing surface is heterogeneous, and is formulated as:

$$q_t = 1/\beta \ln\left(\alpha\beta\right) + 1/\beta \ln t \qquad (15)$$

where α (mg/g·min) is the initial adsorption rate and β is related to the extent of surface coverage and the activation energy involved in chemisorption (g/mg).

The Elovich equation assumes that the active sites of adsorbent are heterogeneous [58], and therefore exhibit different activation energies for chemisorption. Teng and Hsieh [59] proposed that constant α is related to the rate of chemisorptions and β is related to the surface coverage.

The Elovich equation is based on a general second-order reaction mechanism for heterogeneous adsorption processes [58].

Plot of q_t versus $\ln(t)$ should yield a linear relationship if the Elovich is applicable with a slope of $(1/\beta)$ and an intercept of $(1/\beta \ln(\alpha\beta)$ (**Figure 15**). The Elovich constants obtained from the slope and the intercept of the straight line reported in **Table 4**. The correlation coefficients R^2 are very wavy and ranged from low value to high value without definite role (**Table 4**).

3.4.4. Intra-Particle Diffusion Model

The adsorption mechanism of adsorbate onto adsorbent follows three steps: film diffusion, pore diffusion and intra-particle transport. The slowest of three steps controls the overall rate of the process. Generally, intra-particle diffusion is often rate-limiting in a batch reactor, while for a continuous flow system film diffusion is more likely the rate-limiting step. In order to investigate the possibility of intra-particle diffusion resistance affecting the adsorption intra-particle diffusion model [56] was explored

$$q_t = K_i t^{0.5} + I \qquad (16)$$

where K_i (mg·g^{-1}·min$^{-0.5}$) is the intra-particle diffusion rate constant. **Figure 16** represents a plot of q_t vs $t^{0.5}$, it shows two separate regions the initial part is attributed to the bulk diffusion while the final part to the intra-particle diffusion. Values of I give an idea about the thickness of boundary layer (**Table 5**), *i.e.* the larger the intercept the greater is the boundary layer effect [60]. The data indicate that intra-particle diffusion controls the adsorption rate. Simultaneously, external mass transfer resistance cannot be neglected although this resistance is only significant for the initial period of time [61].

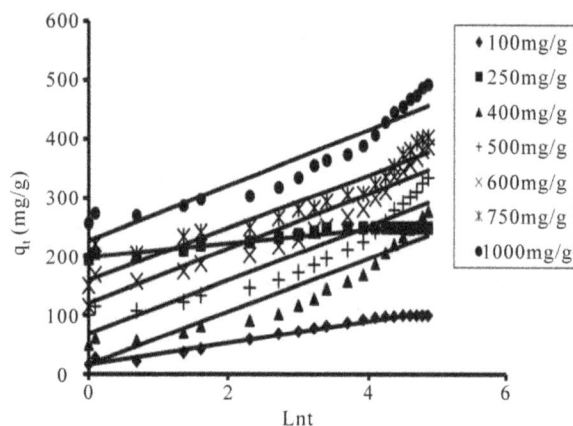

Figure 15. Elovich plots of Cd^{2+} adsorption onto MANC.

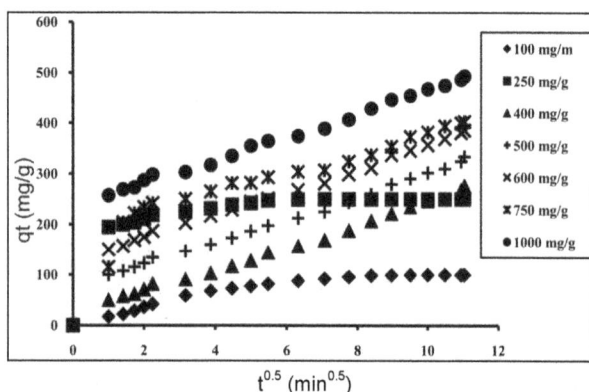

Figure 16. Intra-particle diffusion plots of Cd^{2+} adsorption onto MANC.

Table 5. The intra-particle diffusion parameters for Cd^{2+} at different initial cadmium ions concentration by MANC.

Cd^{2+} (mg/L)	Ki_1	I_1	R_1^2	Ki_2	I_2	R_2^2
100	15.48	3.2	0.979	2.23	77.25	0.78
250	28.08	123.63	0.52	0.11	248.88	0.52
400	22.536	20.15	0.956	24.87	2.84	0.993
500	28.35	53.05	0.853	25.33	48.92	0.993
600	34.168	85.95	0.76	25.38	104	0.994
750	41.52	102.51	0.71	22.97	150.06	0.985
1000	43.73	157.33	0.61	25.45	215.45	0.992

The data exhibit multi-linear plots, revealing that the process is governed by two or more steps (**Figure 16**). The first linear portion (phase I) at all concentrations, can be attributed to the immediate utilization of the most readily available sorbing sites on the sorbent surface. Phase II may be attributed to very slow diffusion of the Cd^{2+} from the MANC surface site into the inner pores. Thus, initial portion of Cd^{2+} ions sorption by MANC may

be governed by the initial intra-particle transport of Cd^{2+} controlled by surface diffusion process and the later part controlled by pore diffusion. The values of ki_1 and ki_2 (diffusion rate constants for phases I and II, respectively) obtained from the slope of linear plots are listed in **Table 5**.

3.5. Thermodynamics Studies

Thermodynamic parameters were evaluated to confirm the adsorption nature of the present study.

The variation in temperature, influencing the distribution of adsorbate between solid and liquid phases, was examined in the range 295 - 328 °K. Moreover the increase in Cd^{2+} sorption with a rise in temperature can be explained on the basis of thermodynamic parameters such as change in enthalpy ($\Delta H°$), entropy ($\Delta S°$) and free energy ($\Delta G°$). The change in enthalpy ($\Delta H°$) and entropy ($\Delta S°$) are calculated by using the van't Hoff equation [62].

$$\ln k_c = \Delta S°/R - \Delta H°/RT \qquad (17)$$

where $k_c = F_e/(1 - F_e)$, and $F_e = (C_o - C_e)/C_o$; is the fraction adsorbed at equilibrium, while T is the temperature in degree K and R is the gas constant [8.314 (J/mol K)].

The plot of $\ln k_c$ vs $1/T$ gives a straight line with acceptable coefficient (R^2) as shown in **Figure 17**. From the slope and the intercept of van't Hoff plots, the values of $\Delta H°$ and $\Delta S°$ were computed, while the Gibbs free enrgy change $\Delta G°$ was calculated using the following equation [63]:

$$\Delta G° = -RT \ln k_c \qquad (18)$$

The thermodynamic parameters for the sorption of cadmium ions onto MANC at various temperatures were calculated and summarized in **Table 6**. The positive value of $\Delta H°$ indicates that the studied sorption processes are endothermic in nature. Furthermore the negative values of $\Delta G°$ demonstrate the spontaneous behavior of the sorption processes [63]. The decrease in the value of $\Delta G°$ with the increase of temperature shows that the reaction

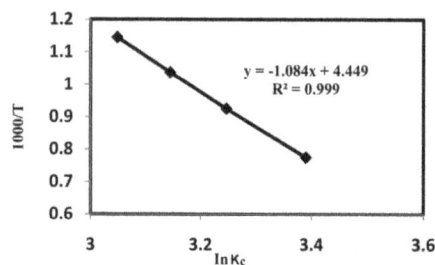

Figure 17. Effect of temperature on Cd^{2+} kinetic sorption for MANC (Initial Cd^{2+} ions concentration = 500 mg·L^{-1} adsorbent dose = 1 g/l, pH = 6, agitation speed = 250 rpm and contact time = 60 min, 240 min.

Table 6. Thermodynamic parameters and activation energy for Cd^{2+} sorption onto MANC.

Temperature (K)	ΔH° (kJ·mol^{-1})	ΔS° (kJ·mol^{-1})	ΔG° (kJ·mol^{-1})	E_a (kJ·mol^{-1})
295			−1.90	11.46
308	9.0156	36.99	−2.37	11.58
318			−2.74	11.66
328			−3.12	11.74

is more spontaneous at higher temperature which indicates that the sorption processes are favored by the increase in temperature [64]. It is noteworthy that adsorption process with ΔG° values between −20 and 0 kJ/mol corresponds to spontaneous physical process, while that with values between −80 and −400 kJ/mol corresponds to chemisorptions [65,66]. From the ΔG° values obtained in this study, it can be deduced that the adsorption mechanism is dominated by physisorption. This also is supported by the fact that $\Delta H^{\circ} < 40$ kJ mol, indicating physical adsorption process [65].

Finally, the positive value of ΔS° suggest that the increased randomness at the solid-solution interface during the sorption process. The adsorbed solvent molecules which are displaced by the adsorbate species gain more translational entropy than ions lost by adsorbate, thus allowing for prevalence of randomness in the system [67]. Normally, adsorption of gases leads to a decrease in entropy due to orderly arrangement of the gas molecules on a solid surface. However, the same may not be true for the complicated system of sorption from solution [68].

Energy of activation was calculated and illustrated in **Table 6** according to a relationship between E_a and ΔH° for reactions in solution by the following equation [69]:

$$E_a = \Delta H^{\circ} + RT \qquad (19)$$

Energies of activation below 42 kJ·mol^{-1} generally indicate diffusion-controlled processes and higher values represent chemical reaction processes [54]. In terms of E_a, diffusion or transport controlled reactions are those governed by mass transfer or diffusion of the adsorbate from the bulk solution to the adsorbent surface and can be described using the parabolic rate law [70]. Conversely, the reaction is surface controlled if the reaction between the adsorbate and adsorbent is slow compared with the transport or diffusion of the adsorbate to the adsorbent. For surface controlled reactions, the concentration of the adsorbate next to the adsorbent surface is equal to the concentration of the adsorbate in the bulk solution and the kinetic relationship between time and adsorbate concentration should be linear [71]. In our study, the small value of the activation energy below 42 kJ·mol^{-1} confirms the fact that the process of the removal Cd^{2+} using MANC is diffusion controlled.

3.6. Single-Stage Batch Adsorber

Adsorption isotherm studies can also be used to predict the design of single stage batch adsorption systems [72-74]. The schematic diagram for a single-stage adsorption process is shown in **Figure 18**. The solution to be treated contains V (L) of water and an initial Cd^{2+} concentration C_o (400 mg/L) which is to be reduced to C_e in the adsorption process. In the treatment stage, the amount of adsorbent W (g) added is added to solution and the Cd^{2+} ions concentration on the solid changes from $q_0 = 0$ to q_e. The mass balance for the dye in the single stage is given by

$$V(C_o - C_e) = W(q_e - q_o) = Wq_e \qquad (20)$$

The Langmuir isotherm data may now be applied to Equation (19) since the Langmuir isotherm gave the best fit to experimental data.

$$W/V = (C_o - C_e)/q_e = C_o - C_e /\left[(Q_m K_L C_e)/(1 + K_L C_e)\right] \qquad (21)$$

Figure 19 shows a series of plots derived from Equation (20) for the adsorption of Cd^{2+} ions on the adsorbent and depicts the amount of effluent which can be treated to reduce the cadmium ions content by 90%, 80%, 70%, 60% and 50% using various masses of the adsorbent.

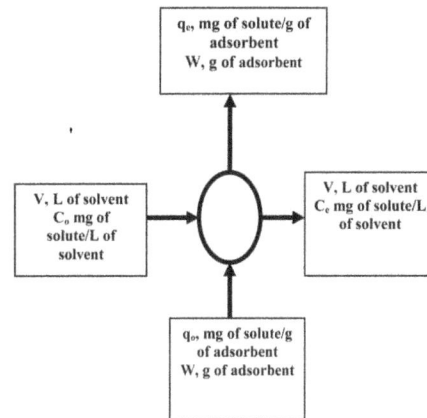

Figure 18. A single-stage batch adsorber.

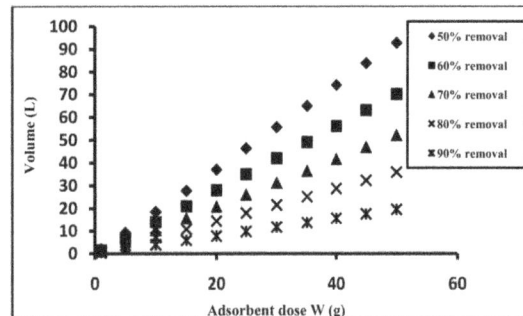

Figure 19. Volume of effluent treated against adsorbent dose for different percentages of Cd^{2+} removal.

4. Conclusions

The prepared Magnetic composite MANC was found to be in nano-scale with high surface area 298 $m^2 \cdot g^{-1}$ and its magnetic behavior was paramagnetic, which makes its removal from treatment media by applying magnetic field possible. These properties made this composite an excellent candidate for removal of cadmium ions from water. Removal of cadmium ions on Magnetic alumina nano composite (MANC) is pH dependent and the maximum removal was attained at pH 6. The equilibrium adsorption is practically achieved through a time of 240 min. It was also a function of initial adsorbent dose and cadmium ions concentration, Also adsorption equilibrium data follows; Langmuir sorption isotherm with maximum adsorption capacity 625 $mg \cdot g^{-1}$. The data indicate that the adsorption kinetics follow the pseudo-second-order model with intraparticle diffusion as one of the rate determining steps. The adsorption process was spontaneous and increased with increase in temperature showing endothermic nature of the adsorption.

REFERENCES

[1] L.-G. Yan, X.-Q. Shan, B. Wen and G. Owensb, "Adsorption of Cadmium onto Al13-Pillared Acid-Activated Montmorillonite," *Journal of Hazardous Materials*, Vol. No. 1-3, 156, 2008, pp. 499-508. doi:10.1016/j.jhazmat.2007.12.045

[2] J. Wase and C. Forster, "Biosorbents for Metal Ions," Taylor & Francis Inc., Bristol, 1997.

[3] E. W. Shin, K. G. Karthikeyan and M. A. Tshabalala, Adsorption Mechanism of Cadmium on Juniper Bark and Wood," *Bioresource Technology*, Vol. 98, No. 3, 2007, pp. 588-594. doi:10.1016/j.biortech.2006.02.024

[4] ATSDR, "Toxicological Profile for Cadmium," Agency for Toxic Substances and Disease Registry, Department of Health and Human Services, USA, 1999.

[5] V. K. Gupta, M. Gupta and S. Sharma, "Process Development for the Removal of Lead and Chromium from Aqueous Solutions Using Red Mud—An Aluminium Industry Waste," Water *Research*, Vol. 35, No. 5, 2001, pp. 1125-1134. doi:10.1016/S0043-1354(00)00389-4

[6] S. J. T. Pollard, G. D. Fowler, C. J. Sollars and R. Perry, "Low-Cost Adsorbents for Waste and Wastewater Treatment," *Science of the Total Environment*, Vol. 116, No. 1-2, 1992, pp. 31-52. doi:10.1016/0048-9697(92)90363-W

[7] E. Erdem, N. Karapinar and R Donat, "The Removal of Heavy Metal Cations by Natural Zeolites," *Journal of Colloid* and *Interface Science*, Vol. 280, No. 2, 2004, pp. 309-314. doi:10.1016/j.jcis.2004.08.028

[8] O. Abollino, M. Aceto, M. Malandrino, C. Sarzanini and E. Mentatsi, "Adsorption of Heavy Metals on Na-Montmorillonite. Effect of pH and Organic Substances," *Water Research*, Vol. 37, No. 7, 2003, pp. 1619-1627. doi:10.1016/S0043-1354(02)00524-9

[9] K. A. Matis, A. I. Zouboulis, G. P. Gallios, T. Erwe and C. Blocher, "Application of Flotation for the Separation of Metal-Loaded Zeolites," *Chemosphere*, Vol. 55, No. 1, 2004, pp. 65-72. doi:10.1016/j.chemosphere.2003.11.030

[10] L. C. A. Oliveira, R. V. R. A. Rios, J. D. Fabris, K. Sapag, V. K. Garg and R. M. Lago, "Clay-Iron Oxide Magnetic Composites for the Adsorption of Contaminants in Water," *Applied Clay Science*, Vol. 22, No. 4, 2003, pp. 169-177. doi:10.1016/S0169-1317(02)00156-4

[11] D. Zamboulis, S. I. Pataroudi, A. I. Zouboulis and K. A. Matis, "The Application of Sorptive Flotation for the Removal of Metal Ions," *Desalination*, Vol. 162. No. 3, 2004, pp. 159-168. doi:10.1016/S0011-9164(04)00039-6

[12] F. Pagnanelli, F. Veglio and L. Toro, "Modelling of the Acid-Base Properties of Natural and Synthetic Adsorbent Materials Used for Heavy Metal Removal from Aqueous Solutions," *Chemospere*, Vol. 54, No. 7, 2004, pp. 905-915. doi:10.1016/j.chemosphere.2003.09.003

[13] Y. Xu and L. Axe, "Synthesis and Characterization of Iron Oxide-Coated Silica and Its Effect on Metal Adsorption," *Journal of Colloid* and *Interface Science*, Vol. 282, No. 1, 2005, pp. 11-19. doi:10.1016/j.jcis.2004.08.057

[14] S. E. Bailey, T. J. Olin, R. M. Bricka and D. D. Adrian, "A Review of Potentially Low-Cost Sorbents for Heavy Metals," *Water Research*, Vol. 33, No. 11, 1999, pp. 2469-2479. doi:10.1016/S0043-1354(98)00475-8

[15] R. Ciccu, M. Ghiani, A. Serci, S. Fadda, R. Peretti and A. Zucca, "Heavy Metal Immobilization in the Mining-Contaminated Soils Using Various Industrial Wastes," Minerals *Engineering*, Vol. 16, No. 3 , 2003, pp. 187-192. doi:10.1016/S0892-6875(03)00003-7

[16] S. V. Dimitrova and D. R. Mehandriev, "Lead Removal from Aqueous Solutions by Granulated Blast-Furnace Slag," *Water Research*, Vol. 32, No. 11, 1998, pp. 3289-3292. doi:10.1016/S0043-1354(98)00119-5

[17] K. C. Sekhar, C. T. Kamala, N. S. Chary, A. R. K. Sastry, T. Nageswara Rao and M. Vairamani, "Removal of Lead from Aqueous Solutions Using an Immobilized Biomaterial Derived from a Plant Biomass," *Journal of Hazardous Materials*, Vol. 108, No. 1-2, 2004, pp. 111-117.

[18] N. Chubar, J. R. Carvalho and M. J. N. Correia, "Cork Biomass as Biosorbent for Cu(II), Zn(II) and Ni(II)," *Colloids Surfaces A*: *Physicochemical and Engineering Aspects*, Vol. 230, No. 1, 2003, pp. 57-65. doi:10.1016/j.colsurfa.2003.09.014

[19] K. T. Park, V. K. Gupta, D. Mohan and S. Sharma, "Removal of Chromium(VI) from Electroplating Industry Wastewater Using Bagasse Fly Ash—A Sugar Industry Waste Material," *The Environmentalist*, Vol. 19, No. 2, 1999, pp. 129-136.

[20] M. A. Karakassides, D. Gournis, A. B. Bourlinos, P. N. Trikalitisb and T. Bakasc, "Magnetic Fe_2O_3-Al_2O_3 Composites Prepared by a Modified Wet Impregnation Method," *Journal of Materials Chemistry*, Vol. 13, No. 4, 2003, pp. 871-876. doi:10.1039/b211330a

[21] C.-F. Changa, P.-H. Lin and W. Holl, "Aluminum-Type Superparamagnetic Adsorbents: Synthesis and Application on Fluoride Removal," *Colloids Surfaces A: Physicochemical and Engineering Aspects*, Vol. 280, No. 3,

2006, pp. 194-202. doi:10.1016/j.colsurfa.2006.02.011

[22] L. M. Sheppard, "Ceramic Transactions," In: K. Ishizaki, L. M. Sheppard, S. Okada, T. Hamasaki and B. Huybrechts, Eds., *Porous Materials*, American Ceramic Society, Westerville, 1993, p. 3.

[23] I. Nettleship, "Applications of Porous Ceramics," *Key Engineering Materials*, Vol. 122-124, No. 1, 1996, pp. 305-324. doi:10.4028/www.scientific.net/KEM.122-124.305

[24] G. R. Doughty and D. Hind, "The Applications of Ion-Conducting Ceramics," *Key Engineering Materials*, Vol. 122-124, No. 1, 1996, pp. 145-162. doi:10.4028/www.scientific.net/KEM.122-124.145

[25] M. Schmidt and F. Schwertfeger, "Applications for Silica Aerogel Products," *Journal of Non-Crystalline*, Vol. 225, No. 1, 1998, pp. 364-368. doi:10.1016/S0022-3093(98)00054-4

[26] L. Ji, J. Lin, K. Ltan and C. Zeng, "Synthesis of High-Surface-Area Alumina Using Aluminum Tri-sec-butoxide-2,4-pentanedione-2-propanol-nitric Acid Precursors," *Journal of Material Chemistry*, Vol. 12, No. 19, 2000, pp. 931-939.

[27] A. M. Ibrahim, M. M. Abd El-Latif and M. M. Mahmoud, "Synthesis and Characterization of Nanosize Cobalt Ferrite Prepared by Convenient Heating Polyol Method and Microwave Heating Technique," *Journal of Alloys and Compounds*, Vol. 506, No. 1 , 2010, pp. 201-204. doi:10.1016/j.jallcom.2010.06.177

[28] C. N. Sawyer, P. L. McCarty and G. F. Parkin, "Chemistry of Environmental Engineering," 5th Edition, McGraw-Hill, New York, 2002.

[29] A. Denizli, G. Ozkan and M. Yakup Arica, "Preparation and Characterization of Magnetic Polymethylmethacrylate Microbeads Carrying Ethylene Diamine for Removal of Cu(II), Cd(II), Pb(II), and Hg(II) from Aqueous Solutions," *Journal of Applied Polymer Science*, Vol. 78, No. 3, 2000, pp. 81-89. doi:10.1002/1097-4628(20001003)78:1<81::AID-APP110>3.0.CO:2-J

[30] F. Y. Wang, H. Wang and J. W. Ma, "Adsorption of Cadmium(II) Ions from Aqueous Solution by a New Low-Cost Adsorbent-Bamboo Charcoal," *Journal of Hazardous Materials*, Vol. 177, No. 1-3, 2010, pp. 300-306. doi:10.1016/j.jhazmat.2009.12.032

[31] C. Niu, W. Wu, Z. Wang, S. Li and J. Wang, "Adsorption of Heavy Metal Ions from Aqueous Solution by Cross-linked Carboxymethyl Konjac Glucomannan," *Journal of Hazardous Materials*, Vol. 141, No. 1, 2007, pp. 209-214. doi:10.1016/j.jhazmat.2006.06.114

[32] A. Akil, M. Mouflith and S. Sebti, "Removal of Heavy Metal Ions from Water by Using Calcined Phosphate as a New Adsorbent," *Journal of Hazardous Materials*, Vol. 112, No. 2, 2004, pp. 183-190. doi:10.1016/j.jhazmat.2004.05.018

[33] G. Wulfsberg, "Principles of Descriptive Chemistry," Brookes/Cole Publishing, Montery, 1987.

[34] N. Khalid, S. A. Chaudhri, M. M. Saeed and J. Ahmed, "Separation and Preconcentration of Lead and Cadmium with 4-(4-Chlorophenyl)-2-phenyl-5-thiazoleacetic Acid

and Its Application in Soil and Seawater," *Separation Science and Technology*, Vol. 31, No. 2, 1996, pp. 229-239. doi:10.1080/01496399608000692

[35] S. Schiewer and B. Volesky, "Modeling of the Proton-Metal Ion Exchange in Biosorption," *Environmental Science and Technology*, Vol. 29, No. 12, 1995, pp. 3049-3058. doi:10.1021/es00012a024

[36] A. Shukla, Y. H. Zhang, P. Dubey, J. L. Margrave and S. S. Shukla, "The Role of Sawdust in the Removal of Unwanted Materials from Water," *Journal of Hazardous Materials*, Vol. 95, No. 1-2, 2002, pp. 137-152. doi:10.1016/S0304-3894(02)00089-4

[37] N. N. Nassar, "Rapid Removal and Recovery of Pb(II) from Wastewater by Magnetic Nanoadsorbents," *Journal of Hazardous Materials*, Vol. 184, No. 1-3, 2010, pp. 538-546. doi:10.1016/j.jhazmat.2010.08.069

[38] Y. H. Huang, C. L. Hsueh, C. P. Huang, L. C. Su and C. Y. Chen, "Adsorption Thermodynamic and Kinetic Studies of Pb(II) Removal from Water onto a Versatile Al_2O_3-Supported Iron Oxide," *Separation and Purification Technology*, Vol. 55, No. 1, 2007, pp. 23-29. doi:10.1016/j.seppur.2006.10.023

[39] A. K. Bhattacharya, T. K. Naiya, S. N. Mandal and S. K. Das, "Adsorption, Kinetics and Equilibrium Studies on Removal of Cr(VI) from Aqueous Solutions Using Different Low-Cost Adsorbents," *Chemical Engineering Journal*, Vol. 137, No. 3, 2008, pp. 529-554.

[40] I. A. W. Tan, A. L. Ahmad and B. H. Hameed, "Adsorption of Basic Dye on High-Surface-Area Activated Carbon Prepared from Coconut Husk: Equilibrium, Kinetic and Thermodynamic Studies," *Journal of Hazardous Materials*, Vol. 154, No. 1-3 , 2008, pp. 337-346. doi:10.1016/j.jhazmat.2007.10.031

[41] I. Langmuir, "The Constitution and Fundamental Properties of Solids and Liquids," *Journal of American Chemical Society*, Vol. 38, No. 11, 1916, pp. 2221-2295. doi:10.1021/ja02268a002

[42] M. M. Abd El-Latif and A. M. Ibrahim, "Removal of Reactive Dye from Aqueous Solutions by Adsorption onto Activated Carbons Prepared from Oak Sawdust," *Desalination and Water Treatment*, Vol. 20, No. 1-3, 2010, pp. 102-113.

[43] H. M. F. Freundlich, "Ueber Die Adsorption in Loesungen," *Zeitschrift für Physikalische Chemie (Leipzig)*, Vol. 57, No. A, 1907, pp. 385-470.

[44] M. M. Abd El-Latif and M. F. Elkady, "Equilibrium Isotherms for Harmful Ions Sorption Using Nano Zirconium Vanadate Ion Exchanger," *Desalination*, Vol. 255, No.1-3, 2010, pp. 21-43. doi:10.1016/j.desal.2010.01.020

[45] M. J. Temkin and V. Pyzhev, "Kinetics of the Synthesis of Ammonia on Promoted Iron Catalysts," *Acta Physiochimica* (URSSR), Vol. 12, 1940, pp. 327-356.

[46] R. Leyva-Ramos, L. A. Bernal-Jacome and I. Acosta-Rodriguez, "Adsorption of Cadmium(II) from Aqueous Solution on Natural and Oxidized Corncob," *Separation and purification Technology*, Vol. 45, No. 1, 2005, pp. 41-49. doi:10.1016/j.seppur.2005.02.005

[47] A. H. Mahvi and L. Diels, "Biological Removal of Cadmium by Alcaligenes Eutrophus CH34," *International*

journal of Environmental Science andTechnology, Vol. 1, No. 3, 2004, pp. 199-204.

[48] A. H. Mahvi, J. Nouri, G. A. Omrani and F. Gholami, "Application of Platanus Orientalis Leaves in Removal of Cadmium from Aqueous Solution," *World Applied Sciences Journal*, Vol. 2, No. 1, 2007, pp. 40-44.

[49] K. A. Bolton and L. J. Evans, "Cadmium Adsorption Capacity of Selected Ontario Soils," *Canadian Journal of Soil Science*, Vol. 5, No. 3, 1996, pp. 183-189. doi:10.4141/cjss96-025

[50] P. Hanzlik, J. Jehlicka, Z. Weishauptova and O. Sebek, "Adsorption of Copper, Cadmium and Silver from Aqueous Solutions onto Natural Carbonaceous Materials," *Plant Soil Environmental*, Vol. 50, No. 6, 2004, pp. 257-264.

[51] S. Lagergren, "About the Theory of So-Called Adsorption of Soluble Substances Zur Theorie der Sogenannten Adsorption Geloster Stoffe," *Kungliga Svenska Vetenskapsakademiens Handlingar*, Vol. 24, No. 4, 1898, pp. 1-39.

[52] Y. S. Ho, G. McKay, D. A. J. Wase and C. F. Foster, "Study of the Sorption of Divalent Metal Ions on to Peat," *Adsorption Science and Technology*, Vol. 18, No. 7, 2000, pp. 639-650. doi:10.1260/0263617001493693

[53] S. H. Chien and W. R. Clayton, "Application of Elovich Equation to the Kinetics of Phosphate Release and Sorption in Soils," *Soil Science Society America Journal*, Vol. 44, No. 2, 1980, pp. 265-268. doi:10.2136/sssaj1980.03615995004400020013x

[54] D. L. Sparks, "Kinetics of Reaction in Pure and Mixed Systems," In: D. L. Sparks, Ed., *Soil Physical Chemistry*, CRC Press, Boca Raton, 1986, pp. 83-145.

[55] J. Zeldowitsch, "Über Den Mechanismus der Katalytischen Oxidation Von CO a MnO_2," *URSS, Acta Physiochim*, Vol. 1, No. 2, 1934, pp. 364-449.

[56] W. J. Weber and J. C. Morris, "Kinetics of Adsorption on Carbon from Solution," *Journal of the Sanitary Engineering Division American Society of Civil Engineering*, Vol. 89, No. 2, 1963, pp. 31-60.

[57] K. Srinivasan, N. Balasubramanian and T. V. Ramakrishan, "Studies on Chromium Removal by Rice Husk Carbon," *Indian Journal Environment and Health*, Vol. 30, No. 4, 1988, pp. 376-387.

[58] C. W. Cheung, J. F. Porter and G. Mckay, "Sorption Kinetic Analysis for the Removal of Cadmium Ions from Effluents Using Bone Char," *Water Research*, Vol. 35, No. 3, 2001, pp. 605-612. doi:10.1016/S0043-1354(00)00306-7

[59] H. Teng and C. Hsieh, "Activation Energy for Oxygen Chemisorption on Carbon at Low Temperatures," *Industrial Engineering and Chemical Research*, Vol. 38, No. 1, 1999, pp. 292-297. doi:10.1021/ie980107j

[60] K. Kannan and M. M. Sundaram, "Kinetics and Mechanism of Removal of Methylene Blue by Adsorption on Various Carbons—A Comparative Study," *Dyes Pigments*, Vol. 51, No. 1, 2001, pp. 25-40. doi:10.1016/S0143-7208(01)00056-0

[61] C. L. Lu, J. G. Lv, L. Xu, X. F. Guo, W. H. Hou, Y. Hu and H. Huang, "Crystalline Nanotubes of γ-AlOOH and γ-Al_2O_3: Hydrothermal Synthesis, Formation Mechanism and Catalytic Performance," *Nanotechnology*, Vol. 20, No. 2, 2009, pp. 1-6. doi:10.1088/0957-4484/20/21/215604

[62] J. M. Murray and J. G. Dillard, "The Oxidation of Cobalt(II) Adsorbed on Manganese Dioxide," *Geochima et Cosmochima Acta*, Vol. 43, No. 2, 1979, pp. 781-787. doi:10.1016/0016-7037(79)90261-8

[63] M. G. Zuhra, M. I. Bhanger, A. Mubeena, N. T. Farah and R. M. Jamil, "Adsorption of Methyl Parathion Pesticide from Water Using Watermelon Peels as a Low Cost Adsorbent," *Chemical Engineering Journal*, Vol. 138, No. 1-3, 2008, pp. 616-621. doi:10.1016/j.cej.2007.09.027

[64] M. Syed, I. Muhammad, G. Rana and K. Sadullah, "Effect of Ni^{2+} Loading on the Mechanism of Phosphate Anion Sorption by Iron Hydroxide," *Separation and Purification Technology*, Vol. 59, No. 1, 2008, pp. 108-114. doi:10.1016/j.seppur.2007.05.033

[65] Y. Seki and K. Yurdakoc, "Adsorption of Promethazine Hydrochloride with KSF Montmorillonite," *Adsorption*, Vol. 12, No. 1, 2006, pp. 89-100. doi:10.1007/s10450-006-0141-4

[66] Y. Yu, Y. Y. Zhuang and Z. H. Wang, "Adsorption of Water-Soluble Dye onto Functionalized Resin," *Journal of Colloid Interface Science*, Vol. 242, No. 2, 2001, pp. 288-293. doi:10.1006/jcis.2001.7780

[67] N. Dizge, C. Aydiner, E. Demirbas, M. Kobya and S. Kara, "Adsorption of Reactive Dyes from Aqueous Solutions by Fly Ash: Kinetic and Equilibrium Studies," *Journal of Hazardous Materials*, Vol. 150, No. 3, 2008, pp. 737-746. doi:10.1016/j.jhazmat.2007.05.027

[68] H. Oualid, S. Fethi, C. Mahdi and N. Emmanuel, "Sorption of Malachite Green by a Novel Sorbent, Dead Leaves of Plane Tree: Equilibrium and Kinetic Modeling," *Chemical Engineering Journal*, Vol. 143, No. 1, 2008, pp. 73-84. doi:10.1016/j.cej.2007.12.018

[69] J. H. Noggle, "Physical Chemistry," 3rd Edition, Vol. 11, Harper Collins Publishers, New York, 1996.

[70] W. Stumm and R. Wollast, "Coordination Chemistry of Weathering. Kinetics of the Surface Controlled Dissolution of Oxide Minerals," *Reviews of Geophysics*, Vol. 28, No. 1, 1990, pp. 53-69. doi:10.1029/RG028i001p00053

[71] K. G. Scheckel and D. L. Sparks, "Temperature Effects on Nickel Sorption Kinetics at the Mineral-Water Interface," *Soil Science Society of America Journal*, Vol. 65, No. 3, 2001, pp. 719-728. doi:10.2136/sssaj2001.653719x

[72] M. M. Abd El-Latif, A. M. Ibrahim and M. F. El-Kady, "Adsorption Equilibrium, Kinetics and Thermodynamics of Methylene Blue from Aqueous Solutions Using Biopolymer Oak Sawdust Composite," *Journal* of American *Science*, Vol. 6, No. 6, 2010, pp. 267-283.

[73] M. Alkan, B. Kalay, M. Dogan and O. Demirbas, "Removal of Copper Ions from Aqueous Solutions by Kaolinite and Batch Design," *Journal of Hazardous Materials*, Vol. 153, No. 1-2, 2008, pp. 867-876. doi:10.1016/j.jhazmat.2007.09.047

[74] B. H. Hameed, D. K. Mahmoud and A. L. Ahmad, "Sorp-

tion Equilibrium and Kinetics of Basic Dye from Aqueous Solution Using Banana Stalk Waste," *Journal of Hazardous Materials*, Vol. 158, No. 2-3, 2008, pp. 499-506. doi:10.1016/j.jhazmat.2008.01.098

12

Transformation Mechanism of Ore Matter in the Weathering

Victor Bragin[1,2], Irina Baksheyeva[1], Margaret Sviridova[2]
[1]Siberian Federal University, Krasnoyarsk, Russia
[2]Institute of Chemistry and Chemical Technology of Siberian Branch
of Russian Academy of Sciences, Krasnoyarsk, Russia
Email: Irina__igorevna@mail.ru

ABSTRACT

Grain-size class redistribution of non-ferrous, precious metals and iron in copper-nickel ores tailings from Norilsk industrial region was after artificial weathering investigated. Possible mechanisms of metal redistribution were suggested.

Keywords: Technogenic Raw Materials; Cu-Ni Ores; Nonferrous and Precious Metals; Iron; Platinum Group Elements; Grain-Size Class Redistribution of Metals; Redistribution Mechanisms

1. Introduction

The process of non-ferrous and precious metals extracting from the tails is an essential reserve for non-ferrous industry in old mining areas. Ecologically, development of technology is important too. Solution of the problem is particularly urgent in reference to highly oxidized tails, as metal extraction technology for them is quite undeveloped, and, besides, their ecological effect is significant.

Current approaches to the problem of old tails are usually considered technologically and ecologically. There are a great number of researches and practical works describing old tails as a potential source of metal. In this case, the major scientific and engineering problem is the optimization of reextraction methods. One of the latest researches in the field has been done by V. A. Chanturiya and V. Ye. Vigdergauz [1,2]. On the whole, available results show that a non-ferrous, precious metals extraction rate reduces significantly due to the increase of weathering. The principal mechanisms which affect technological properties are the following: oxidation of sulphide minerals surfaces, formation of secondary mineral phases (which are normally slime with a low contrast of precious metals distribution) and adverse movement in the composition of pulp liquid phase. In some cases, when the increase of weathering is considerable, it is possible to extract metals-zinc, copper, nickel-by using hydrometallurgical methods. It can be done only if sulphide matrix is nearly completely decomposed and, thus, soluble forms of non-ferrous metals and a favorable composition of barren rock generate.

During the recent 10 - 15 years much attention has been paid to the ecological aspect in the old tails research. The major studies have been carried out on the materials from old mining areas in Europe and North America [3-7]. In Russia research in the field was done by D. V. Makarov [8,9]. The research was mainly devoted to the problems of transformation of primary minerals (both ore and non-metallic ones) into secondary mobile forms, which can be especially dangerous for the environment. There have been developed a few methods in order to reduce the ecological impact of oxidized tails. They include measures aimed at oxidation preventing or slowing down and building up in drainage lines which transfer toxic mineralized waters. There are also designs of geochemical barriers made of natural calcium carbonate for getting commercial products out of drainage waters.

The problems of precious metals reextraction have been given much less consideration due to a fewer number of such objects, which thus resulted in a fewer ecological problems and the less total reextraction value. Besides, technically the objects are difficult to research. The most well-known work, researching gold behavior in anthropogenic formations on the territory of Krasnoyarsk region and Khakassia, were carried out by V. A. Makarov Reextraction of platinum and platinum metals from anthropogenic products of Norilsk region were studied and described by a group of scientists, Yu. V. Blagodatin and D. A. Dodin among them [10]. The data concerning behavior and technological properties of platinoids are quite important as the platinum extraction anthropogenic products within Norilsk industrial area is expected to increase and reach 10% out of total output. In general, the research results prove a particular complexity of the problem. Modern technologies make it possible to extract free forms of precious metals of gravity size. However, the most common metal forms are unexposed impregnation, isomorphic component of both metallic and non-metallic minerals, loose grains of micron and submicron grade-size. Thus the traditional approach cannot be used. Moreover, the process becomes more complicated due to the presence of various secondary forms of non-ferrous metals in the tails. Consequently, there is no universal method for simultaneous extraction of non-ferrous and precious metals from old tails.

The solution of the problem can be found in terms of the approach, based on the mechanism of ore material transformation when being weathered. The approach could use the mechanism in order to control a hypergene process for the formation of material constitution, which could be used optimally while reextracting valuable constituent from tails. We suppose that low efficiency of process solutions of the problem is caused by the lack of knowledge about how precious metals are redistributed in different phases during weathering and also about the correlation between the redistribution and phase transformations of non-ferrous and rock-forming metals. Detailed analysis of the processes could control their rate, choose conditions, when characteristics of precious metals distribution are optimal for their simultaneous extraction with non ferrous metals.

The most important process, which determines formation of technological properties of weathered copper-nickel tails, is redistribution of platinum, platinum metals and gold within different mineral phases and grain-size classes, caused by sulphide oxidation. The phenolmenon is described generally, considering the case of oxidation zones in sulphide deposits. However, specific mechanisms, concerning raw materials from Norilsk industrial hub, are unknown. In order to study the mechanisms, there have been conducted experiments with artificial weathering of anthropogenic materials samples.

2. Materials and Methods

The study conducted in the Institute of Chemistry and Chemical Technology of SB RAS was founded in 2004. Tails from the Norilsk region were put to weathering simulate in the laboratory in an evaporating mode. The material was being kept humid by adding distilled water regularly. The experiment was determining the tails granulometric composition and grain-size class redistribution. Solid samples were taken in 6 months, 1 year and 4 years since the experiment beginning. Screen test was done according to a wet scheme, followed by check dry screening. The content of precious and non-ferrous metals in the samples was determined by mass-spectrometric [11] method after opening with the help of chloroazotic

acid according to the standard scheme. As for the non-ferrous metals, in addition, atomic absorption method was used after the nitric opening. The application of mass spectrometric analysis in the research made it possible to obtain the information about the distribution of all periodic table elements according to both grain-size class of weathered tails.

3. Results and Discussion

The carried out research enabled to reveal a number of characteristics in the precious metals behavior after the weathering of copper-nickel tails. It is determined that during weathering of sulphide minerals of copper nickel ores tails in Norilsk industrial hub, dispersive impregnation of precious metals opens up at the rate which is higher than the one of sulphide matrix oxidation. Besides it is followed by transition of platinum and palladium into slime with the grain-size of less than 0.044 mm (**Figure 1**). There have been found the association of opened up dispersive particles of precious metals with secondary iron-oxide materials.

Concentration of non-ferrous metals (copper and iron) after weathering is situated in class –0.044 mm, in which oxide phase dominate according to X-ray phase analysis and microscopic observation. Furthermore, as **Figure 1** demonstrates, the growth of the metals can be seen in grain-size classes of about 0.1 mm, which are typical of sulphide minerals, having been subjected to partial oxidation on the surface. The most interesting observations can be done when comparing the behavior of precious metals (platinum and palladium). While palladium is being redistributed similarly to iron (with the growth both in slime and sand fraction of 0.1 mm), platinum contrariwise is transferring from the materials into slime. Moreover, platinum redistribution from sand classes into slime is much more considerable, than that of iron, copper and palladium and tails material as well generally: platinum density of distribution grows by as 5 times as much as compared to initial density (**Figure 2**). Thus, after weathering there is separation of platinum and palladium flows, platinum looses faster, being drawn to iron-oxide slime phases, whereas palladium tends to have the primary associations with iron and non-ferrous metals.

The revealed features can be explained considering the differences in platinum and palladium forms which occur in primary sulphides. It is well known, that in Norilsk deposits ores platinum and palladium associate closely with sulphide minerals. However, platinum is represented mainly as thin impregnations of independent phases, whereas palladium occurs as isomorphic admixture. Regarding the latter, the revealed feature is easy to explain. Sulphide matrix corrosion on the surface of intergrowth with platinum mineral causes loosing and pitting of plati

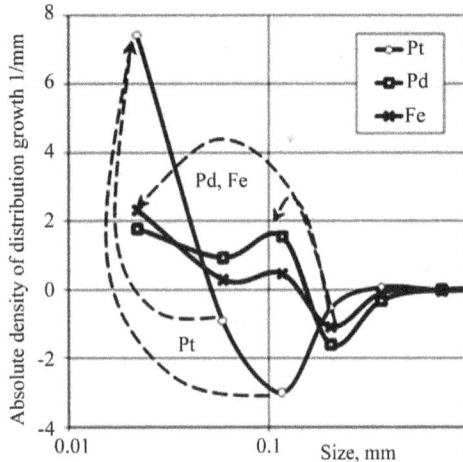

Figure 1. Grain-size class redistribution of platinum, palladium and iron after a storage period of 4 years.

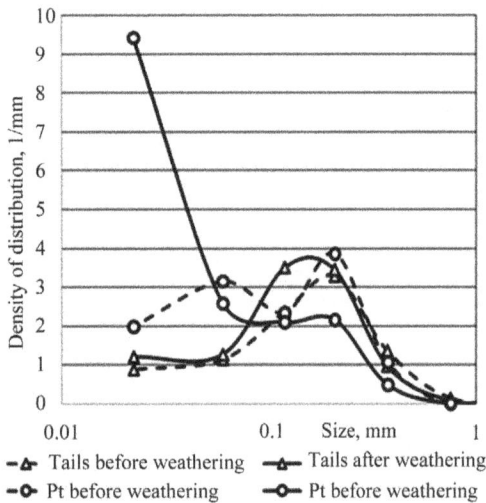

Figure 2. Grain-size class redistribution of tails material and platinum.

num grains. In this case, catalytic properties of platinum could are important. If impregnation size is quite big (less than 5 - 10 microns), it is slime, where platinum accumulates. At the same loosing of isomorphic associated palladium is impossible without compete matrix destruction, Therefore, the metal growth after weathering can be observed both in slimes with iron-oxides and nonferrous metals oxides and in grain-size class of about 0.1 mm (where sulfide grains, having been partially oxidized on the surface, concentrate, they still contain palladium as isomorphic admixture. Revealed features of platinum and palladium redistribution after weathering are of significant interest for establishing the technology of precious metals extraction from old nickel-copper tails. Separation of platinum and palladium flows, with platinum being loosed and separated from the bulk of sulphides, gives interesting opportunities in developing com-

bined schemes of anthropogenic material processing on the territory of Norilsk region.

4. Conclusions

The redistribution of non-ferrous and precious metals in copper-nickel ores tailings after artificial weathering was investigated. Possible mechanisms of metal redistribution have been discussed. On basis of obtained data, processes mechanisms were suggested.

There were found essential distinctions in the behavior of platinum and palladium caused by the difference in their forms occurring in primary sulphides. It should be noted, that redistribution of palladium (occurring mainly as isomorphic admixture in primary sulphides) after weathering is simultaneous to the redistribution of iron and non-ferrous metals, which are associated with sulphide and iron-oxide phases. In contrast to that, redistribution of platinum (occurring mainly as dispersive impregnation in primary sulphides) goes at a higher rate; platinum accumulates in slime fractions in association with ironoxide secondary phases. It is assumed that accelerated platinum loosing can be controlled by catalytic reactions on the surface of platinum impregnations.

REFERENCES

[1] V. E. Vigdergauz and V. A. Chanturia, "Alteration of Ion-Molecular Composition of Sulfide Ore Slurry Depending on the Development of Redox Processes in It," *Mining Journal*, No. 6, 2008, pp. 71-75.

[2] V. E. Vigdergauz, D. V. Makarov, I. V. Zorenko, E. V. Belogub, M. N. Maljarenko, E. A. Shrader and I. N. Kuznetsova, "Effect Exerted by Structural Features of Copper-Zinc Ores on Their Oxidation and Technological Properties," *Journal of Mining Science*, Vol. 44, No. 4, 2008, pp. 101-110.

[3] S. Suárez, H. M. Prichard, F. Velasco, P. C. Fisher and I. McDonald, "Alteration of Platinum-Group Minerals and Dispersion of Platinum-Group Elements during Progressive Weathering of the Aguablanca Ni-Cu Deposit, SW Spain," *Mineralium Deposita*, Vol. 45, No. 4, 2010, pp. 331- 350.

[4] D. A. Holwell and I. McDonald, "Distribution of PGE in the Platreef at Overysel, Northerm Bushveld Complex: A Combined PGM and LA-ICP-MS Study," *Contrib Mineral Petrol*, Vol. 154, No. 2, 2007, pp. 171-190. doi:10.1007/s00410-007-0185-9

[5] D. Hutchinson and I. McDonald, "Laser Ablation ICP-MS Studu of Platinum-Group Elements in Sulphides from the Platreef at Turfspruit Norther Limb of the Bushveld Complex, South Africa," *Miner Deposita*, Vol. 43, No. 6, 2008, pp. 695-711. doi:10.1007/s00126-008-0190-6

[6] A. J. H. Newell, D. J. Bradshaw and P. J. Harris, "The Effect of Heavy Oxidation upon Flotation and Potential Remedies for Merensky Type Sulfides," *International Journal of Mineral Processing*, Vol. 19, 2006, pp. 675-686.

[7]   C. J. Moraza, N. Iglesias and I. Palencia, "Application of Sugar Foam to a Pyrite-Contaminated Soil," *International Journal of Mineral Processing*, Vol. 19, No. 5, 2006, pp. 399-406.

[8]   V. T. Kalinnikov, V. N. Makarov, S. I. Mazuhina, D. V. Makarov and V. A. Masloboev, "The Study of Supergene Processes in the Tailings of Sulfide Copper-Nickel Ore," *Chemistry for Sustainable Development*, Vol. 13, No. 4, 2005, pp. 515-519.

[9]   D. V. Makarov, V. N. Makarov, S. V. Drogobuzhskaja, A. A. Alkatseva, E. R. Farvazova and M. V. Tunina, "Envi-

ronmental Pollution. The Content of Ni, Cu, Co, Fe, MgO in the Tailings of Pore Fluids of Copper-Nickel Ore, after Long-Term Storage," *Environmental Geoscience*, No. 2, 2006, pp. 135-142.

[10]   D. A. Dodin, D. V. Lenchuk and V. M. Izoitko, "Technogenetics Deposits of Norilsk Region", *International Symposium of Mineral Resources of Russia*, St. Petersburg, 10 - 13 November 1993, pp. 10-12.

[11]   A. T. Lebedev, "Mass Spectrometry in Organic Chemistry," Publishing House BINOM, Knowledge Laboratory, Moscow, 2003, p. 493.

13

Influence of Ferric and Ferrous Iron on Chemical and Bacterial Leaching of Copper Flotation Concentrates

Ali Ahmadi

Department of Mining Engineering, Isfahan University of Technology, Isfahan, Iran

Email: a.ahmadi@cc.iut.ac.ir

ABSTRACT

The effects of ferrous and ferric iron as well as redox potential on copper and iron extraction from the copper flotation concentrate of Sarcheshmeh, Kerman, Iran, were evaluated using shake flask leaching examinations. Experiments were carried out in the presence and absence of a mixed culture of moderately thermophile microorganisms at 50°C. Chemical leaching experiments were performed in the absence and presence of 0.15 M iron (ferric added medium, ferrous added medium and a mixture medium regulated at 420 mV, Pt. vs. Ag/AgCl). In addition, bioleaching experiments were carried out in the presence and absence of 0.1 M iron (ferric and ferrous added mediua) at pulp density 10% (w/v), inoculated bacteria 20% (v/v), initial pH 1.6, nutrient medium Norris and yeast extract addition 0.02% (w/w). Abiotic leaching tests showed that the addition of iron at low solution redox potentials significantly increased the rate and extent of copper dissolution but when ferric iron was added, despite a higher initial rate of copper dissolution, leaching process stopped. Addition of both ferrous and ferric iron to the bioleaching medium levelled off the copper extraction and had an inhibitory effect which decreased the final redox potential. The monitoring of ferrous iron, ferric iron and copper extraction in leach solutions gave helpful results to understand the behaviour of iron cations during chemical and bacterial leaching processes.

Keywords: Ferric Iron; Ferrous Iron; Copper Concentrate; Leaching; Redox Potential

1. Introduction

During the last recent decades, copper leaching has been extensively studied as an alternative route to treat chalcopyrite concentrates.

The driving force in the leaching of metallic sulfides such as chalcopyrite, chalcocite and covellite is determined by the difference in the rest potential of the minerals present, and the redox potential of the leaching solution which mainly depends on the ferric to ferrous iron ratio as described by the Nernst equation (Equation (1)) [1].

$$Eh = Eh^\circ + \frac{R \cdot T}{n \cdot F} \ln \frac{\left[Fe^{3+} \right]}{\left[Fe^{2+} \right]} \qquad (1)$$

where Eh is the solution redox potential in mV (with respect to hydrogen electrode), Eh° is the standard equilibrium potential in mV, R is the gas constant, T is temperature in K, n is the charge number, F is Faraday's constant, $[Fe^{3+}]$ and $[Fe^{2+}]$ are the concentrations of the ferric and ferrous iron, respectively, in mol/l.

In bioleaching of copper concentrates, redox potential is one of the most important environmental parameters which affect biological and chemical subsystems. It in-

creases by the activity of iron-oxidizing microorganisms. This increase is beneficial for the dissolution of most of the metallic sulfides such as chalcocite, covellite and pyrite, but, in the case of chalcopyrite, the leaching behavior is more complicated in which its dissolution rate is maximal at low redox potentials [2-11].

Furthermore, the growth and activity of iron-oxidizing microorganisms are affected by the concentrations of ferrous and ferric ions. Das et al. [12] reported that a low concentration of ferric iron enhances the oxygen uptaken by the acidophilic microorganisms, but at higher concentrations, ferric iron inhibits ferrous iron oxidation. In addition, increasing the ferrous iron to a certain critical concentration (3 g/l) enhanced the oxidation rate while higher ferrous iron concentrations had an inhibitory effect [13].

Howard and Crundwell [14] found that increasing concentration of ferric iron in the range of 0.05 to 0.5 M decreased the rate of chalcopyrite dissolution. Third et al. [6] also studied the effect of ferrous and ferric iron on the chemical and bacterial leaching of chalcopyrite at 37°C and found that high ferric concentration inhibits the bioleaching of chalcopyrite. They also confirmed the results obtained by Hiroyoshi et al. [2] and found that chemical leaching of chalcopyrite was increased 4-fold in the presence of 0.1 M ferrous iron compared with 0.1 M ferric iron. Cordoba et al. [8] studied the influence of ferric iron on chalcopyrite dissolution at low and high solution potentials and found that although ferric iron was responsible for the oxidation of chalcopyrite, ferrous iron had an important role in controlling the precipitation and nucleation of jarosite (Equation (2)).

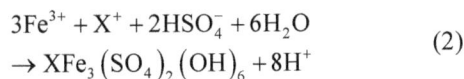

$$3Fe^{3+} + X^+ + 2HSO_4^- + 6H_2O \rightarrow XFe_3(SO_4)_2(OH)_6 + 8H^+ \quad (2)$$

The effect of ferric and ferrous iron on the extraction of copper and iron from flotation concentrates including both primary and secondary copper sulfides has not been well studied as yet. Hence, this research was done to investigate the effect of these cations on chemical and bacterial leaching of Sarcheshmeh copper concentrate in the presence and absence of mixed cultures of moderately thermophile microorganisms. The monitoring of the mentioned cations during the leaching processes is helpful to understand their behaviour during chemical and bacterial leaching processes.

2. Materials and Methods

2.1. Minerals

A flotation concentrate was obtained from Sarcheshmeh Copper Mine, Kerman, Iran. Mineralogical investigation on the representative sample was performed by optical microscopy using a Leica phase contrast microscope (DMLP). It showed that the concentrate contained 44.0% chalcopyrite (CuFeS$_2$), 24.0% pyrite (FeS$_2$), 6.9% covellite (CuS), 5.8% chalcocite (CuS$_2$), 13.6% non-metallic minerals and 4.8% copper oxide minerals. X-ray fluorescence (XRF) analysis showed that the concentrate included 27.5% Cu, 23.0% Fe, 14.8% S, 3.9% Si and 1.0% Zn. The particle size distribution of the concentrate was determined by wet sieving and cyclosizer and showed that 80% was passing 76 μm.

2.2. Microorganisms

A mixed culture of moderately thermophilic iron- and sulfur oxidizing bacteria mainly containing *Acidithiobacillus caldus*, *Solfobacillus* and *Thermosulfidooxidans* obtained from Biohydrometallurgy Laboratory of Sarcheshmeh Copper Complex was used as inoculum in the bioleaching tests. Experiments were carried out at initial pH 1.5, 50°C and Norris nutient medium [15] modified without iron with the following com-position: 0.4 g/l (NH$_4$)$_2$SO$_4$, 0.4 g/l K$_2$HPO$_4$, 0.5 g/l MgSO$_4$·7H$_2$O.

2.3. Chemical Leaching Experiments

To investigate the effect of ferric and ferrous iron as well

as redox potential on the chemical leaching of Sarcheshmeh copper concentrate, four slurries with different initial redox potentials were prepared (**Table 1**) in 500 ml-Erlenmeyer flasks containing 200 ml suspension. These shake flask experiments were carried out in the following conditions: initial iron concentration 0.15 M (except control test R1), pulp density 1% (w/v), initial pH 1.8, stirring rate 150 rpm and 50°C. To prevent microbial growth, the medium was sterilized with 2% (v/v) bactericide (2% (w/w) thymol in ethanol) added. The solution of test R4 was prepared to obtain the redox potential of 420 mV (Pt vs. Ag/AgCl) by mixing ferric and ferrous sulfates in which total iron concentration was maintained at 0.15 M. The low solid to liquid ratio was chosen to avoid sharp changes of the redox potential of the medium during the first instant of leaching.

2.4. Bacterial Leaching Experiments

In order to determine the influence of ferrous and ferric iron addition on the bioleaching of the copper concentrate, shake flasks experiments were carried out in the presence and absence of 0.1 M of the mentioned cations. These experiments were carried out in 500 ml-Erlenmeyer flasks containing 200 ml of suspension of the concentrate at a pulp density of 10% (w/v) in the modified Norris's medium supplemented with 0.02% (w/v) yeast extract and having an initial pH of 1.5. Each flask was inoculated with a bacterial suspension (20% v/v) and then incubated at 50°C and 150 rpm on a rotary shaker. To inoculate bacteria to a fresh medium, the bacterial solution was added to an Erlenmeyer flask containing the required fresh nutrient solution (at the desired pH). Then, the desired amount of concentrate and ferric and/or ferrous iron when necessary, was added to the flask. After mixing the resulting slurry, the pH was regulated and the redox potential was recorded. These biotic experiments were inoculated with an active culture (as solution) with the cell density of about 2.7 × 10^8 cells/ml, which had been previously adapted to a 10% pulp density of the concentrate.

Evaporation loss was measured by weighting flasks and then was compensated by adding distilled water to the slurry before sampling.

Table 1. Initial concentrations of ferrous and ferric ions and the values of initial redox potential in the solutions.

Test No. \ Parameter	R1	R2	R3	R4
Iron Concentration	0	0.15 M Fe (II)	0.15 M Fe (III)	0.15 M Fe total
Initial redox potential (mV, vs. Ag/AgCl)	359	345	535	420

2.5. Analyses

The concentration of total iron and copper in the leach solutions was determined by atomic absorption spectroscopy (AAS) (model: Varian 240). After each test, the solid residue was filtered, rinsed with distilled water, left to air dry and sent for analyzing by AAS. The concentration of ferrous iron in the solution was determined by titration with potassium dichromate 0.01 N using sodium diphenylamine sulfonate as indicator. The difference in the concentration of total iron and ferrous iron was considered as the concentration of ferric iron in the solution. The pH and redox potential values in the leach solutions were measured with a pH meter (model: Jenway 3540) and a Pt electrode in reference to an Ag/AgCl electrode (+207 mV vs. SHE at 25°C), respectively. For all experiments, analytical grade reagents and distilled water were used.

3. Results and Discussion

3.1. Effect of Ferric and Ferrous Iron on Chemical Leaching

Variation of redox potential with time is shown in **Figure 1**. It can be seen that the redox potential in the ferrous medium rose from 345 to 449 mV over 8 days while in the ferric medium its value decreased from 535 to 472 mV in the same period of time. On the other hand, redox potential in both media regulated at 420 mV and the sterile medium did not change remarkably during the leaching experiments. Variation of redox potential shows that in the ferrous medium, ferric ions were produced during leaching of the concentrate, while in the ferric medium, ferrous ions were entered into the solution. This result was confirmed by analyzing the ferrous and total iron in the final solutions (**Figure 2**). The increase of the redox potential in the ferrous medium was probably related to the oxidation of ferrous iron by oxygen present in the solution which can surpass ferric reduction as a result of low solid content. It may also be as a consequence of the activity of bacteria present in the concentrate which had not been killed by the bactericide. While, the decrease of redox potential in the ferric medium was mainly attributed as a result of reducing ferric iron to ferrous iron on the surface of the minerals (dissolution process). The redox potential in the control test remained around 326 to 360 mV. This low increase may be as a result of natural ferrous oxidation at the relatively high temperature. **Figure 3** shows that the values of solution pH at the end of experiments remained at low levels (from 1.17 to 1.52) in which after the initial two days, decreasing the pH by H_2SO_4 was not necessary. The minimum and maximum pH values were related to ferric (without any acid addition) and control media (acid was added up to day 4), respectively. pH decrease could be related to the forma-

tion of H_2S as a result of acidic leaching of copper sulfides according to Equations (3) to (5), which can then be converted to acid according to Equations (6) to (8).

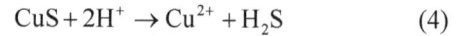

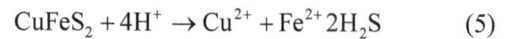

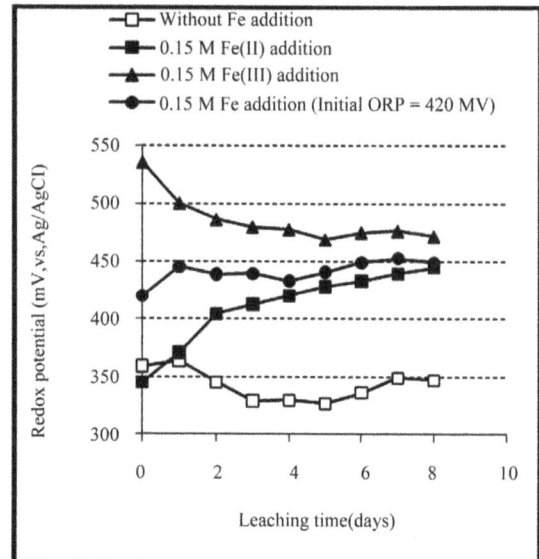

$$Cu_2S + 2H^+ \rightarrow Cu^{2+} + Cu° + H_2S \qquad (3)$$

$$CuS + 2H^+ \rightarrow Cu^{2+} + H_2S \qquad (4)$$

$$CuFeS_2 + 4H^+ \rightarrow Cu^{2+} + Fe^{2+} 2H_2S \qquad (5)$$

Figure 1. Variation of redox potential during the chemical leaching of the copper concentrate at 0.15 M iron addition (except the control test), pulp density 1% (w/v), 50°C, nutrient medium Norris and initial pH 1.8.

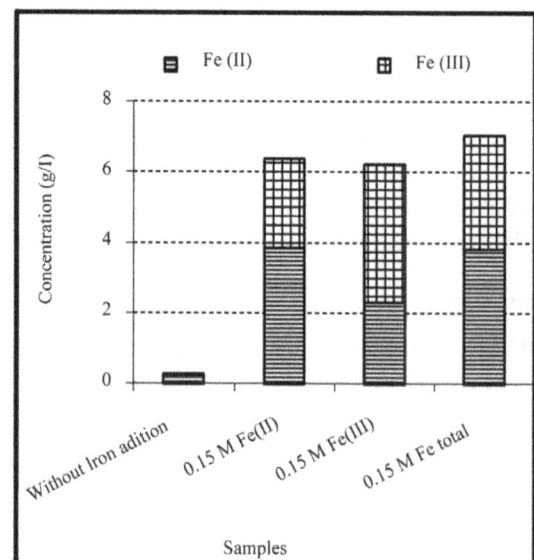

Figure 2. Concentration of ferrous and ferric iron in the final solutions of chemical leaching of the copper concentrate at 0.15 M iron addition (except the control test), pulp density 1% (w/v), 50°C, nutrient medium Norris and initial pH 1.8.

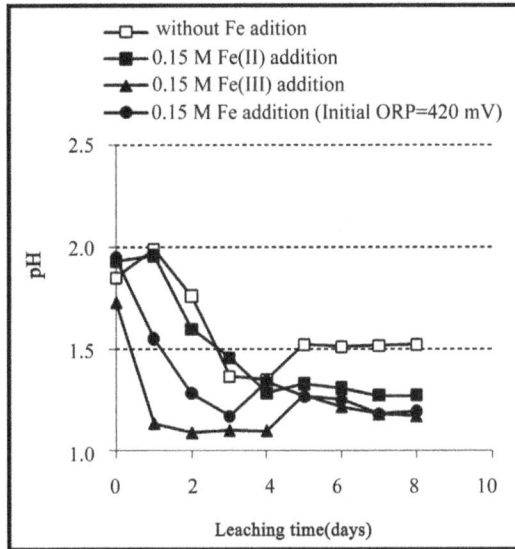

Figure 3. Variation of pH during the chemical leaching of the copper concentrate at 0.15 M iron addition (except the control test), pulp density 1% (w/v), 50°C, nutrient medium Norris and initial pH 1.8.

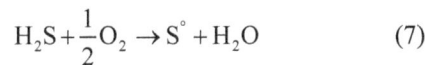

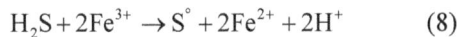

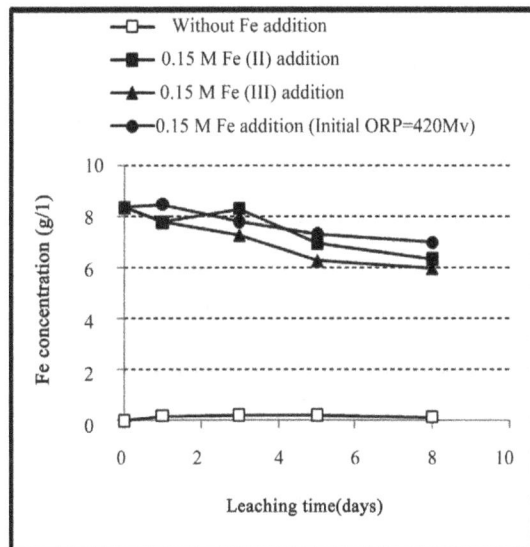

Figure 4. Extraction of copper during the chemical leaching of the copper concentrate at 0.15 M iron addition (except the control test), pulp density 1% (w/v), 50°C, nutrient medium Norris and initial pH 1.8.

$$H_2S + Cu^{2+} \rightarrow CuS + 2H^+ \qquad (6)$$

$$H_2S + \frac{1}{2}O_2 \rightarrow S^° + H_2O \qquad (7)$$

$$H_2S + 2Fe^{3+} \rightarrow S^° + 2Fe^{2+} + 2H^+ \qquad (8)$$

Figure 4 presents the results of copper extraction during leaching of the concentrate. It can be seen that the initial rates of copper extraction in the media with iron addition (especially ferric medium and the medium regulated at 420 mV) were significantly higher than those in the test with no iron addition. However, the final extraction in the test regulated at redox potential of 420 mV and the test with ferrous addition are maximal. The initial increases were followed by a parabolic behaviour attributed to the chalcopyrite passivation. The high initial rates of copper dissolution were mainly related to the dissolution of copper oxides and secondary copper sulfides. The final copper recoveries were 21.9%, 39.6%, 44.6% and 46.2% for control test, ferric added medium, ferrous added medium and the medium with the initial redox potential of 420 mV, respectively. This result is in agreement with those obtained by Hiroyoshi et al. [2] and Third et al. [6] who reported that ferrous iron is more effective than ferric iron to dissolve chalcopyrite. Higher copper extraction rate in the first day of ferric added medium was related to the higher efficiency of chalcocite and covellite in ferric media.

On the other hand, as shown in **Figure 5**, in the experiments that iron has been added, the iron concentration in the solution was reduced during the process. This

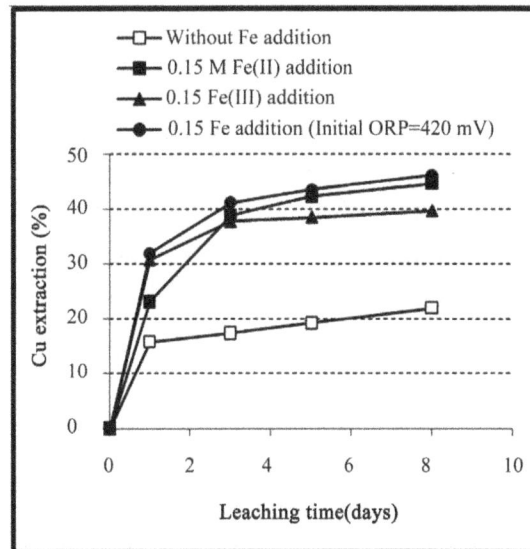

Figure 5. Variation of iron concentration in solution during the chemical leaching of the copper concentrate at 0.15 M iron addition (except the control test), pulp density 1% (w/v), 50°C, nutrient medium Norris and initial pH 1.8.

decrease was occurred in both ferrous and ferric media especially in the latter medium. The reason for this decrease was ascribed to the precipitation of a part of iron present in the solution as iron-hydroxy precipitates such as jarosite in the ferric medium and goethite and/or hematite in the ferrous medium. These precipitates were considered as one of the most probable causes to slow down the leaching rate.

3.2. Effect of Ferric and Ferrous Iron on Bacterial Leaching

Figures 6 and **7** present the variation of redox potential and pH during the bioleaching experiments. **Figure 6** shows that the addition of ferric iron caused a high initial redox potential (490 mV) but it did not remain constant in the following days. At the first day of the leaching process, the redox potential decreased from 490 to 370 mV, after that, it began to increase, but its final values remained lower than those in the test with no iron addition. This indicates that the high concentration of ferric iron reduced the ability of bacteria to oxidize ferrous iron. This inhibition effect had previously been reported by Lin *et al.* [16] in bioleaching of pyrite using *At. ferrooxidans* and *Leptospirillum ferriphilum* bacteria, and the research conducted by Nyavor *et al.* [17] who reported that the high concentration of ferric iron competitively inhibits ferrous iron oxidation by *At. ferrooxidans*.

For the case of ferrous addition, the redox potential began to increase rapidly from 319 mV to about 490 mV after 5 days (**Figure 6**), this was then followed by a slight increase to the maximum redox potential of 515 mV over the next 11 days. This result shows that the high level of initial ferrous iron had also an inhibitory effect on the yield of bacterial oxidation of ferrous iron. However, this inverse effect may also be related to the high concentration of ferric iron (~4.12 g/l in day 6) which had been produced by the bacterial oxidation of added ferrous iron. As a result, both solutions which iron had been added to them would have a similar composition after several days. Lower values of redox potential were also attributed to the formation of ferric hydroxides such as jarosite (Equation (1)), causing a decrease in ferric iron in the solution, which in turn led to a decrease in the value of redox potential. Essential nutrients needed for the activity and growth of bacteria were also precipitated with iron hydroxides which hamper the activity and growth of bacteria. These results were in agreement with the research conducted by Kawabe *et al.* [18] who reported that the complete inhibitory effect of ferric iron on bacterial growth was achieved at a high level of ferric iron (>0.3 M).

The variations of pH during bioleaching experiments were shown in **Figure 7**. It can be seen that, for the experiment with no iron addition, after a transient increase, pH value decreased gradually to about 1.5 (after 9 days) mainly due to the activity of sulfur-oxidizing bacteria to produce acid (Equation (9)). After this time, pH remained constant around 1.5 over the next 7 days.

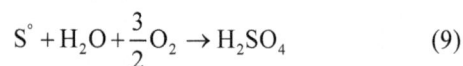

$$S^{\circ} + H_2O + \frac{3}{2}O_2 \rightarrow H_2SO_4 \qquad (9)$$

For the experiments with iron addition, a similar pattern was observed in which after a transient increase in the first few days, pH was gradually decreased to about

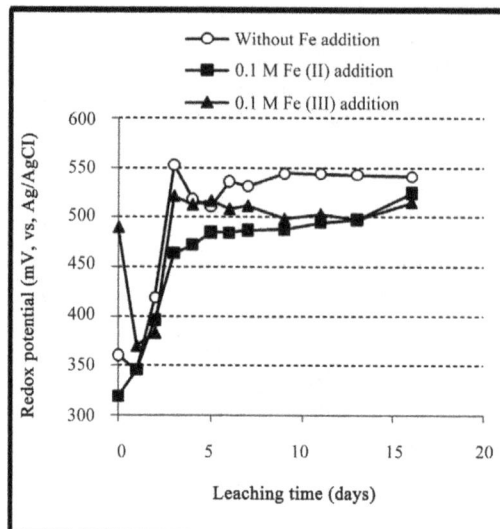

Figure 6. Variation of redox potential during the bioleaching of the copper concentrate with and without 0.1 M iron addition at pulp density 10% (w/v), 50°C, nutrient medium Norris and initial pH 1.6.

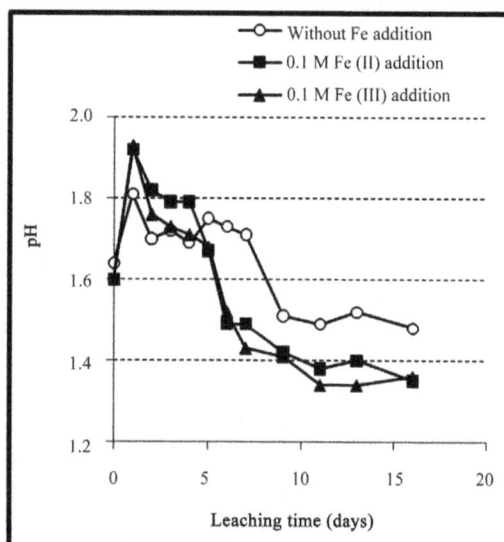

Figure 7. Variation of pH during the bioleaching of the copper concentrate with and without 0.1 M iron addition at pulp density 10%, 50°C, nutrient medium Norris and initial pH 1.6.

1.4 after 9 days. The lower pH value in the experiments with iron addition was attributed to the formation of jarosite which is an acid producing reaction (Equation (2)).

Figures 8 and **9** show the results of copper and iron extraction during the bioleaching experiments. **Figure 8** shows that the addition of iron into the solution increased the initial rates of copper extraction (~14%/day in the iron added medium versus 10.5%/day in the non added iron medium) which were followed by leveling off the extraction curves. The flat regions in the last 10 days of iron added cultures were attributed to the passivation of chalcopyrite surface. It can be seen that the yield of copper extraction in the ferric medium (36.7%) was significantly lower than that obtained in the experiment with no iron addition (43.1%) and one that ferrous iron was added (41.5%). Copper extraction rates in the last 10 days were 0.02%, 0.18% and 1.1% day in the ferric, ferrous and non added iron media, respectively.

Figure 9 shows that in the ferric added medium the amount of dissolved iron was lower than the amount of

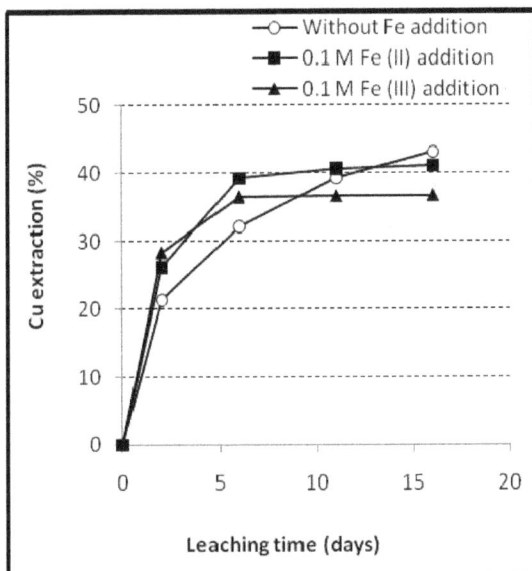

Figure 8. Extraction of copper during the bioleaching of the copper concentrate with and without 0.1 M iron addition at pulp density 10% (w/v), 50°C, nutrient medium Norris and initial pH 1.6.

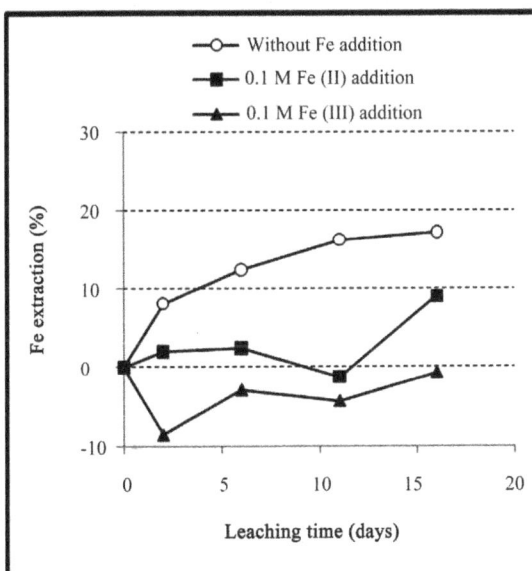

Figure 9. Extraction of iron during the bioleaching of the copper concentrate with and without 0.1 M iron addition at pulp density of 10% (w/v), 50°C, nutrient medium Norris and initial pH 1.6.

precipitated iron (negative extraction). The drop of iron extraction in the 2nd day of ferric added medium and 11th day of ferrous added medium were probably due to the precipitation of jarosite. The final iron extraction in the experiment with no iron addition reached 17.2%.

To explain the role of ferrous iron in chalcopyrite leaching two models have been proposed. The first model expressed that ferrous iron acts as a catalyst for converting chalcopyrite to chalcocite as a more soluble mineral (Hiroyoshi et al., [2]) and the second one related that the dissolved oxygen converted ferrous to ferric iron at the chalcopyrite surface and the reaction proceeds as expected [19].

4. Conclusions

The following results were obtained from chemical and bacterial leaching experiments for Sarcheshmeh chalcopyrite concentrate:

In the abiotic leaching tests, it was found that the addition of iron at low solution redox potentials (around 420 mV) significantly increased the rate and extent of copper dissolution but when only ferric iron was added, despite a higher initial rate of copper dissolution, leaching process stopped after several days. Decreasing the concentration of dissolved iron was occurred in both ferrous and ferric media especially in the latter medium. The reason for this decrease could be related to the removal of a part of iron present in the solution as iron-hydroxy precipitates. It can be jarosite in the ferric medium and goethite and/or hematite in the ferrous medium.

Addition of both ferrous and ferric iron to the bioleaching medium leveled off the copper extraction and had an inhibitory effect which decreased the final redox potential. It indicates the lower ability of bacteria to oxidize ferrous iron in the presence of these cations.

5. Acknowledgements

The support of the National Iranian Copper Industry Company for providing the materials is gratefully acknowledged.

REFERENCES

[1] G. Rossi, "Biohydrometallurgy," McGraw Hill, Hamburg, 1990.

[2] N. Hiroyoshi, M. Hirota, T. Hirajima and M. Tsunekawa, "A Case of Ferrous Sulfate Addition Enhancing Chalcopyrite Leaching," *Hydrometallurgy*, Vol. 47, No. 1, 1997, pp. 37-45. doi:10.1016/S0304-386X(97)00032-7

[3] N. Hiroyoshi, H. Miki, T. Hirajima and M. Tsunekawa, "Enhancement of Chalcopyrite Leaching by Ferrous Ions in Acidic Ferric Sulfate Solutions," *Hydrometallurgy*, Vol. 60, No. 3, 2001, pp. 185-197. doi:10.1016/S0304-386X(00)00155-9

[4]   N. Hiroyoshi, H. Kitagawa and M. Tsunekawa, "Effect of Solution Composition on the Optimum Redox Potential for Chalcopyrite Leaching in Sulfuric Acid Solutions," *Hydrometallurgy*, Vol. 91, No. 1-4, 2008, pp. 144-149. doi:10.1016/j.hydromet.2007.12.005

[5]   A. Pinches, P. J. Myburgh and C. Merwe, "Process for the Rapid Leaching of Chalcopyrite in the Absence of Catalysis," US patent No. 6277341 B1, 2001.

[6]   K. A. Third, R. Cord-Ruwisch and H. R. Watling, "Control of the Redox Potential by Oxygen Limitation Improves Bacterial Leaching of Chalcopyrite," *Biotechnology and Bioengineering*, Vol. 78, No. 4, 2002, pp. 433-441. doi:10.1002/bit.10184

[7]   A. Sandström, A. Shchukarev and J. Paul, "XPS Characterisation of Chalcopyrite Chemically and Bio-Leached at High and Low Redox Potential," *Minerals Engineering*, Vol. 18, No. 5, 2005, pp. 505-515. doi:10.1016/j.mineng.2004.08.004

[8]   E. M. Cordoba, J. A. Munoz, M. L. Blázquez, F. González and A. Ballester, "Leaching of Chalcopyrite with Ferric Ion. Part IV: The Role of Redox Potential in the Presence of Mesophilic and Thermophilic Bacteria," *Hydrometallurgy*, Vol. 93, No. 3-4, 2008. pp. 106-115. doi:10.1016/j.hydromet.2007.11.005

[9]   M. Gericke, Y. Govender and A. Pinches, "Tank Bioleaching of Low-Grade Chalcopyrite Concentrates Using Redox Control," *Hydrometallurgy*, Vol. 104, No. 3-4, 2010, pp. 414-419. doi:10.1016/j.hydromet.2010.02.024

[10]   A. Ahmadi, M. Schaffie, Z. Manafi and M. Ranjbar, "Electrochemical Bioleaching of High Grade Chalcopyrite Flotation Concentrate in a Stirred Tank Reactor," *Hydrometallurgy*, Vol. 104, No. 1, 2010, pp. 99-105. doi:10.1016/j.hydromet.2010.05.001

[11]   A. Ahmadi, M. Schaffie, J. Petersen, A. Schippers and M. Ranjbar, "Conventional and Electrochemical Bioleaching of Chalcopyrite Concentrates by Moderately Thermophile Bacteria at High Pulp Density," *Hydrometallurgy*, Vol. 106, No. 1-2, 2011, pp. 84-92.

doi:10.1016/j.hydromet.2010.12.007

[12]   T. Das, S. Ayyappan and G. R. Chaudhury, "Factors Affecting Bioleaching Kinetics of Sulfide Ores Using Acidophilic Micro-Organisms," *BioMetals*, Vol. 12, No. 1, 1999, pp. 1-10. doi:10.1023/A:1009228210654

[13]   J. L. Barron and D. R. Lueking, "Growth and Maintenance of *Thiobacillus ferrooxidans* Cells," *Applied and Environmental Microbiology*, Vol. 56, 1990, pp. 2801-2806.

[14]   D. Howard and F. K. Crundwell, "A Kinetic Study of the Leaching of Chalcopyrite with Sulfolobus Metallicus," *Biohydrometallurgy and the Environment toward the Mining of the 21st Century*, 1999, pp. 209-217.

[15]   P. R. Norris and D. W. Barr, "Growth and Iron Oxidation by Acidophilic Moderate Thermophiles," *FEMS Microbiology Letters*, Vol. 28, No. 3, 1985, pp. 221-224. doi:10.1111/j.1574-6968.1985.tb00795.x

[16]   L. Zhang, G.-Z. Qiu, Y.-H. Hu, X.-J. Sun, J.-H. Li and G.-H. Gu, "Bioleaching of Pyrite by *A. ferrooxidans* and *L. ferriphilum*," *Transactions of Nonferrous Metals Society of China*, Vol. 18, No. 6, 2008, pp. 1415-1420. doi:10.1016/S1003-6326(09)60018-2

[17]   K. Nyavor, N. O. Egiebor and P. M. Fedorak, "The Effect of Ferric Ion on the Rate of Ferrous Oxidation by *Thiobacillus ferrooxidans*," *Applied Microbiology and Biotechnology*, Vol. 45, No. 5, 1996, pp. 688-691. doi:10.1007/s002530050749

[18]   Y. Kawabe, C. Inoue, K. Suto and T. Chida, "Inhibitory Effect of High Concentrations of Ferric Ions on the Activity of *Acidithiobacillus ferrooxidans*," *Journal of Bioscience and Bioengineering*, Vol. 96, No. 4, 2003, pp. 375-379.

[19]   C. Klauber, "A Critical Review of the Surface Chemistry of Acidic Ferric Sulfate Dissolution of Chalcopyrite with Regards to Hindered Dissolution," *International Journal of Mineral Processing*, Vol. 86, No. 1-4, 2008, pp. 1-17. doi:10.1016/j.minpro.2007.09.003

14

Removal of Chromium(III) from the Waste Solution of an Indian Tannery by Amberlite IR 120 Resin

Pratima Meshram[1], Sushanta Kumar Sahu[1]*, Banshi Dhar Pandey[1],
Vinay Kumar[1], Tilak Raj Mankhand[2]

[1]Metal Extraction & Forming Division, CSIR-National Metallurgical Laboratory, Jamshedpur, India
[2]Department of Metallurgical Engineering, IIT BHU, Varanasi, India
Email: *sushanta_sk@yahoo.com

ABSTRACT

The extraction of chromium(III) from a model waste solution and also from a waste solution of an Indian tannery with Amberlite IR 120 resin is described, and the performance of this resin is compared with other similar resins. The parameters that were optimized include effect of mixing time, pH, loading and elution behaviours of chromium(III) for this resin. Sorption of chromium(III) on Amberlite IR 120 followed Freundlich isotherm and Langmuir isotherm model, and the maximum sorption capacity was determined to be 142.86 mg Cr(III)/g of the resin. Higher Freundlich constant (K_f) values (6.30 and 13.46 for aqueous feed of 500 and 1000 ppm Cr(III)) indicated strong chemical interaction through ion exchange mechanism of the metal ion with the resin. The kinetic data showed good fit to the Lagergren first order model for extraction of chromium(III). Desorption of chromium(III) from the loaded resin increased with the increase in concentration of eluent (5% - 20% H_2SO_4). With 20% (v/v) sulphuric acid solution 94% chromium(III) was eluted in three stages. Elution of the Cr(III) in the column experiments was however, found to be lower (82%) than that of the shake flask data. In case of Indian tannery's waste solution, it was observed that almost total chromium was extracted in four stages with Amberlite IR 120.

Keywords: Chromium(III); Ion exchange; Amberlite IR 120; Tannery Waste Solution

1. Introduction

In recent years, chromium has received considerable attention owing to uses of its compounds in pigments and paints, leather tanning, oxidative dying, electroplating, fungicides, catalysis, refractory materials, glass Industries and various other industrial applications. These industrial processes discharge large quantities of chromium into the environment. Chromium occurs in aqueous systems in the trivalent and hexavalent forms. Out of the two forms, hexavalent chromium is more hazardous to living organisms than the chromium(III). Rapid oxidation of chromium(III) to chromium(VI) state in aquatic and solid wastes situations accounts for mobility of chromium. Therefore, removal and recovery of chromium(III) from industrial wastewater and effluents are critical from both ecological and economic point of view. It may reduce the risk of polluting environment while the recovered compounds of chromium(III) can be reused.

There are around 2500 tanneries in India including Tamilnadu (50%), West Bengal (20%) and Uttar Pradesh (15%). The other important states for the leather processing are Maharashtra, Andhra Pradesh and Punjab. India annually produces around 180 million m^2 of leather, which accounts for about 10% of global production [1]. As leather tanning industry effluent is one of the main sources of chromium pollution in aquatic system and the production of leather is increasing in India, development of a process for the removal of chromium ions is very important for the country. Basic Cr(III) sulphate is the main chemical used in the tanning process after which the spent tanning solution is discharged with a high concentration of chromium, causing harmful effects to the environment. The consumption of basic chromium salts by the Indian leather industry is about 24,000 tons per annum. Thus, about 2000 - 3200 tons of elemental chromium escape into the environment annually from these industries, with a chromium concentration ranging between 2000 - 5000 mg/L [2]. These waste solutions are generally diluted and treated to convert Cr(III) to a hydroxide sludge for disposal [3]. Whilst the treated waste water is allowed to enter to land stream with still higher level of chromium compared to the recommended permissible limit of 2 mg/L. This trend, if not arrested, will certainly lead to a huge material loss as well as creating an ecological imbalance.

A number of methods *viz.* chemical precipitation [4],

---

coagulation [5], adsorption [6], solvent extraction [7], ion exchange [8], biosorption [9], membrane separation [10, 11], etc. are available for removal of metal ions from liquid waste streams. Literature survey further shows that ion exchange is one of the most frequently studied and widely applied techniques for the treatment of metal-contaminated wastewater, recovery of metallic substances from such streams, and the regeneration of solutions for recycling [12-16]. This is a promising technique based on adsorption of cations or anions on synthetic resins with essential characteristics of its regeneration after elution. Further, this has two main advantages over its competitor like solvent extraction: mixing and settling arrangements are not required and organic phase losses are completely avoided. Besides, ion exchange is appropriately suited to treat even the low metal ion containing solutions particularly those of waste streams obtained from the processing of low grade materials and effluents.

Several studies considering the chromium removal by ion exchange resins have been reported in the literature. Agrawal et al. [17] have recently reviewed the remediation options for the treatment of chromium containing waste solutions and summarized the possibilities of using different methods including ion exchange. Petruzzelli et al. [18,19] have developed a process, known as IERE-CHROME (Ion Exchange Recovery of Chromium) for removal, recovery and reuse of chromium(III) from segregated tannery wastewater. The process has the advantage of recovery of almost pure chromium (>99%) from other interfering metals and organic compounds. Kocaoba and Akcin [20], have studied the removal of chromium(III) and cadmium(II) from aqueous solution using Amberlite IR 120, a strong cation exchange resin. Both batch and continuous ion exchange process for the recovery of chromium(III) from the aqueous chloride solution by using Amberlite IR 120 has been described by Alguacil et al. [21]. Kocaoba and Akcin [22] have also compared the performance of Amberlite IR 120 with two weakly acidic resins, Amberlite IRC 76 and Amberlite IRC 718, and observed that both the weakly acidic resins exhibited better performance than Amberlite IR 120. Recent studies [23-25] have shown that chelating resins could be used for the selective removal and recovery of trivalent chromium. Three chelating ion-exchange resins (Amberlite IRC 748, Diaion CR 11 and Diphonix) were tested [26] for separation of Cr(III) from industrial effluents produced in hard and decorative electroplating. Efforts are also made to develop newer chelating sorbents with new organic functionalities for recovery of chromium(III) from wastewaters [27-29].

Recently, Sahu et al. [30] studied the removal of chromium(III) from tannery waste solution using a new strongly acidic cation exchange resin, Indion 790. Kinetics and loading capacity of Indion 790 were found to be

much better than other reported resins viz. Lewatit S100 [31] and IRN 77 [32], although extraction efficiency suffered after pH 3.5. In view of the presence of high salt content and various organic substances in the waste tanning solution [33] a detailed investigation on the applicable of Amberlite IR 120 for the removal of chromium(III) from model and actual waste tanning solution of an Indian industry has been reported in the present study. The performance of Amberlite IR 120 has also been compared with other commercially available cation exchange resins of similar type.

2. Experimental

In the present study, recovery of chromium from the model tanning solutions containing 500 and 1000 ppm chromium(III) by ion exchange using a macro-porous strongly acidic cation exchange resin derived from sulfonated polystyrene group, Amberlite IR 120 has been used. Amberlite IR 120 manufactured from Rohm and Haas Company Limited (USA), has the physical properties and specifications as given in **Table 1**. Before experiments, about 20 g of the resin was washed properly with 200 mL of distilled water and pre-treated with 50 mL of 5% (v/v) hydrochloric acid for 10 min. The treated resin was again washed properly with distilled water to remove excess Cl^- and dried at room temperature for 24 h. The resin was stored in dry condition for experimental work. A synthetic stock solution of 1000 ppm of chromium(III) was prepared by dissolving required amount of $CrCl_3 \cdot 6H_2O$ in distilled water, and solution of desired concentration was prepared by diluting the stock solution.

To study the extraction of chromium(III) by ion exchange, 50 mL of the aqueous solution of known concentration and pH (2.7 which is similar to the waste tannery solution) was equilibrated with weighed amount

Table 1. Properties of Amberlite IR 120 ion exchange resin.

Parameter	Remarks
Ionic form	H^+
Functional group	SO_3^-
Matrix type	Styrene DVB
Resin type	Macroporous strong acidic cation
Particle size (mm)	0.6 - 0.8
Moisture (%)	45 - 50
Max. operating temp. (°C)	120 - 150
pH range	0 - 14
Total exchange capacity (meq/mL)	2.0

(1.0 g) of dry resin so as to maintain the A/R ratio of 50 (A = volume of the aqueous feed in mL and R = amount of resin in g) unless stated otherwise. The loaded resin was washed thoroughly with distilled water and then chromium was eluted with 50 mL of sulphuric acid solution of a known strength. All the extraction and elution studies were carried out at 303 K. The raffinate and the eluted solutions were analyzed for chromium concentration by Atomic Absorption Spectrometer (ECIL, India) and the material balance was checked. Sorption isotherms were determined by repeatedly loading the resin with 500 or 1000 ppm chromium solutions for eight and six times, respectively. The extraction data were fitted into different isotherms in order to determine the sorption performance and loading capacity of the resin for chromium(III) ion.

Ion exchange behaviour of Amberlite IR 120 was also investigated by passing chromium(III) solution at a flow rate of 2 mL/min over 1.0 g of this resin packed in a column. Industrial application of the resin was examined and compared by studying the recovery of chromium(III) from the actual tanning waste solution. The tanning waste solution was obtained through Central Leather Research Institute (CLRI), Chennai from a plant situated in Chennai, India. The composition of the waste solution is given in **Table 2**. The tannery waste solution contains biogenic matter and a large variety of inorganic chemicals and organic contaminants [33,34] such as aliphatic sulfonates, sulphates, aromatic and aliphatic ethoxylates, sulfonated poly-phenols, acrylic acid, fatty acids, dye, proteins, soluble carbohydrates etc.; the organic load of the solution is presented as COD of 4.4 g/L and BOD of 2.2 g/L (**Table 2**). The waste tannery solution was diluted 10 times and used for extraction of chromium(III) with Amberlite IR 120.

3. Results and Discussion

3.1. Effect of Mixing Time

The effect of mixing time on extraction of chromium

Table 2. Composition of the tanning bath solution collected from CLRI, Chennai.

Constituent	g/L
Cr(III)	4.57
Fe(III)	0.05
Al(III)	0.12
COD	4.4
BOD	2.2
SO_4^{2-}	12.0
NaCl	60
pH	2.5

from aqueous feeds containing 500 or 1000 ppm Cr(III) was investigated using Amberlite IR 120 at an initial pH of 2.7 and A/R = 50. The extraction of chromium(III) increased as the contact time increased and a maximum of 99% Cr(III) was extracted in 15 min from 500 ppm chromium(III) solution, although the major extraction (93%) was achieved in 5 min time (**Figure 1**). On the other hand 96% chromium was extracted in 15 min and major extraction (90%) was achieved in 8 min for 1000 ppm Cr(III) solution. The metal extraction versus time curve is smooth and continuous, leading to saturation indicating that equilibrium is attained on the surface of the resin.

Extraction of chromium is also dependent on initial concentration of the metal in solution. **Figure 1** shows that in 15 min extraction of chromium with Amberlite IR 120 was higher (99%) for the feed concentration of 500 ppm Cr(III) as compared to the concentrated solution containing 1000 ppm chromium where extraction was 96%. This is due to the fact that in dilute solution the amount of active sites available in the resin is higher as compared to that of concentrated solutions, and therefore higher level of extraction is observed from the dilute solution

3.2. Effect of pH

In order to optimize the pH for maximum removal efficiency, the experiments were conducted with 500 and 1000 ppm Cr(III) at A/R of 50 and 15 min contact time using Amberlite IR 120 in the pH range 1.0 - 5.0. The effect of pH on the extraction of Cr(III) is presented in **Figure 2**. Extraction was found to be almost constant in the pH range 1.0 - 4.5 and further increase in pH caused decrease in chromium extraction. The pH value of the solution is an important factor that controls the sorption of chromium(III), affecting its uptake on the resins; in general the uptake decreases at higher pH values [32].

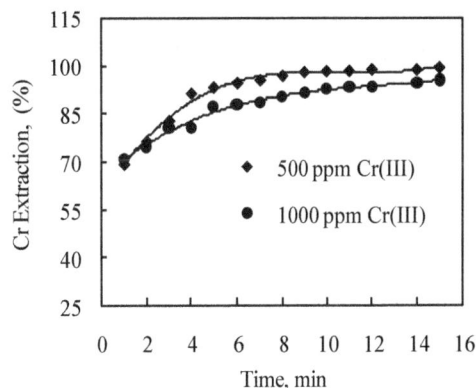

Figure 1. Effect of mixing time on extraction of chromium (III) with Amberlite IR 120. [Cr(III)] = 500 ppm, pH = 2.7, A/R = 50.

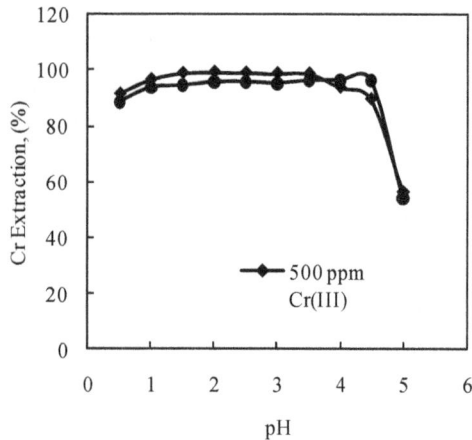

Figure 2. Effect of pH on extraction of chromium(III) with Amberlite IR 120. [Cr(III)] = 500 ppm, A/R = 50, Contact time = 15 min.

High removal efficiency of chromium by Amberlite IR 120 in the pH range 1.0 - 4.5 can likely to be ascribed to the effect of competitive binding between chromium(III) and hydrogen ions on the surface of the resin. At pH < 1, an excess of hydrogen ions can compete effectively with chromium(III) for binding sites, resulting in a lower level of chromium(III) uptake. At pH values above 4.5, chromium(III) ions might hydrolyse and precipitate forming the hydroxyl complexes of chromium, $Cr(OH)_3$. Chromium(III) thus showed strong affinity towards the resin in the pH range 1.0 to 4.5.

3.3. Sorption Isotherm

An adsorption isotherm is used to characterize the equilibria between the amount of adsorbate that accumulates on the adsorbent and the concentration of the dissolved adsorbate. In order to understand the nature of adsorption isotherm of chromium(III) on Amberlite IR 120, 1.0 g of the resin was repeatedly contacted with fresh 500 ppm and also with 1000 ppm chromium(III) solutions. The trend of extraction of chromium(III) by the resin is shown in **Figure 3**, which indicates that in eight contacts 100 mg chromium(III) is transferred to the resin with 500 ppm chromium(III) solution. Whereas, in case of 1000 ppm Cr(III) solution, in six contacts 136.7 mg chromium(III) is transferred to the resin. The data so obtained while repeatedly loading Amberlite IR 120 with chromium(III) were fitted into Freundlich and Langmuir isotherms.

Freundlich model assumes that the uptake or adsorption of metal ions occurs on a heterogeneous surface by monolayer adsorption. The model is described by the following equation:

$$q = K_f \left(C_e\right)^{1/n} \qquad (1)$$

taking log of Equation (1)

$$\log q = \frac{1}{n}\log C_e + \log K_f \qquad (2)$$

The plot (**Figure 4**) of $\log q$ versus $\log C_e$ gave a straight line indicating that the amount of chromium(III) adsorbed on the surface of Amberlite IR 120 depends linearly on the equilibrium chromium(III) concentration in the aqueous solution with a correlation coefficient (R^2) of 0.988 and 0.985 with a feed concentration of 500 and 1000 ppm Cr(III) solution respectively. The Freundlich constants (K_f) were found to be 6.30 and 13.46 for 500 and 1000 ppm chromium(III) solution respectively.

According to Langmuir model, uptake of metal ions occurs on a homogeneous surface by monolayer sorption without any interaction with the sorbed ionic species. The model is described by:

$$q = \frac{\left(q_m K_l C_e\right)}{\left(1 + K_l C_e\right)} \qquad (3)$$

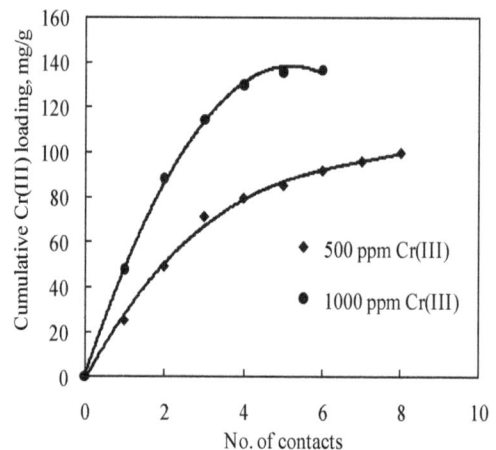

Figure 3. Saturation loading of Amberlite IR 120 with 500 ppm chromium(III) solution. pH = 2.7, Contact time = 15 min, A/R = 50.

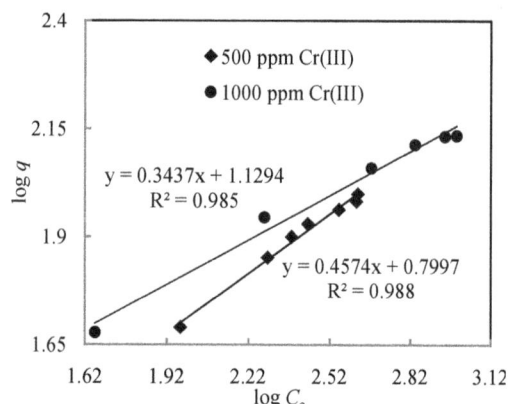

Figure 4. Freundlich isotherm for sorption of chromium(III) with Amberlite IR 120.

rearranging Equation (3)

$$\frac{1}{q} = \frac{1}{(q_m K_l C_e)} + \frac{1}{q_m} \qquad (4)$$

The plot (**Figure 5**) of $1/q$ versus $1/C_e$ shows that $1/q$ depends linearly with $1/C_e$ with a correlation coefficient of 0.996 and 0.990 for 500 and 1000 ppm Cr(III) solution respectively. Langmuir constants (K_l) were found to be 0.006 and 0.011 for aqueous feed of 500 ppm and 1000 ppm chromium(III) solution respectively (**Table 3**). From the Langmuir model, loading capacity of Amberlite IR 120 for chromium(III) was also determined and found to be 142.86 mg/g of resin. Loading capacity of Amberlite IR 120 was found to be higher in the present investigation than that of other resins of similar characteristics (**Table 4**) because of higher metal concentration in the aqueous solution.

The plots depicted in **Figures 4** and **5** show that the equilibrium adsorption data fitted well into the linear Langmuir and Freundlich isotherms. However, higher Freundlich constant values (K_f = 6.30 and 13.46) and values of n (2.18 and 2.92) between 1 and 10 indicated strong chemical interaction between the resin and chromium(III) [35], this also showed that the loading of metal on to the resin surface followed ion exchange mechanism rather than the adsorption mode.

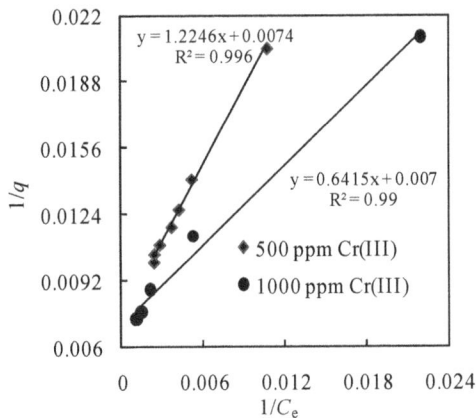

Figure 5. Langmuir isotherm for sorption of chromium(III) with Amberlite IR 120.

Table 3. Physical constants obtained for Freundlich and Langmuir isotherms.

Feed concentration	Freundlich isotherm			Langmuir Isotherm	
	R^2	K_f	n	R^2	K_l
500 ppm Cr(III)	0.988	6.30	2.18	0.996	0.006
1000 ppm Cr(III)	0.985	13.46	2.92	0.990	0.011

3.4. Kinetics of Extraction

To understand the sorption mechanism of chromium(III) on Amberlite IR 120, the kinetic data were fitted into Lagergren first order model [36]:

$$\log(q_e - q_t) = \log q_e - \frac{K_l}{2.303}t \qquad (5)$$

where all the terms have usual meaning as defined in the nomenclature section. From **Figure 6** on $\log(q_e - q_t)$ versus time, the first order rate constant value determined from the slope of straight line, for exchange of chromium(III) ions from an aqueous feed of 500 ppm to Amberlite IR 120 was found to be 0.297 min^{-1} with correlation coefficient (R^2) of around 0.978, whilst with 1000 ppm Cr(III) feed solution, the first order rate constant for exchange of chromium(III) ions was calculated to be 0.129 min^{-1} with correlation coefficient (R^2) of about 0.970. The low values of rate constants further indicate that the extraction process is controlled by internal particle diffusion of sorbed chromium(III) species on the resins as discussed below.

Different mechanisms and steps in ion-exchange phenomena can control the kinetics. Four major rate limiting steps are generally cited. Out of these first two are mass transfer resistance steps comprising of the mass transfer of solute from solution to the boundary film initially and then mass transfer of metal ions from boundary film to the surface. After mass transfer steps sorption and ion exchange of ions onto sites takes place and finally internal diffusion of solute takes place. The external mass transfer resistance steps are neutralized with the homogeneity of the solution, and the third step is fast. Therefore, internal particle diffusion of sorbed chromium(III) limits the process.

As reported by Ho and McKay [37], the sorption will follow the first order kinetics unless the experimental data are much scattered or deviated with time. In the

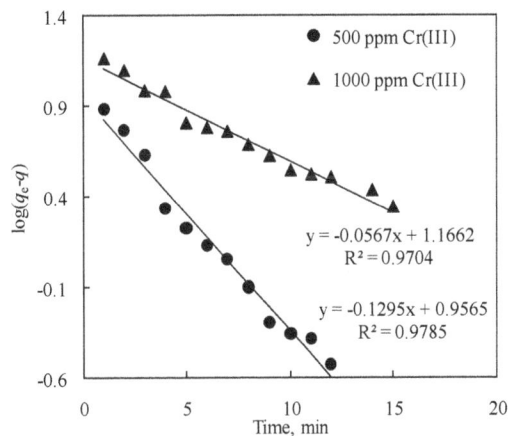

Figure 6. First order kinetics for extraction of chromium (III) by ion exchange with Amberlite IR 120.

Table 4. Comparisons of performances of different resins used for chromium(III) extraction.

Ion Exchanger	Nature of resin	A/R (mL/g)	Initial Conc. (mg/L)	pH	Cr(III) removal efficiency	Wastewater	Loading Capacity (mg/g)	Highlights	References
Lewatit S 100	Sulfonic acid group with cross linked polystyrene matrix	300	52	3.5	99%	$CrCl_3$	20.3	Sorption followed Langmuir isotherm and first-order reversible kinetics.	[31]
Amberlite IR 120	Sulfonic acid group with Styrene-DVB polymer matrix	100,000	10	5	94.27%, 87.61% in (H⁺) form	$[Cr_4(SO_4)_5 (OH)_2]$	2.53 (Na⁺), 2.19 (H⁺)	Amberlite IRC 76 and 718 weakly acidic resins showed better performance than Amberlite IR 120 for recovery of Cr(III) in Na⁺ form. Elution was achieved with 10% HCl at 5 BV h⁻¹ flow rate.	[22]
Amberlite IR 120	Sulfonic acid group with Styrene-DVB polymer matrix	10,000	20	5.5	68.24% Cr, 90.8% Cd in Na⁺ form and 72.41% Cr, 81.1% Cd in H⁺ form	$[Cr_4(SO_4)_5 (OH)_2]$, $[Cd(NO_3)_2. 4H_2O]$	2.53 (Na⁺) 2.19 (H⁺)	Cadmium adsorption was much higher than chromium in Na⁺ form.	[20]
Amberlite IR 120	Sulfonic acid group with Styrene-DVB polymer matrix	200	500	0.92	-	$CrCl_3$	67.7	Followed Langmuir isotherm, and film-diffusion controlled kinetics.	[21]
IRN 77	Sulfonic acid group with polystyrene DVB matrix	200	100	2.7	>95%	$Cr(NO_3)_3$/chromium containing synthetic primary coolant water	35.4	Followed Freundlich adsorption isotherm. Adsorption of Co, Cr and Ni followed the first order kinetic model.	[32]
SKN 1	Sulfonic acid group	200	100	2.7	>95%	$Cr(NO_3)_3$	46.3	Followed Freundlich adsorption isotherm. Adsorption of Co, Cr and Ni followed the first order kinetic model.	[32]
Indion 790	Styrene DVB with sulfonic acid groupmacro-porous strongly acidic cation exchange resin of sulfonated polystyrene group	50	500	2.74	92%	$CrCl_3$	86.9	Followed Freundlich isotherm and first order kinetics. With 20% sulfuric acid solution 89% Cr(III) was eluted in two stages.	[30]
Amberlite IR 120	Sulfonic acid group with Styrene DVB matrix	50	500	2.7	99%	$CrCl_3$	142.86	Sorption followed Freundlich isotherm and first-order kinetics. With 20% sulphuric acid solution 94% Cr(III) was eluted in three stages.	Present work

present study, the kinetic plot (**Figure 6**) with time had no deviation and showed good fit (correlation coefficient ≥ 0.97) to the Lagergren first order model for extraction of chromium(III).

3.5. Desorption of Chromium(III)

Desorption studies help to recover the metals from the loaded resin. The reliability of the ion exchange process strongly depends on the regeneration efficiency of the exchanger. In fact, if a resin is characterized by high selectivity but low desorption capacity, the overall process is then considered less attractive. Desorption of chromium(III) from the loaded Amberlite IR 120 with different concentration of sulfuric acid solution was also studied and the results are given in **Table 5**. As can be seen the elution of Cr(III) showed increasing trend with increase in acid concentration from 5% to 15% (v/v) in the first stage, although the overall elution in 3-stages was almost the same. A maximum of 94% chromium(III) elution was achieved with 20% (v/v) sulfuric acid solution in three stages. The incomplete elution of chromium(III) from the loaded resins may be attributed to the tendency of formation of hydrolysis product such as Cr(OH)SO$_4$ which is difficult to elute under normal conditions [32].

3.6. Extraction of Chromium(III) in Column

Extraction of chromium(III) by Amberlite IR 120 in a continuous mode was also studied. The aqueous solution of 500 ppm Cr(III) at pH 2.7 was passed through a small column containing 1.0 g of resin (0.9 mL volume in the column) at a flow rate of 2.0 mL/min. The plot (**Figure 7**) of C_r/C_o versus bed volume shows a breakthrough of Cr(III) after 100 bed volumes for the resin. Amberlite IR 120 gets saturated after 300 bed volume i.e. after passage of 270 mL of the solution. Saturation loading of 110 mg Cr(III)/g Amberlite IR 120 was obtained, which is only slightly lower than the maximum loading capacity determined from the isotherm. Larger scale column test with slower flow rates would achieve proper raffinate concentration, breakthrough points and maximum metal loading.

Table 5. Elution of chromium(III) from loaded Amberlite IR 120 resin.

H$_2$SO$_4$ (%)	Chromium(III) elution (%)			
	Stage-I	Stage-II	Stage-III	Total
5	57.61	20.0	9.53	87.14
10	66.84	15.11	3.85	85.8
15	71.56	12.42	2.76	86.74
20	73.53	16.19	4.61	94.33

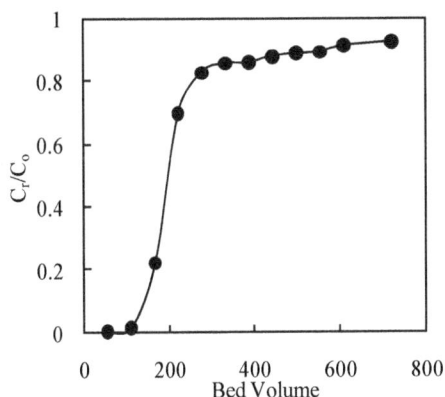

Figure 7. Extraction of chromium(III) with Amberlite IR 120 in column. Resin = 1.0 g, [Cr(III)] = 500 ppm, pH = 2.7, flow rate 2.0 mL/min.

When the metal loaded Amberlite IR 120 in the column was eluted with 20% (v/v) H$_2$SO$_4$ at a flow rate of 2.0 mL/min, only 82% chromium(III) was recovered after passing 200 mL of eluent. However, immediate elution at a slow flow rate after loading may improve the recovery.

3.7. Recovery of Chromium(III) from Tannery Waste Solution

A tannery waste solution containing 4.58 g/L chromium(III) and small quantities of aluminium and iron, and other constituents obtained through Central Leather Research Institute (CLRI), Chennai, India, was used to establish the process for industrial application of Amberlite IR 120. The discharge limit of chromium, prescribed by Central Pollution Control Board (CPCB), India is 2 ppm for streams [38]. In order to achieve this discharge level for chromium(III), 10 times diluted waste tannery solution was contacted with Amberlite IR 120 at pH 2.7, A/R ratio of 50 and a mixing time of 15 min. The raffinate obtained was again contacted with fresh resin. About 83% chromium(III) was extracted (**Figure 8**) in the first stage itself and in fourth stage the raffinate with trace amount (0.022 ppm) of chromium(III) was obtained, which was found to be safe for discharge. The efficiency of the resin towards the removal of chromium from the model chromium solution is found to be higher than that of actual tannery waste solution. This may be attributed to the presence of other species/impurities including metal ions present in tannery waste water which may interfere in the ion exchange process. Elution of chromium(III) from loaded resins with 20% H$_2$SO$_4$ was also studied and 90% chromium(III) elution was observed from the loaded Amberlite IR 120 in two stages.

3.8. A Comparative Performance of Amberlite IR 120 and Other Resins for Removal of Chromium(III) from Waste Solutions

The extraction behavior of Amberlite IR 120 for chromium(III) from a model waste solution and also from a waste tanning solution has been compared with that of another resin, Indion 790, which was used in our previous work [30]. Both the resins showed similar extraction behaviour towards chromium(III) from a model solution containing 500 ppm chromium(III) at pH 2.7. A maximum of 99% and 92% chromium(III) was recovered with Amberlite IR 120 and Indion 790, respectively in 12 min time.

In comparison, the kinetics of adsorption of chromium(III) with Amberlite IR 120 was faster than that of Indion 790. Both the resins showed constant chromium(III) extraction in the pH range 1.0 - 3.5. Beyond pH 3.5 extraction of chromium(III) with Indion 790 drastically

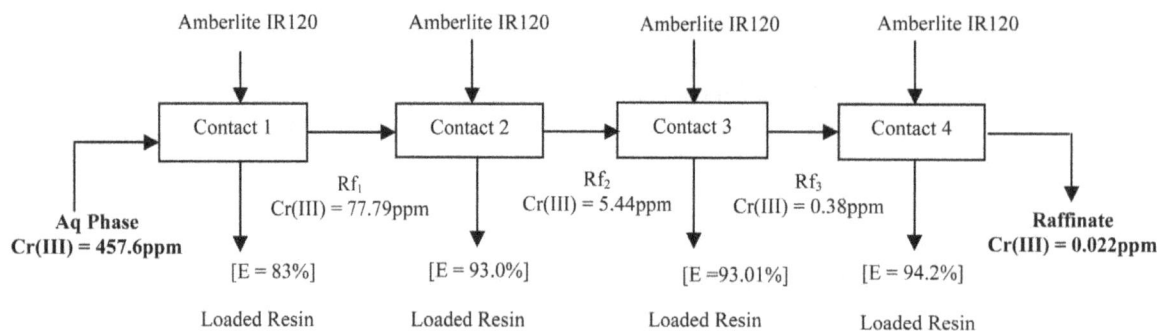

Figure 8. Extraction of chromium(III) from the diluted waste solution of tannery by Amberlite IR 120. Aq. Feed = 457.6 ppm Cr(III), Initial pH = 2.7, A/R = 50, Rf_{1-4} = Raffinate in different contacts, E = % extraction of Cr(III).

decreased from 92% to 76% whereas, Amberlite IR 120 was effective even up to 4.5 pH. Extraction of Cr(III) from the tannery waste solution (458 ppm Cr^{3+}) was found to be 99.9% and 95% with Amberlite IR 120 and Indion 790 in 3 stages at the aqueous feed pH of 2.7.

Sorption of chromium(III) on Amberlite IR 120 and Indion 790 followed Freundlich isotherm. The higher Freundlich constant (K_f) values such as 6.30 and 8.57 respectively indicated strong chemical interaction of the metal ion with the resins. The Amberlite IR 120 showed higher loading capacity than that of Indion 790 with 500 ppm chromium(III) feed. Both the resin follows the first order kinetic model for extraction of chromium(III). Desorption of chromium(III) from the loaded resins increased with the increase in concentration of eluant (5% - 20% H_2SO_4). With 20% (v/v) sulfuric acid solution 94% chromium(III) was eluted from loaded Indion 790 in two stages, and from loaded Amberlite IR 120, 94% chromium(III) was eluted in three stages.

Comparative data from some ion exchange studies for chromium(III) are further summarised in **Table 4**. A comparison between these resins showed that all the resins effectively removed Cr(III) from the wastewaters. Lewatit S 100 also showed higher chromium removal efficiency (99%) from a chloride solution due to the low metal ion concentration used. The removal efficiency as well as loading capacity of Amberlite IR 120 used in the present work even at lower pH of 2.7 was found to be higher than that of the other similar resins as given in **Table 4**.

4. Conclusion

Amberlite IR 120—a strong acidic cation exchanger has been used for the extraction of chromium(III) from the tanning waste solution. Different process parameters such as pH of the feed solution, sorption of metal, kinetics of extraction, elution of metal etc. have been optimized using this resin. When a model solution containing 500 ppm chromium(III) was contacted with Amberlite IR

120, about 99% chromium was extracted in 15 min at pH 2.7 and A/R ratio of 50. Sorption of chromium(III) on Amberlite IR 120 followed Freundlich isotherm model, and the maximum sorption capacity determined was 142.86 mg/g of the resin. Higher Freundlich constant (K_f) values (6.30 and 13.46) indicated strong chemical interaction of the metal ion with the resin which also showed extraction of Cr(III) by ion exchange mechanism rather than surface adsorption. The kinetics of the sorption of chromium by the resin followed the first order kinetics. A 20% (v/v) sulphuric acid solution gave maximum elution (94%) of Cr(III) from the loaded resin in three stages when treated for 15 min in each case. Study in column showed breakthrough at 100 bed volume from an aqueous feed of 500 ppm Cr(III) with a loading of 110 mg Cr^{3+}/g of the resin. Incomplete desorption of the loaded metal may be attributed to the formation of hydrolysis products of chromium which are difficult to elute under normal conditions. Almost all chromium can be extracted from a diluted tannery waste solution in four stages, though 90% Cr(III) was eluted with 20% (v/v) sulphuric acid in two stages. The process has a potential for large scale trials to treat the waste tannery solution.

5. Acknowledgements

Authors are thankful to the Director, CSIR-National Metallurgical Laboratory, Jamshedpur, India for giving permission to publish the paper. Thanks are also due to Dr. B. U. Nair, CLRI, Chennai for providing solution from a tannery. The financial support received from Planning Commission, Govt. of India through Council of Scientific & Industrial Research (CSIR) New Delhi under 10th five year plan is gratefully acknowledged.

REFERENCES

[1]   CPCB, "Recovery of Better Quality Reusable Salt from Soak Liquor of Tanneries in Solar Evaporation Pans," Central Pollution Control Board (CPCB) Ministry of Environment & Forests Control of Urban Pollution Series:

Cups/2009-10, 2009.

[2] P. Chandra, S. Sinha and U. N. Rai, "Bioremediation of Cr from Water and Soil by Vascular Aquatic Plants," In: E. L. Kruger, T. A. Anderson and J. R. Coats, Eds., *Phytoremediation of Soil and Water Contaminants*, ACS Symposium Series #664, American Chemical Society, Washington DC, 1997, pp. 274-282.

[3] B. D. Pandey, G. Cote and D. Bauer, "Extraction of Chromium(III) from Spent Tanning Baths," *Hydrometallurgy*, Vol. 40, No. 3, 1996, pp. 343-357. doi:10.1016/0304-386X(95)00006-3

[4] Q. Chen, Z. Luo, C. Hills, G. Xue and M. Tyrer, "Precipitation of Heavy Metals from Wastewater Using Simulated Flue Gas: Sequent Additions of Fly Ash, Lime and Carbon Dioxide," *Water Research*, Vol. 43, No. 10, 2009, pp. 2605-2614. doi:10.1016/j.watres.2009.03.007

[5] P. D. Johnson, P. Girinathannair, K. N. Ohlinger, S. Ritchie, L. Teuber and J. Kirby, "Enhanced Removal of Heavy Metals in Primary Treatment Using Coagulation and Flocculation," *Water Environment Research*, Vol. 80, No. 5, 2008, pp. 472-479.

[6] S. Santosa, D. Siswanta, S. Sudiono and R. Utarianingrum, "Chitin-Humic Acid Hybrid as Adsorbent for Cr(III) in Effluent of Tannery Wastewater Treatment," *Applied Surface Science*, Vol. 254, No. 23, 2008, pp. 7846-7850. doi:10.1016/j.apsusc.2008.02.102

[7] S. K. Sahu, V. K. Verma, D. Bagchi, V. Kumar and B. D. Pandey, "Recovery of Chromium(VI) from Electroplating Effluent by Solvent Extraction with Tri-n-Butyl Phosphate," *Indian Journal of Chemical Technology*, Vol. 15, No. 4, 2008, pp. 397-402.

[8] S. Verbych, N. Hilal, G. Sorokin and M. C. Leaper, "Ion Exchange Extraction of Heavy Metal Ions from Wastewater," *Separation Science and Technology*, Vol. 39, No. 9, 2004, pp. 2031-2040. doi:10.1081/SS-120039317

[9] G. Zupancic and A. Jemec, "Anaerobic Digestion of Tannery Waste: Semi-Continuous and Anaerobic Sequencing Batch Reactor Processes," *Bioresource Technology*, Vol. 101, No. 1, 2010, pp. 26-33. doi:10.1016/j.biortech.2009.07.028

[10] M. Pazouki and A. Moheb, "An Innovative Membrane Method for the Separation of Chromium Ions from Solutions Containing Obstructive Copper Ions," *Desalination*, Vol. 274, No. 1-3, 2011, pp. 246-254.

[11] C. Justina, M. Elsa, P. Ana, L. Ana, S. Luis and N. Maria, "Membrane-Based Treatment for Tanning Wastewaters," *Canadian Journal of Civil Engineering*, Vol. 36, No. 2, 2009, pp. 356-362. doi:10.1139/S08-053

[12] T. A. Kurniawan, Y. S. C. Gilbert, W. H. Lo and S. Babel, "Physico-Chemical Treatment Techniques for Wastewater Laden with Heavy Metals," *Chemical Engineering Journal*, Vol. 118, No. 1-2, 2006, pp. 83-98. doi:10.1016/j.cej.2006.01.015

[13] I. Lee, Y. Kuan and J. Chern, "Equilibrium and Kinetics of Heavy Metal Ion Exchange," *Journal of Chinese Institute of Chemical Engineers*, Vol. 38, No. 1, 2007, pp. 71-84. doi:10.1016/j.jcice.2006.11.001

[14] J. S. Kentish and G. W. Stevens, "Innovations in Separa-tion Technology for the Recycling and Reuse of Liquid Waste Streams," *Chemical Engineering Journal*, Vol. 84, No. 2, 2001, pp. 149-159. doi:10.1016/S1385-8947(01)00199-1

[15] S. Yalcin, R. Apak, J. Hizal and H. Afsar, "Recovery of Copper(II) and Chromium(III) from Electroplating Industry Wastewater by Ion Exchange," *Separation Science and Technology*, Vol. 36, No. 10, 2001, pp. 2181-2196. doi:10.1081/SS-100105912

[16] J. A. S. Tenorio and D. C. R. Espinosa, "Treatment of Chromium Plating Process Effluents with Ion Exchange Resins," *Waste Management*, Vol. 21, No. 7, 2001, pp. 637-642. doi:10.1016/S0956-053X(00)00118-5

[17] A. Agrawal, V. Kumar and B. D. Pandey, "Remediation Options for the Treatment of Electroplating and Leather Tanning Effluent Containing Chromium: A Review," *Mineral Processing and Extractive Metallurgy Review*, Vol. 27, No. 2, 2006, pp. 99-130. doi:10.1080/08827500600563319

[18] D. Petruzzelli, R. Passino, M. Santori and G. Tiravanti, "Industrial Waste Management, the Case of the Tannery Industry in Chemical Water and Wastewater Treatment III," Springer-Verlag, Berlin, 1994.

[19] D. Petruzzelli, R. Passino and G. Tiravanti, "Ion Exchange Process for Chromium Removal and Recovery from Tannery Wastes," *Industrial Engineering and Chemical Research*, Vol. 34, 1995, pp. 2612-2617. doi:10.1021/ie00047a009

[20] S. Kocaoba and G. Akcin, "Removal of Chromium(III) and Cadmium(II) from Aqueous Solutions," *Desalination*, Vol. 180, No. 1-3, 2005, pp. 151-156. doi:10.1016/j.desal.2004.12.034

[21] F. J. Alguacil, M. Alonso and L. J. Lozano, "Chromium(III) Recovery from Waste Acid Solution by Ion Exchange Processing Using Amberlite-IR 120 Resin: Batch and Continuous Ion Exchange Modeling," *Chemosphere*, Vol. 57, No. 8, 2004, pp. 789-793. doi:10.1016/j.chemosphere.2004.08.085

[22] S. Kocaoba and G. Akcin, "Removal and Recovery of Chromium and Chromium Speciation with MINTEQA2," *Talanta*, Vol. 57, No. 1, 2002, pp. 23-30. doi:10.1016/S0039-9140(01)00677-4

[23] N. Kabay, N. Gizli, M. Demircioglu, M. Yuksel, A. Jyo, K. Yamabe and T. Shuto, "Cr(III) Removal by Macroreticular Chelating Ion Exchange Resins," *Chemical Engineering Communications*, Vol. 190, No. 5-8, 2003, pp. 813-822. doi:10.1080/00986440302114

[24] S. A. Cavaco, S. L. Fernandes, M. M. Quina and L. M. G. Ferreira, "Removal of Chromium from Electroplating Industry Effluents by Ion-Exchange Resins," *Journal of Hazardous Materials*, Vol. 144, No. 3, 2007, pp. 634-638. doi:10.1016/j.jhazmat.2007.01.087

[25] S. Fernandes, S. A. Cavaco, M. J. Quina and L. M. G. Ferreira, "Selective Separation of Chromium(III) from Electroplating Effluents by Ion-Exchange Processes," *Proceedings of European Congress of Chemical Engineering* (ECCE-6)," Copenhagen, 2007.

[26] S. A. Cavaco, S. Fernandes, C. M. Augusto, M. J. Quina and L. M. G. Ferreira, "Evaluation of Chelating Ion-Ex-

change Resins for Separating Cr(III) from Industrial Effluents," *Journal of Hazardous Materials*, Vol. 169, No. 1-3, 2009, pp. 516-523. doi:10.1016/j.jhazmat.2009.03.129

[27] S. Pramanik, S. Dey and P. Chattopadhyay, "A New Chelating Resin Containing Azophenolcarboxylate Functionality: Synthesis, Characterization and Application to Chromium Speciation in Wastewater," *Analytica Chimica Acta*, Vol. 584, No. 2, 2007, pp. 469-476. doi:10.1016/j.aca.2006.11.041

[28] P. Chattopadhyay, C. Sinha and D. K. Pal, "Preparation and Properties of a New Chelating Resin Containing Imadazolylazo Groups," *Fresenius' Journal of Analytical Chemistry*, Vol. 357, No. 4, 1997, pp. 368-372.

[29] F. Gode and E. Pehlivan, "A Comparative Study of Two Chelating Ion-Exchange Resins for the Removal of Chromium (III) from Aqueous Solution," *Journal of Hazardous Materials*, Vol. 100, No. 1-3, 2003, pp. 231-243. doi:10.1016/S0304-3894(03)00110-9

[30] S. K. Sahu, P. Meshram, B. D. Pandey, V. Kumar and T. R. Mankhand, "Removal of Chromium(III) by Cation Exchange Resin, Indion 790 for Tannery Waste Treatment," *Hydrometallurgy*, Vol. 99, No. 3-4, 2009, pp. 170-174. doi:10.1016/j.hydromet.2009.08.002

[31] F. Gode and E. Pehlivan, "Removal of Chromium (III) from Aqueous Solutions Using Lewatit S 100: The Effect of pH, Time, Metal Concentration and Temperature," *Journal of Hazardous Materials*, Vol. 136, No. 2, 2006, pp. 330-337. doi:10.1016/j.jhazmat.2005.12.021

[32] S. Rengaraj, K. H. Yeon and S. H. Moon, "Removal of

Chromium from Water and Wastewater by Ion Exchange Resins," *Journal of Hazardous Materials*, Vol. 87, No. 1-3, 2001, pp. 273-287. doi:10.1016/S0304-3894(01)00291-6

[33] M. Muruganathan, G. BhaskarRaju and S. Prabhakar, "Separation of Pollutants from Tannery Effluents by Electro Flotation," *Separation and Purification Technology*, Vol. 40, No. 1, 2004, pp. 69-75. doi:10.1016/j.seppur.2004.01.005

[34] T. Reemtsma and M. Jekel, "Dissolved Organics in Tannery Wastewaters and Their Alteration by a Combined Anaerobic and Aerobic Treatment," *Water Research*, Vol. 31, No. 5, 1997, pp. 1035-1046. doi:10.1016/S0043-1354(96)00382-X

[35] G. McKay, H. S. Blair and J. R. Gardner, "Adsorption of Dyes on Chitin. I. Equilibrium Studies," *Journal of Applied Polymer Science*, Vol. 27, No. 8, 1982, pp. 3043-3057. doi:10.1002/app.1982.070270827

[36] S. Lagergren, "About the Theory of So-Called Adsorption of Soluble Substances, the Royal Swedish Academy of Sciences," *Handlingar*, Vol. 24, 1898, pp. 1-39.

[37] Y. S. Ho and G. McKay, "Pseudo-Second Order Model for Sorption Processes," *Process Biochemistry*, Vol. 34, No. 5, 1999, pp. 451-465. doi:10.1016/S0032-9592(98)00112-5

[38] P. W. Ramteke, S. Awasthi, T. Srinath and B. Joseph, "Efficiency Assessment of Common Effluent Treatment Plant (CETP) Treating Tannery Effluents," *Environmental Monitoring and Assessment*, Vol. 169, No. 1-4, 2010, pp. 125-131. doi:10.1007/s10661-009-1156-6

Nomenclature

q = Cr(III) loaded on the resin (mg/g);
C_e = Equilibrium concentration of Cr(III) (mg/L);
q_m = Loading capacity of the resin (mg/g);
K_f = Freundlich Constant;
n = Sorption intensity for Freundlich model;
K_l = Langmuir constant;
q_t = Adsorption capacity of the resin at time t(mg/g);
q_e = Adsorption capacity of the resin at equilibrium (mg/g);
k_l = First order rate constant (min^{-1}).

An Exploratory Study of Tridentate Amine Extractants: Solvent Extraction and Coordination Chemistry of Base Metals with *Bis*((1*R*-benzimidazol-2-yl)methyl)amine

Nomampondo P. Magwa[1], Eric Hosten[2], Gareth M. Watkins[1], Zenixole R. Tshentu[1*]

[1]Department of Chemistry, Rhodes University, Grahamstown, South Africa
[2]Department of Chemistry, Nelson Mandela Metropolitan University, Port Elizabeth, South Africa
Email: z.tshentu@ru.ac.za

ABSTRACT

Solvent extraction of base metals using *bis*((1-decylbenzimidazol-2-yl)methyl)amine (BDNNN) showed a lack of pH-metric separation of the metals. The extraction system was described quantitatively using the equilibria involved to derive the mathematical explanation for the two linear log D vs pH_e plots for each metal ion extraction curve, and coordination numbers could also be extracted from the two slopes. The lack of separation was attributed to the absence of stereochemical "tailor making" since the complexes isolated from the reaction of the ligand, *bis*((1*H*-benzimidazol-2-yl)methyl)amine (NNN), with base metals suggested the formation of similar octahedral complex species from spectral and crystal structure evidence. The *bis* tridentate coordination observed was in agreement with information extracted from the extraction data. This investigation opens up an opportunity and an approach for the evaluation of amines as extractants but cautions against tridentate ligands.

Keywords: Base Metals; Tridentate; *Bis*((1*H*-benzimidazol-2-yl)methyl)amine; Extractive and Coordination Chemistry

1. Introduction

There is an increasing demand for substantial production of metals of high purity in the metallurgical industries in an attempt to meet the corresponding demand for both industrial and domestic applications. This has been the catalyst for the development of simpler and more economical routes for purification of metal ions from their ore solutions. The scale for example of nickel operations is limited, and therefore one has to maximise profit *via* processing efficiency rather than volume. The ultimate goal is to design processes that are environmentally friendly, cost-effective, time-saving and selective. Solvent extraction, a widely applied technique for the recovery of base metals [1,2], meets some of these principles since high boiling point and high flashpoint solvents are used and are recycled together with an extractant after the stripping step, and the system can be tuned through appropriate extractant design to achieve selectivity [3]. The basis of base metal ion separation is a unique property of the coordination chemistry of the particular metal ion. In the development of a metal ion specific extractant, it is necessary to consider the charac-

teristics of the metal ions from which the desired metal ion must be removed as well as its own [3]. Improvements of the chemical processes in the solvent extraction system require a thorough knowledge and investigation of the chemistry involved in order to achieve a meaningful advancement of this technology.

There has been an envisaged shift towards amine-based extractants, and this is motivated by the favourable properties that are offered by the nitrogenous ligands especially the aromatic amines as compared with oxygen-based extractants [3-5]. Some of these properties include the intermediate pK_a values as well as σ and π bonding capabilities resulting in extractions in the low pH range and in a possibility of separation through bonding preferences respectively. The latter property has been dubbed "stereochemical tailor-making" by du Preez [5]. On the other hand, the strong aliphatic amine ligands with σ-donor only character show lack of relative preference for the metal ions but this can be improved if chelates are used, and also tend to form metal complexes at relatively high pH values which is undesirable [5]. The chelate effect has also been exploited effectively in a bidentate aromatic system (1-octyl-2,2'-pyridylimidazole) providing for effective separation of nickel from other base

*Corresponding author.

metals in strong acidic sulfate solutions with a possibility of back-extraction [3].

An extension of these systems to tridentate ligands would have an additional advantage of increasing the extraction equilibrium constants [6] due to the high complex formation constants for reactions of base metals and tridentate ligands [7,8], thereby requiring relatively low extractant-to-metal ratios to achieve quantitative extractions [6]. The only example of a tridentate amine extractant in the literature is that of a derivative of diethylenetriamine [9] but extractions occurred at relatively high pH values as expected due to high pK_a values of aliphatic amines [8], and the small $\Delta pH_{0.5}$ values implied a lack of pH-metric separation of the later 3d metal ions [9]. It would be hoped that good separation factors of the metal chelates would be achieved through stereochemical considerations since nickel(II) is known to form the most stable spin free octahedral (O_h) complexes of all base metal ions [10] while the copper(II) and cobalt(II) ions tend to form stable tetrahedral (T_d) complexes [4,11]. This study, therefore, also interrogates the coordination chemistry aspects of base metals, that infuence the extractions, with tridentate amine-based ligands.

In this account, we present the extractive and coordination chemistry of an aromatic tridentate ligand, bis((1H-benzimidazol-2-yl)methyl)amine (NNN) (**Figure 1(a)**), towards base metals in a sulfate/sulfonate medium. Dinonylnaphthalene sulfonic acid (DNNSA) (**Figure 1(b)**) was used as a bulky anion to ion-pair and transfer the cationic complexes formed in this extraction system (ion-association system) to the organic phase since the sulfate ion is known to have high hydration energies leading to lack of phase transferability [12]. The sulfate medium has become particularly important to explore since it is encountered in liquors produced in sulfate-based high nickel matte leach processes or those produced in sulfuric acid pressure leaching of laterites [13].

2. Experimental

2.1. Materials

o-Phenylenediamine (99.5%, Sigma-Aldrich), iminodiacetic acid (98%, Sigma-Aldrich), hydrochloric acid

(32%, Merck Chemicals), ammonia (28%, Merck Chemicals), methanol (99%, Sigma-Aldrich), and ethyl acetate (98%, Sigma-Aldrich) were reagent grade chemicals used as received for the synthesis of bis((1H-benzimidazol-2-yl)methyl)amine (NNN). The reagent grade octylbromide (98%) was also obtained from Sigma-Aldrich and used as received for the synthesis of the extractant. $NiSO_4 \cdot 6H_2O$ (98%), $CuSO_4 \cdot 5H_2O$ (99%) were obtained from Merck chemicals. $CoSO_4 \cdot 7H_2O$ (97.5%) was obtained from Fluka, while $ZnSO_4 \cdot 7H_2O$ (99.5%) was obtained from BDH Chemicals. The copper(II), nickel(II), cobalt(II) and zinc(II) perchlorate hexahydrate salts which were used to prepare the metal(II) sulfonate salts using toluene-4-sulfonic acid (98%, Sigma-Aldrich) were obtained from Sigma-Aldrich. Analytical grade reagents were used without further purification in the preparations of the 0.10 M metal ion stock solutions in 3 M H_2SO_4 solution. The ICP/AAS 1000 ppm metal standards, dissolved in 0.5 N nitric acid, were used to prepare standard solutions for the construction of calibration curves using distilled, deionized, milliQ water for the dilutions. Dinonylnaphthalene sulfonic acid (50 wt% in heptane), Shellsol 2325 (17 - 22 v/v% aromatic content) and 2-octanol (98%) were obtained from Sigma-Aldrich, Shell Chemicals (SA) and Merck Chemicals, respectively.

2.2. Instrumentation

^{1}H NMR spectrometry was carried out on a Bruker AMX 400 MHz NMR spectrometer and reported relative to tetramethylsilane (δ 0.00). A Vario Elementary ELIII Microcube CHNS analyser was used for elemental analyses. A Thermo Electron (iCAP 6000 Series) inductively coupled plasma (ICP) spectrometer equipped with an OES detector was used for metal ion analysis. The Labcon micro-processor controlled orbital platform shaker model SPO-MP 15 was used for contacting the two phases of extraction. The pH measurements were performed on a Metrohm 827 pH meter using a combination electrode with 3 M KCl as electrolyte. The metal-complexes were characterised using infrared spectrometry on both Perkin Elmer 400 FTIR and 100 FTIR-ATR

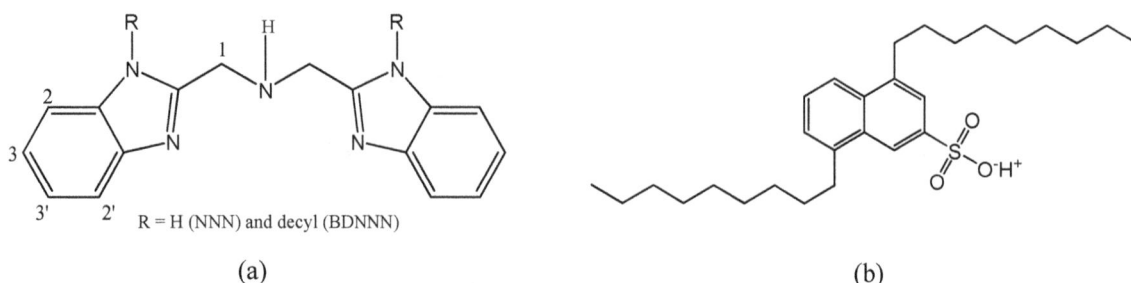

(a)   (b)

Figure 1. The chemical structures of (a) bis((1R-benzimidazol-2-yl)methyl)amine (NNN for R = H, and BDNNN for R = Decyl) and (b) dinonylnaphthalene sulfonic acid (DNNSA).

spectrometers. The solid reflectance spectra of the metal complexes were recorded on a Shimadzu UV-VIS-NIR Spectrophotometer UV-3100 with a MPCF-3100 sample compartment with samples mounted between two quartz discs which fit into a sample holder coated with barium sulfate. The spectra were recorded over the wavelength range of 2000 - 250 nm, and the scans were conducted at a medium speed using a 20 nm slit width. The Gallenkamp melting point apparatus (temperature range, 0°C - 350°C) was used to measure the melting points. The conductivity measurements were carried out on a A.W.R. Smith Process Instrumentation cc Laboratory Bench Meter Model AZ 86555 with ABS graphite cell probe using an aqueous standard which has a conductivity value of 135 ohm^{-1}·cm^2·mole^{-1} at 20°C for the calibration. All the complexes were prepared in DMF as a solvent to a concentration of 10^{-3} M for the conductivity measurements.

2.3. Synthesis of the Ligand and Extractant

2.3.1. Bis((1H-benzimidazol-2-yl)methyl)amine (NNN)

The bis((1H-benzimidazol-2-yl)methyl)amine (NNN) was synthesized as reported elsewhere [14], except that iminodiacetic acid was used and the decolorization step using activated charcoal in methanol was necessary. The characterization data for the white precipitate of the free base was as follows: Yield: 84%, m.p., 268°C - 270°C. Anal. Calcd. for C$_{16}$H$_{16}$N$_4$OS (%): C, 65.10; H, 5.80; N, 23.70. Found: C, 65.78; H, 5.80; N, 23.07. ^{1}H·NMR (CDCl$_3$) δ (ppm): 4.06 (4H, s, H1), 7.15 (4H, m, H3,H3'), 7.52 (4H, m, H2,H2'). IR (cm^{-1}): 3208 v(N-H), 3049 v(sec N-H), 1592 v(C = N).

2.3.2. Bis((1-decylbenzimidazol-2-yl)methyl)amine (BDNNN)

The alkylated derivative of the ligand was prepared according to a literature method [15]. However, the purification step was carried out as follows: The resulting solution after the removal of the KBr salt was concentrated via rotary evaporation, and purified using a silica gel chromatographic column with ethyl acetate/methanol (4:1) solvent system. After the removal of the solvent by rotary evaporation the product was obtained as brown oil. Yield = 67%. Anal. Calcd. for C$_{36}$H$_{59}$N$_5$O$_2$ (%): C, 72.81; H, 10.01; N, 11.79. Found: C, 72.95; H, 10.61; N, 11.86. ^{1}H NMR (CDCl$_3$) δ (ppm): δ 0.89 (6H, t, CH$_3$), 1.13 (24H, m, CH$_3$(CH$_2$)$_6$), 1.01(4H, t, CH$_2$-CH$_3$), 1.55(4H, t, CH$_2$-CH$_2$N), 3.96 (4H, t, CH$_2$-N), 4.02 (4H, s, H1), 7.26 (4H, q, H3 & H3'), 7.28 (2H, d, H2), 7.75 (2H, d, H2'). IR (cm^{-1}): 1522 v(C = N).

2.4. Syntheses of Metal Complexes

All the reactions for the formation of coordination complexes (sulfonate and sulfate compounds) were conducted in absolute ethanol using the toluene-4-sulfonate salts of the metals, and inert conditions were adopted for the synthesis. 10 mL of hot ethanolic solution (60°C) containing 2 mmol of the ligand was added dropwise to 10 mL of the metal ion solution (1 mmol, respectively for each metal ion). The mixture was heated under reflux overnight and the precipitate that formed was filtered off, washed with ethanol and dried.

2.4.1. Sulfate Complexes

[Co(NNN)$_2$]SO$_4$·4H$_2$O: Color: pink. Yield = 51%, m.p., 252°C - 254°C. Anal. Calcd. for C$_{32}$H$_{38}$N$_{10}$CoO$_8$S (%):C, 49.17; H, 4.90; N, 17.92; S, 4.10. Found: C, 49.20; H, 4.46; N, 17.85; S, 4.86. IR (cm^{-1}): 3242 v(N-H), 1545 v(C = N), 1037 v$_3$(SO$_4$), 226 v(M-N). Conductivity (10^{-3} M, ohm^{-1}·cm^2·mole^{-1}): 66.

[Ni(NNN)$_2$]SO$_4$·3H$_2$O: Color: purple. Yield = 57%, m.p., 253°C - 255°C. Anal. Calcd. for C$_{32}$H$_{36}$N$_{10}$NiO$_7$S (%): C, 50.34; H, 4.75; N, 18.35; S, 4.20. Found: C, 50.64; H, 4.74; N, 18.28; S, 4.10. IR (cm^{-1}): 3234 v(N-H), 1538 v(C=N), 1037 v$_3$(SO$_4$), 230 v(M-N). Conductivity (10^{-3} M, ohm^{-1}·cm^2·mole^{-1}): 69.

[Cu(NNN)$_2$]SO$_4$·7H$_2$O: Color: green. Yield = 65%, m.p., 228°C - 229°C. Anal. Calcd. for C$_{32}$H$_{44}$N$_{10}$CuO$_{11}$S (%): C, 45.74; H, 5.28; N, 16.67; S, 3.82. Found: C, 45.81; H, 5.11; N, 16.41; S, 3.44. IR (cm^{-1}): 3233 v(N-H), 1565 v(C = N), 1060 - 1088 v$_3$(SO$_4$), 224 v(M-N). Conductivity (10^{-3} M, ohm^{-1}·cm^2·mole^{-1}): 71.

[Zn(NNN)$_2$]SO$_4$·11H$_2$O: Color: white. Yield = 64%, m.p., 221°C - 222°C. Anal.Calcd. for C$_{32}$H$_{52}$N$_{10}$ZnO$_{15}$S (%): C, 42.04; H, 5.73; N, 15.31; S, 3.51. Found: C, 41.96; H, 5.71; N, 15.00; S, 3.59. IR (cm^{-1}): 3227 v(N-H), 1548 v(C = N), 1037 v$_3$(SO$_4$), 222 v(M-N). Conductivity (10^{-3} M, ohm^{-1}·cm^2·mole^{-1}): 82.

2.4.2. Sulfonate Complexes

[Co(NNN)$_2$](RSO$_3$)$_2$·4H$_2$O·2EtOH: Color: red. Yield = 71%, m.p., 225°C - 226°C. Anal. Calcd. for C$_{50}$H$_{64}$N$_{10}$CoO$_{12}$S$_2$ (%): C, 53.66; H, 5.67; N, 12.52; S, 5.73. Found: C, 53.60; H, 5.34; N, 12.67; S, 5.14. IR (cm^{-1}): 3311 v(N-H), 1551 v(C = N), 1150-1161 v$_3$(RSO$_3$), 279 v(M-N). Conductivity (10^{-3} M, ohm^{-1}·cm^2·mole^{-1}): 136.

[Ni(NNN)$_2$](RSO$_3$)$_2$·3H$_2$O·2EtOH: Color: purple. Yield = 58%, m.p., 246°C - 248°C. Anal. Calcd. for C$_{50}$H$_{62}$N$_{10}$NiO$_{11}$S$_2$ (%): C, 54.50; H, 5.67; N, 12.71; S, 5.82. Found: C, 54.43; H, 5.58; N, 12.47; S, 5.70. IR (cm^{-1}): 3313 v(N-H), 1550 v(C = N), 1151 - 1164 v$_3$(RSO$_3$), 246 v(M-N). Conductivity (10^{-3} M, ohm^{-1}·cm^2·mole^{-1}): 139.

[Cu(NNN)$_2$](RSO$_3$)$_2$·12H$_2$O: Color: blue. Yield = 68%, m.p., 201°C - 202°C. Anal. Calc. for C$_{46}$H$_{68}$N$_{10}$CuO$_{18}$S$_2$ (%): C, 46.95; H, 5.82 N, 11.90; S, 5.45. Found: C, 46.48; H, 5.47; N, 11.76; S, 5.85. IR (cm^{-1}): 3213 v(N-H), 1550 v(C = N), 1147 - 1172 v$_3$(RSO$_3$), 262 v(M-N). Conductivity (10^{-3} M, ohm^{-1}·cm^2·mole^{-1}): 141.

[Zn(NNN)₂](RSO₃)₂·3H₂O·2EtOH: Color: white. Yield = 62%, m.p., 201°C - 202°C. Anal. Calc. for $C_{50}H_{62}N_{10}$-$ZnO_{11}S_2$ (%): C, 54.17; H, 5.64; N, 12.63; S, 5.78. Found: C, 54.50; H, 5.64; N, 12.30; S, 5.53. IR (cm⁻¹): 3304 ν(N-H), 1549 ν(C=N), 1172 - 1187 ν_3(RSO₃), 258 ν(M-N). Conductivity (10⁻³ M, ohm⁻¹·cm²·mole⁻¹): 147.

2.5. Solvent Extraction Procedure

All the extractions were carried out at 25°C (±1°C) in a temperature controlled laboratory. Equal volumes (10 mL) of 0.001 M metal ion solution (aqueous layer) and 80% 2-octanol/shellsol solution (organic layer containing the extractant BDNNN and the counterion DNNSA) were pipetted into 50 mL conical separating funnels. The contents in the funnel were shaken using an automated orbital platform shaker for 30 minutes at an optimised speed of 200 rpm. A minimum period of 60 minutes was observed before harvesting the raffinates. The raffinates were filtered through a 33 mm millex-HV Millipore filter (0.45 μm) and diluted appropriately for analysis by ICP. The percentage extractions (%E) of the metal ions were calculated from the concentrations of the metal ions in the aqueous phase using equation 1 below:

$$\%E = \left(\frac{C_i - C_s}{C_i} \right) \times 100 \tag{1}$$

where C_i is the initial solution concentration (mg/L) and C_s is the solution concentration after extraction.

The extraction efficiencies were investigated as a function of pH, and all the extraction curves were plotted with Sigma Plot 11.0.

2.6. X-Ray Structure Determination and Refinement

Single crystals of [Cu(NNN)₂](RSO₃)₂·12H₂O, suitable for X-ray diffraction, were obtained by slow evaporation of the ethanolic mother liquor of this complex at room temperature. X-ray diffraction studies were performed at 200 K using a Bruker Kappa Apex II diffractometer with graphite monochromated Mo Kα radiation (λ = 0.71073 Å). The crystal structures were solved by direct methods using *SHELXTL* [16]. All non-hydrogen atoms were refined anisotropically. C-bound H atoms were placed in calculated positions and refined as riding atoms, with bond lengths 0.95 (aromatic CH), 0.99 (CH₂), 0.98 (CH₃) Å and with U_{iso}(H) = 1.2 (1.5 for methyl) U_{eq}(C). Hydrogens bonded to nitrogen were located on a Fourier map and allowed to refine freely. Hydrogens on water molecules were restrained to an O-H bond length of 0.84 Å and H-O-H angle of 110°. Diagrams and publication material were generated using *SHELXTL*, *PLATON* [17] and *ORTEP*-3 [18].

3. Results and Discussion

3.1. Solvent Extraction Studies

These studies were carried out in dilute synthetic sulfate solutions. The extractant (BDNNN) was used for the extraction studies while the ligand (NNN) was used to study the coordination chemistry involved. The complexes of the extractant were oily and not easily isolated hence we used the NNN ligand for the coordination chemistry studies. The decyl groups of the extractant (BDNNN) would be positioned away from the coordination sphere and therefore the use of the ligand (NNN) would not change the coordination chemistry of the extractant from a steric hindrance point of view.

The use of low pK_a groups (benzimidazoles) on the ligand was expected to result in metal extractions from highly acidic solutions in comparison with tridentate aliphatic amine extractants based on diethylenetriamine [9]. This was well exploited but also resulted in lack of room for back-extraction of some metals (like Cu(II), Co(II) and possibly Ni(II)) at the "left legs" of their extraction curves. The lack of pH-metric separation of the base metal ions, however, was evident from the small $\Delta pH_{0.5}$ values (**Figure 2**). Interestingly, the order of the extraction of the metal ions somewhat followed the Irving-Williams stability order [19] in the shifting of the curves towards the acidic region with the exception of Co(II) and Ni(II) extractions, which could be influenced by kinetic effects [20]. It is also noteworthy to report on the hard ions (Mn^{2+} and Mg^{2+}) extracting ability of this tridentate ligand but with rejection of Fe(III) in the pH range 0 - 2.6, which is attractive for the latter but not the former.

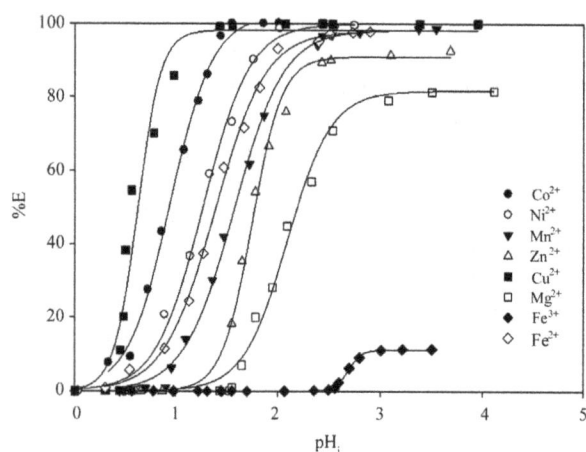

Figure 2. A plot of %E vs initial pH of equimolar concentrations (0.001 M) of Mg^{2+}, Mn^{2+}, Fe^{3+}, Fe^{2+}, Co^{2+}, Ni^{2+}, Cu^{2+} and Zn^{2+}, extracted with BDNNN (at M:L ratio 1:40) and 0.02 M DNNSA in 2-octanol/Shellsol 2325 (8:2) from a dilute sulfate medium.

It is clear, therefore, that the move towards tridentate ligands, even with low pK_a aromatic nitrogen groups and a strong aliphatic amine group, is not sufficient to successfully exploit the low pH range interactions with metal ions since extraction isotherms are pushed further into the low pH range thereby compromising back-extraction. These ligands also lack selectivity for the important borderline metal ions which can possibly be tuned through bonding preferences. The coordination chemistry studies (Section 3.2) were, therefore, conducted to try and elucidate the underlying aspects of bonding that influenced the extraction isotherms observed.

In a quantitative treatment for this solvent extraction system, similar to that applied for a chelating system (HL) [6], the protonation, complexation and phase distribution equilibria can be used to describe the system mathematically with respect to the distribution ratio of a metal ion (M^{n+}), and also give insight into the coordination numbers involved in the extraction reaction. The protonation equilibria which were studied using potentiometry in the pH range of 2 - 10 by Hay *et al.* [7] showed two constants for *bis*((1*H*-benzimidazol-2-yl)methyl)amine (NNN, L), and they were 5.64 and 10.12 respectively for the cumulative protonation steps (LH^+ and LH_2^{2+}). The species distribution plot, constructed from the above constants using the computer program HYSS [21], is given in **Figure 3** for the pH-metric speciation involved for L in the aqueous phase.

The chelating agent (L) must distribute between the organic and aqueous phases to effect coordination in the aqueous phase, and that distribution coefficient is represented by $K_D(L)$:

$$(L)_a \; \square \; (L)_o \quad \text{and} \quad K_D(L) = \frac{[L]_o}{[L]_a} \qquad (2)$$

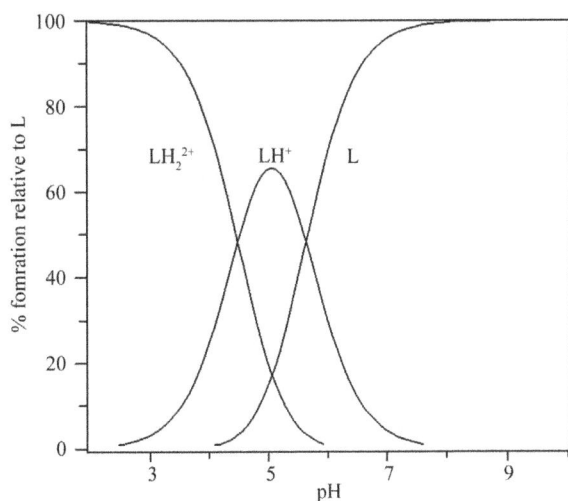

Figure 3. Protonation species distribution diagram for *bis*((1*H*- benzimidazol-2-yl)methyl)amine (NNN, L).

However, in the aqueous phase the following two protonation equilibria may exist depending on pH:

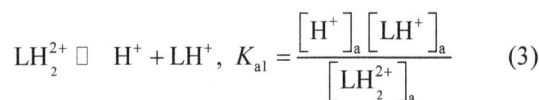

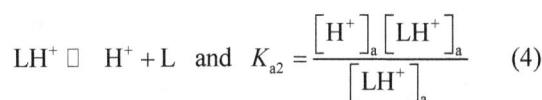

$$LH_2^{2+} \; \square \; H^+ + LH^+, \quad K_{a1} = \frac{[H^+]_a [LH^+]_a}{[LH_2^{2+}]_a} \qquad (3)$$

$$LH^+ \; \square \; H^+ + L \quad \text{and} \quad K_{a2} = \frac{[H^+]_a [LH^+]_a}{[LH^+]_a} \qquad (4)$$

Then, the metal ion chelates with the neutral ligand to form a cationic complex:

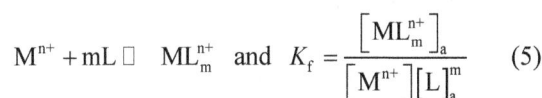

$$M^{n+} + mL \; \square \; ML_m^{n+} \quad \text{and} \quad K_f = \frac{[ML_m^{n+}]_a}{[M^{n+}][L]_a^m} \qquad (5)$$

It must be borne in mind, however, that the metal ion will replace proton(s) in the pH ranges under investigation but the protonation equilibria will accommodate this in the mathematical treatment. Finally, the chelate which is ion-paired by an anion (in our case two sulfonate anions represented by X^{n-}) to form an extractible species, $[ML_m]X$, distributes itself between the organic and aqueous phases:

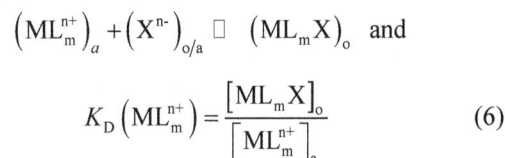

$$(ML_m^{n+})_a + (X^{n-})_{o/a} \; \square \; (ML_mX)_o \quad \text{and}$$

$$K_D(ML_m^{n+}) = \frac{[ML_mX]_o}{[ML_m^{n+}]_a} \qquad (6)$$

The distribution ratio (D), defined as the ratio of the concentration of the total metal species in the organic phase to that in the aqueous (regardless of its mode), is given by expression 7, on the assumption that the metal chelate distributes largely in the organic phase and that the metal ion does not hydrolyse in the aqueous phase.

$$D \approx \frac{[ML_mX]_o}{(M^{n+})_a} \qquad (7)$$

Substituting Equations (6) and (5) respectively into Equation (7) yields Equation (8), depicting the formation constant and the concentration of the ligand in the aqueous phase as important parameters as well as the distribution coefficient of the chelate:

$$D = K_D(ML_m^{n+}) K_f [L]_a^m \qquad (8)$$

which can be transformed to Equation (9) if Equation (2) is substituted into Equation (8), indicating that the concentration of L in the aqueous phase is dependent on its concentration in the organic phase and that its distribution between the two phases affects the distribution ratio of the complex formed:

$$D = \frac{K_D \left(ML_m^{n+} \right) K_f}{K_D \left(L \right)^m} \left[L \right]_o^m \qquad (9)$$

However, since the extractions are carried out at low pH, it is necessary to consider the two protonation equilibria respectively because these species occur over wide pH ranges (**Figure 3**), and competition of metal ions with protons for the ligand occurs early with pH due to the higher formation constants [7] compared with protonation constants thereby resulting in release of the protons from the ligand. Now, substituting Equation (4), and Equations (3) and (4), respectively, into Equation (8) yields the following respective Equations (10) and (11):

$$D = K_D \left(ML_m^{n+} \right) K_f K_{a2}^m \frac{\left[LH^+ \right]^m}{\left[H^+ \right]_a^m} \qquad (10)$$

And $$D = K_D \left(ML_m^{n+} \right) K_f K_{a2}^m K_{a1}^m \frac{\left[LH_2^{2+} \right]_a^m}{\left[H^+ \right]_a^{2m}} \qquad (11)$$

Therefore, in the pH range where only one protonation equilibrium (Equation (4)) is involved then a plot of log D vs pH (from taking the logarithms of both sides in equation 10) should yield a straight line with slope m (number of ligands bonded to the metal ion M^{n+}). But in the highly acidic region where the second proton equilibrium (Equation (3)) is also active, then a plot of log D vs pH should yield a straight line with slope (2 m).

A plot of log D vs pHe (since the extraction isotherms relate to the equilibrium condition) for this extraction system is presented in **Figure 4**. There is a clear change in the slope of each curve and respectively become steeper with an increase in the order of stabilization of metal ions [19]. The higher pH range for Mg^{2+}, Zn^{2+}, Mn^{2+}, Fe^{2+} and Ni^{2+} is represented by a slope in the range 1 - 2 with nickel at m = 1.97 while it peaks close to 3 - 4 ($\approx$2 m) respectively at the lower pH range with the exception of nickel (slope = 15) which gets seriously affected by the extremely acidic medium. This observation is somewhat in agreement with the mathematical model described here for this complex equilibria system suggesting two ligands per metal ion are involved in the coordination. These observations (as well as those discussed below) are also in line with the protonated ligand species observed in **Figure 3** (for the protonation equilibria). The extremely acidic region (affecting the copper and cobalt curves, and to a minor extent the nickel curve in the acidic end) is only characterised by a slope of ca. (3 m - 2 m) for copper and cobalt at the higher pH end (around the pH where other metal ions also have a slope close to 4). This occurred mainly because our mathematical modelling does not take into account the third protonation equilibrium which is possible under those

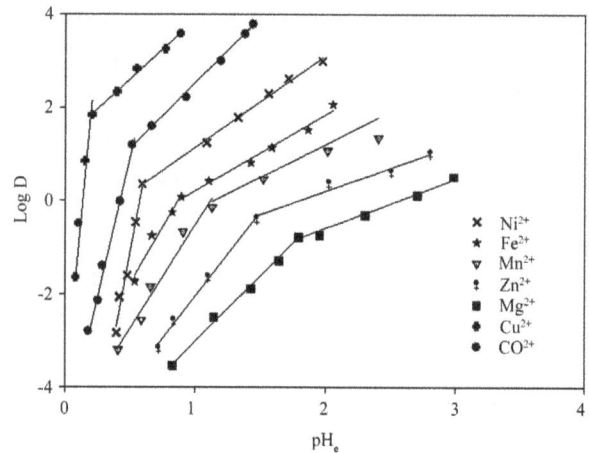

Figure 4. A plot of log D vs equilibrium pH (pHe) for the extraction of 0.001 M M^{2+} (M = Mg^{2+}, Zn^{2+}, Mn^{2+}, Fe^{2+}, Ni^{2+}, Co^{2+} and Cu^{2+}) with 0.002 M BDNNN and 0.02 M DNNSA from sulfate medium.

highly acidic conditions where cobalt and copper are extracted. This third protonation constant was also not determined by potentiometry by Hay *et al.* [7] due to the inaccessibility of the measurement in that pH range by potentiometry. According to our mathematical treatment, involving the third equilibrium should result in a slope 3 m = 6 at the lower pH end for cobalt and copper (and to some extent nickel), however, it is much steeper with slopes of 12, 15 and 26 for cobalt, nickel and copper respectively.

It is possible, however, that not only the double (slope = 2 m) and the triple (slope = 3 m) protonation equilibria dominate in the pH range where extractions of copper and cobalt ions occur but complex multiple protonations (*i* m, *i* = 1, 2, 3, $\cdots$) with *i* exceeding 3 from the hydrogen bonding with the rings' π electrons [22], hence the coordination number (m) cannot be calculated accurately from data in that pH range. It seems though that a linear plot of the points (calculated at the intersections of the two lines from each metal ion extraction data) at the vertices of the two linear plots, respectively for each metal ion, gives a negative slope $\approx$ 1.6 which is in agreement with the coordination numbers of 2 involved for all the metals (discussed in Section 3.2), but a mathematical treatment has not been provided herein. This observation may be coincidental, and therefore one cannot conclude on an isolated study. Once it is verified for other extraction studies with extractants similar to the one studied here, a mathematical function may be derived.

3.2. Coordination Chemistry Studies

The elemental analyses data suggested the following empirical formulae; $[M(NNN)_2]X \cdot xH_2O \cdot yEtOH$ (M = Co, Ni, Cu and Zn; X = SO_4^{2-} or $(RSO_3^-)_2$, x = 3 - 12 and y = 0 - 2). The involvement of two ligands per metal ion is in agreement with what was observed in the extraction studies. Complexes of Fe(II), Fe(III) and other hard ions could not be isolated with a good level of purity but *bis*-coordination is implied by the slopes of the extraction data. The molar conductivity data in DMF showed that the sulfate and sulfonate complexes have molar conductance values of 66 - 82 and 136 - 147 $\Omega^{-1}\cdot cm^2\cdot mol^{-1}$, respectively. This indicated that all the sulfate complexes behaved as 1:1 electrolytes while the sulfonate complexes were 1:2 electrolytes [23]. This behavior in solution suggested the non-coordinated nature of the counteranions. Both the sulfate and sulfonate complexes were prepared in order to elucidate the nature of these anions respectively with respect to their innocence to coordination, and the full structural analyses are discussed in Sections 3.2.1 and 3.2.2 below. Both the sulfate and sulfonate complexes were prepared since the extraction studies were carried out in a sulfate medium and a bulky sulfonate anion was also used to replace the sulfate ion.

3.2.1. Spectral Analysis

The C=N stretching vibration of the benzimidazole rings of the free ligand (NNN) appeared at 1592 cm^{-1} [24,25], and coordination-induced frequencies were observed in the range 1548 - 1565 cm^{-1} upon complex formation. The lowering in the double bond character of C=N is perhaps due to the influence of the benzimidazole groups being *trans* to each other. The far infrared spectra of the sulfate and sulfonate complexes displayed bands in the range 222 - 279 cm^{-1} which were assigned to the ν(M-N) [26]. A strong and broad peak in the range 1137 - 1188 cm^{-1} and 1050 - 1090 cm^{-1} was present in the spectra of sulfate and sulfonate complexes respectively, and this is typical of the uncoordinated sulfate and sulfonate ions [27]. The infrared spectra of these complexes suggested that all the three donor atoms of the ligand are involved in the coordination sphere and that both the sulfate and sulfonate anions are non-coordinating. The geometry of the complexes was confirmed by UV-Vis solid reflectance electronic studies as well as by single crystal X-ray crystallography (Section 3.2.2).

Three d-d transitions that are expected in the visible region of the spectrum for an octahedral Co(II) complex are; $^4T_{1g}(F) \rightarrow \, ^4T_{2g}(F)$ (ν_1), $^4T_{1g}(F) \rightarrow \, ^4A_{2g}(F)$ (ν_2) and $^4T_{1g}(F) \rightarrow \, ^4T_{1g}(P)$ (ν_3) [28]. These absorption bands were observed at 1095, 544 and 482 nm respectively (**Figure 5**). For the nickel complexes the bands were observed at 919, 585 and 555 nm which may be assigned to $^3A_{2g}(F) \rightarrow \, ^3T_{2g}(F)$ (ν_1), $^3A_{1g}(F) \rightarrow \, ^3T_{2g}(F)$ (ν_2) and ($^3A_{2g}(F) \rightarrow \, ^3T_{1g}(P)$ (ν_3) transitions respectively for an octahedral symmetry [28]. The electronic spectrum of the Cu(II) complex showed one broad band at 619 nm which was ascribed to the $^2B_{1g} \rightarrow \, ^2B_{2g}$ and, assuming that the second transition ($^2B_{1g} \rightarrow \, ^2A_{1g}$) is masked by the intraligand

transition, this is consistent with a distorted octahedral geometry [28].

3.2.2. X-Ray Structural Analysis

An ORTEP diagram of the crystal structure of $[Cu(NNN)_2](RSO_3)_2\cdot 12H_2O$ is presented in **Figure 6**. The selected crystallographic data is presented in **Table 1**, and selected bond lengths and angles in **Table 2**.

The crystal structure of $[Cu(NNN)_2](RSO_3)_2\cdot 12H_2O$ conclusively depicted that the complex is cationic with relatively isolated sulfonate anions (**Figure 6**). The closest contact that the sulfonate ion has with the cationic molecule is H(23)-O(31) = 2.560(1). The two ligands are

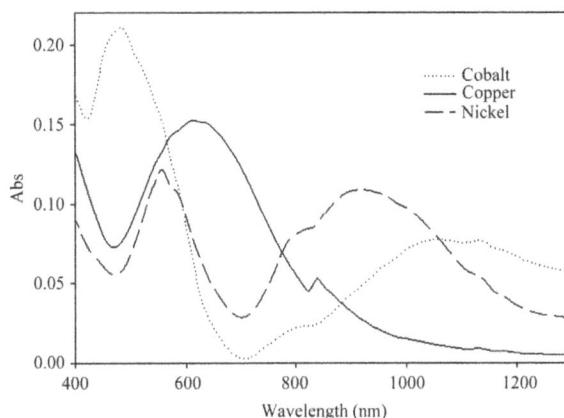

Figure 5. The UV-Vis solid reflectance spectra of [M(NNN)$_2$]SO$_4\cdot x$H$_2$O (M=Co, Ni and Cu; x = 3 - 12).

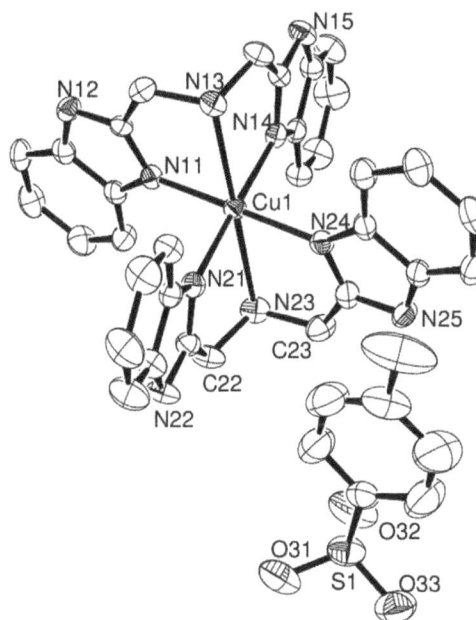

Figure 6. ORTEP diagram of [Cu(NNN)$_2$](RSO$_3$)$_2$·12H$_2$O showing the atom labeling scheme and ellipsoids drawn at 50% probability level. One toluene-4-sulfonate anion and twelve water molecules have been omitted for clarity.

Table 1. Selected crystallographic data for [Cu(NNN)₂](RSO₃)₂·12H₂O.

Compound	[Cu(NNN)$_2$](RSO$_3$)$_2$·12H$_2$O
Chemical formula	C$_{54}$H$_{86}$CuN$_{10}$O$_{18}$S$_2$
Formulae weight	4514.66
Crystal color	Blue
Crystal system	Tetragonal
Space group	$I4$
Temperature (K)	200
Crystal size (mm^3)	0.05 × 0.11 × 0.50
a (Å)	34.4165(11)
b (Å)	34.4165(11)
c (Å)	9.3238(4)
α (°)	90
β (°)	90
γ (°)	90
V (Å^3)	11044.0(9)
Z	2
D_{calc} (g/cm^{-3})	1.358
μ/mm^{-1}	0.545
F(000)	4696
Wavelength (Å)	0.71073
Theta Min-Max (°)	1.9, 28.3
S	0.98
Tot., Uniq. Data, R(int)	91906, 13572, 0.095
Observed data [I > 2.0 sigma(I)]	8997
R	0.0507
R_w	0.1218

Table 2. Selected bond lengths (Å) and angles (°) for [Cu(NNN)₂](RSO₃)₂·12H₂O.

Bond lengths			
Cu1–N11	1.998(14)	Cu1-N21	2.045(14)
Cu1–N13	2.509(14)	Cu1-N23	2.547(14)
Cu1–N14	2.040(14)	Cu1-N24	2.012(15)

Bond angles			
N11-Cu1-N13	77.0(5)	N13-Cu1-N24	104.0(5)
N11-Cu1-N14	86.5(6)	N14-Cu1-N21	179.9(8)
N11-Cu1-N21	93.6(6)	N14-Cu1-N23	105.3(5)
N11-Cu1-N23	103.1(5)	N14-Cu1-N24	93.9(6)
N11-Cu1-N24	179.0(6)	N21-Cu1-N23	74.8(5)
N13-Cu1-N14	75.9(5)	N21-Cu1-N24	86.0(6)
N13-Cu1-N21	104.0(5)	N23-Cu1-N24	75.9(5)
N13-Cu1-N23	178.9(4)	C22-N23-C23	114.2(14)

rameters were reported because the complete refinement of the model was unsuccessful, and the ORTEP diagram presented was based on parameters obtained in "best" model. A similar nickel complex crystal structure has also been reported [30,31] but had a different arrangement of the donor groups, for example the aliphatic amine groups from each ligand were bonded *trans* to the benzimidazole group of another ligand. The similarity in the geometry of these base metal complexes afforded us to conclude that the lack of pH-metric separation with the use of *bis*((1-decylbenzimidazol-2-yl)methyl)amine as extractant (Section 3.1) is influenced by the lack of stereochemical "tailor making".

4. Conclusion

The combination of two low pK_a aromatic nitrogenous groups of benzimidazole with a strong aliphatic amine, in the design of a tridentate ligand, resulted in extraction curves that are pushed deep as a function of pH possibly due to the high complex formation constants of tridentate coordination and the relatively low protonation constants of the benzimidazole groups. This may compromise the stripping of the metal ions from the loaded organic phase through pH adjustment to lower pH. The Fe(III) rejection ability of this tridentate ligand in the range where the other later 3d metal ions extract is remarkable. It can be concluded that the exploitation of the subtle stereochemical aspects of coordination for the extraction of base metals is lacking with tridentate ligands (at least those of the nature presented here). This leads us to propose the evaluation of a bidentate derivative, [2-methyl-aminomethyl-(1-*R*-benzimidazole), *R* = octyl or decyl],

tridentately coordinated to the copper(II) ion in the formation of a tetragonally distorted O$_h$ geometry with the four benzimidazoles in the square plane while the two aliphatic amines occupy the apical positions. The five-membered chelate rings have bite angles in the range 74.8° - 77.0°. The aliphatic amines are bonded *trans* to each other and the Cu-N lengths of 2.509 and 2.547 Å are rather long possibly due to Jahn-Teller distortion. However, the Cu-N (benzimidazole) lengths are rather short (average of 2.014 Å) due to the π-acidity character of the benzimidazole group.

An X-ray crystal structure of the copper complex similar to [Cu(NNN)₂](RSO₃)₂·12H₂O was presented by Berends and Stephan [29] but no crystallographic pa-as a potential extractant of base metal ions. We have also demonstrated, for the first time, that the ion-association solvent extraction system studied here can be interpreted quantitatively for the complexation aspect of the formation of the cationic complex species to describe the two linear log D vs pH$_e$ plots, and to extract information on coordination numbers from the extraction data using the respective slopes for each extraction curve.

5. Supporting Information

CCDC 891865 contains the supplementary crystallographic data for [Cu(NNN)$_2$](RSO$_3$)$_2$·12H$_2$O. These data can be obtained free of charge via www.ccdc.cam.ac.uk/ data_request/cif [or from the Cambridge Crystallographic Data Centre (CCDC), 12 Union Road, Cambridge CB2 1EZ, UK; fax: +44(0)1223 - 336033; e-mail: deposit@c- cdc.cam.ac.uk].

6. Acknowledgements

The authors thank Mr F. Chindeka (DST/Mintek-NIC, Rhodes University Chemistry Department) for the microanalysis results. We also acknowledge Shell Chemicals (SA) (Pty) Ltd for supplying Shellsol 2325. We would also like to thank Mr A. S. Ogunlaja and Mr P. Kleyi for their assistance. For financial support, we thank the National Research Foundation (NRF-CPRR grant).

REFERENCES

[1] C. Kumar, S. K. Sahu and B. D. Pandey, "Prospects for Solvent Extraction Process in the Indian Context for Recovery of Base Metals. A Review," *Hydrometallurgy*, Vol. 103, No. 1-4, 2010, pp. 45-53. doi:10.1016/j.hydromet.2010.02.016

[2] K. C. Sole, A. M. Feather and P. M. Cole, "Solvent Extraction in Southern Africa: An Update of Some Recent Hydrometallurgical Developments," *Hydrometallurgy*, Vol. 78, No. 1-2, 2005, pp. 52-78. doi:10.1016/j.hydromet.2004.11.012

[3] A. I. Okewole, N. P. Magwa and Z. R. Tshentu, "The Separation of Nickel(II) from Base Metal Ions Using 1-Octyl-2-(2′-pyridyl)imidazole as Extractant in a Highly Acidic Sulfate Medium," *Hydrometallurgy*, Vol. 121-124, 2012, pp. 81-89. doi:10.1016/j.hydromet.2012.04.002

[4] J. G. H. du Preez, J. Postma, S. Ravindran and B. J. A. M. van Brecht, "Nitrogen Reagents in Metal Ion Separation. Part VI. 2-(1′-Octylthiomethyl)pyridine as Extractant for Later 3d Transition Metal Ions," *Solvent Extraction and Ion Exchange*, Vol. 15, No. 1, 1997, pp. 79-96. doi:10.1080/07366299708934467

[5] J. G. H. du Preez, "Recent Advances in Amines as Separating Agents for Metal Ions," *Solvent Extraction and Ion Exchange*, Vol. 18, No. 4, 2000, pp. 679-701. doi:10.1080/07366290008934703

[6] G. D. Christian, "Analytical Chemistry," 6th Edition, John Wiley and Sons Inc, Hoboken, 2003, pp. 444-445.

[7] R. W. Hay, T. C. Clifford and P. Lightfoot, "Copper(II) and Zinc(II) Complexes of N,N-*Bis*(benzimidazole-2-yl-methyl)-amine. Synthesis, Formation Constants and the Crystal Structure of [ZnLCl]$_2$]·MeOH. Catalytic Activity of the Complexes in the Hydrolysis of the Phosphotriester 2,4-Dinitrophenyl Diethyl Phosphate," *Polyhedron*, Vol. 17, No. 20, 1998, pp. 3575-3581. doi:10.1016/S0277-5387(98)00152-1

[8] B. Kurzak, D. Kroczewska and J. Jezierska, "Ternary Copper(II) Complexes with Diethylenetriamine and α-(or β-) Alaninehydroxamic Acids in Water Solution," *Polyhedron*, Vol. 17, No. 11, 1998, pp. 1831-1841. doi:10.1016/S0277-5387(97)00528-7

[9] S. O. Bondareva, Y. I. Murinov and V. V. Lisitskii, "Extraction of Non-Ferrous Metals by Bisacylated Diethylenetriamine," *Russian Journal of Inorganic Chemistry*, Vol. 52, No. 5, 2007, pp. 796-799. doi:10.1134/S0036023607050257

[10] J. G. H. du Preez, T. I. A. Gerber, W. Edge, V. L. V. Mtotywa and B. J. A. M. van Brecht, "Nitrogen Reagents in Metal Ion Separation. Part XI. The Synthesis and Extraction Behavior of a New Imidazole Derivative," *Solvent Extraction and Ion Exchange*, Vol. 19, No. 1, 2001, pp. 143-154. doi:10.1081/SEI-100001379

[11] G. Wilkinson, J. A. McCleverty, and R. D. Gillard, Eds., Comprehensive Coordination Chemistry, Late Transition Elements, Vol. 5, Pergamon Press, 1987, pp. 596-681.

[12] K. A. Allen, "Equilibrium between Didecylamine and Sulphuric Acid," *Journal of Physical Chemistry*, Vol. 60, No. 7, 1956, pp. 943-946. doi:10.1021/j150541a027

[13] B. R. Reddy, S. V. Rao and K. H. Park, "Solvent Extraction Separation and Recovery of Cobalt and Nickel from Sulphate Medium using Mixtures of Tops 99 and TIBPS Extractants," *Minerals Engineering*, Vol. 22, No. 5, 2009, pp. 500-505. doi:10.1016/j.mineng.2009.01.002

[14] J. V. Dagdigian and C. A. Reed, "A New Series of Imidazole-Thioether Chelating Ligands for Bioinorganic Copper," *Inorganic Chemistry*, Vol. 18, No. 9, 1979, pp. 2623-2626. doi:10.1021/ic50199a058

[15] M. Haring, "A Novel Route to N-Substituted Heterocycles," *Helvetica Chimica Acta*, Vol. 42, 1957, pp. 1845-1850.

[16] Bruker SHELXTL Version 5.1. (Includes XS, XL, XP, XSHELL), Bruker AXS Inc., Madison, Wisconsin, USA, 1999.

[17] A. L. Spek, "Single-Crystal Structure Validation with the Program PLATON," *Journal of Applied Crystallography*, Vol. 36, 2003, pp. 7-13. doi:10.1107/S0021889802022112

[18] L. J. Farrugia, "*ORTEP*-3 for Windows: A Version of *ORTEP*-III with a Graphical User Interface (GUI)," *Journal of Applied Crystallography*, Vol. 30, 1997, p. 565. doi:10.1107/S0021889897003117

[19] H. M. Irving and R. J. P. Williams, "The Stability of Transition Metal Complexes," *Journal of the Chemical Society*, 1953, pp. 3192-3210. doi:10.1039/jr9530003192

[20] D. S. Flett, "Cobalt-Nickel Separation in Hydrometallurgy: A Review," *Chemistry for Sustainable Develop-

ment, Vol. 12, 2004, pp. 81-91.

[21] L. Alderighi, P. Gans, A. Ienco, D. Peters, A. Sabatini and A. Vacca, "Hyperquad Simulation and Speciation (HySS): A Utility Program for the Investigation of Equilibria Involving Soluble and Partially Soluble Species," *Coordination Chemistry Reviews*, Vol. 184, No. 1, 1999, pp. 311-318. doi:10.1016/S0010-8545(98)00260-4

[22] M. F. Perutz, "The Role of Aromatic Rings as Hydrogen-Bond Acceptors in Molecular Recognition," *Philosophical Transactions: Physical Sciences and Engineering*, Vol. 345, No. 1674, 1993, pp. 105-112. doi:10.1098/rsta.1993.0122

[23] W. J. Geary, "The Use of Conductivity Measurements in Organic Solvents for the Characterisation of Coordination Compounds," *Coordination Chemistry Reviews*, Vol. 7, No. 1, 1971, pp. 81-122.

[24] T. J. Lane, I. Nakagawa, J. L. Walker and A. J. Kandathil, "Infrared Investigation of Certain Imidazole Derivatives and Their Metal Chelates," *Inorganic Chemistry*, Vol. 1, No. 2, 1962, pp. 267-276. doi:10.1021/ic50002a014

[25] J. Reedijk, "Pyrazoles and Imidazoles as Ligands. Part VI. Coordination Compounds of Metal(II) Perchlorates, Tetrafluoroborates and Nitrates Containing the Ligand *N-n*-Butylimidazole," *Journal of Inorganic and Nuclear Chemistry*, Vol. 33, No. 1, 1971, pp. 179-188. doi:10.1016/0022-1902(71)80020-9

[26] E. S. Raper and J. L. Brooks, "Complexes of 1-Methylimidazoline-2-thione with Co(II) and Zn(II) Halides and Perchlorates," *Journal of Inorganic and Nuclear Chemistry*, Vol. 39, No. 12, 1977, pp. 2163-2166. doi:10.1016/0022-1902(77)80387-4

[27] Z. Nakamoto, "Infrared and Raman Spectra of Inorganic and Coordination Compounds," 3rd Edition, John Wiley and Sons, New York, 1978, p. 239.

[28] A. B. P. Lever, "Inorganic Electronic Spectroscopy," 2nd Edition, Elsevier, New York, 1984, pp. 554-557.

[29] H. P. Berends and D. W. Stephan, "Copper(I) and Copper(II) Complexes of Biologically Relevant Tridentate Ligands," *Inorganica Chimica Acta*, Vol. 93, No. 4, 1984, pp. 173-178. doi:10.1016/S0020-1693(00)88159-1

[30] J.-Y. Xu, W. Gu, L. Li, S.-P. Yan, P. Cheng, D.-Z. Liao and Z.-H. Jiang, "Synthesis and Crystal Structure of Nickel Complex of *N,N*-Bis(benzimidazol-2-yl-methyl)amine," *Journal of Molecular Structure*, Vol. 644, No. 1-3, 2003, pp. 23-27. doi:10.1016/S0022-2860(02)00281-8

[31] P. Thangarasu, S. Bernès and C. Durán de Bazúa, "*Bis* [*bis*(benzimidazol-2-ylmethyl-N³)amine-*N*]nickel(II) Dichloride", *Acta Crystallographica Section C-Crystal Structure Communications*, Vol. 53, No. 11, 1997, pp. 1607-1609. doi:10.1107/S0108270197006513

Kinetic Analysis of Isothermal Leaching of Zinc from Zinc Plant Residue

Ali Reza Eivazi Hollagh[1], Eskandar Keshavarz Alamdari[1,2], Davooud Moradkhani[3], Ali Akbar Salardini[4]

[1]Mining & Metallurgical Engineering, Amirkabir University of Technology, Tehran, Iran
[2]Research Center for Materials and Mining Industries Technology, Amirkabir University of Technology, Tehran, Iran
[3]Faculty of Engineering, Zanjan University, Zanjan, Iran
[4]Materials and Energy Research Center, Tehran, Iran.
Email: alamdari@aut.ac.ir

ABSTRACT

The sulfuric acid leaching of zinc plant residues was studied in an attempt to find a suitable hydrometallurgical method for zinc recovery. The parameters evaluated consist of reaction time, Solid-to-liquid-ratio, reaction temperature, agitation rate and pH. The results of kinetic analysis of the leaching data under various experimental conditions indicated that there is a reaction controlled by the solution transport of protons through the porous product layer with activation energy of about 1 kJ/mol for different constant solid to liquid ratios. Based on the shrinking core model (SCM), the following semi-empirical rate equation was established:

$$1 - 3(1-\alpha)^{\frac{2}{3}} + 2(1-\alpha) = 0.001187 \times \left[H^+\right]^{0.016} \times \left[\left(\frac{S}{L}\right)\right]^{-1.34} \times \exp\left(-\frac{1}{RT}\right) \times t \,.$$ On the other hand, activation energy was

obtained from a model-free method using isothermal measurements. Values for activation energy were calculated as a result of the conversion function with an average of 2.9 kJ/mol. This value is close to that determined previously, using shrinking core model (SCM).

Keywords: Zinc; Hydrometallurgy; Kinetic Model; Zinc Plant Residue; Model-Free Method

1. Introduction

Zinc is an important base metal needed for different applications in metallurgical, chemical, textile [1], agricultural, painting, and rubber industries. Identified zinc resources of the world estimated by the US. Geological Survey to be about 1.9 billion tons of zinc. Zinc primary resources include zinc sulfide, carbonate, silicates and oxide minerals. Moreover, a part of zinc is recovered by different secondary resources such as zinc ash, zinc dross, flue dusts of electric arc furnace and brass smelting, automobile shredder scrap, rayon industry sludge [1] and zinc plant residues (ZPR) etc.

It was obviously observed that Pyrometallurgical and Hydrometallurgical methods or their combination could be applied for the treatment of primary and secondary zinc containing materials (ZCM). Nevertheless, when these methods are compared, the hydrometallurgical processes are more suitable for the materials with low zinc content because of their higher zinc recoveries. In addition to some other operational advantages hydrometallurgical processes are more environmentally safe and economically feasible.

Sphalerite, a zinc sulfide ore is considered as the main and primary source to produce zinc metal. Oxidative leaching is one of the major processes applied for the zinc recovery from sphalerite and many researchers have investigated its kinetic analyses [2-8].

Extensive investigations have been carried out on the treatment of zinc oxide ores by hydrometallurgical and pyrometallurgical methods [9]. For instance, Thomas and Fray [10] investigated the kinetics of leaching of zinc oxide materials by using chlorine and chlorine hydrate. In all cases studied, lead was also leached out with zinc; however, iron oxides remained almost undissolved. They applied shrinking core diffusion model to describe the kinetic analysis and found that the rate of leaching of Adrar Turkish ore was controlled by surface reaction. In another study, Frenay [11] examined the leaching of oxidized zinc ores in different solution media and gained the best leaching results by using sulfuric acid and caustic soda. Other workers studied the dissolution of zinc from zinc silicate ore [12-14] and carbonate ore [15].

Abdel-Aal [12] studied the kinetic leaching in sulfuric acid solution from an Egyptian zinc silicate ore and demonstrated that the diffusion through the product layer was a rate controlling process during the reaction. The kinetic dissolution of zinc from zinc silicate calcine was studied and a kinetic model of porous solids was applied to describe the rate controlling step and the rate of leaching controlled by chemical reaction and diffusion in porous solids [13].

As a result of limited availability of primary high grade zinc ores or concentrates, the recovery of zinc from secondary resources seems to be quite economical and because of their environmental issues, avoidable. First, the disposal of secondary resources is now becoming expensive because of increasingly severe environmental protection regulations. Second, due to the chemical nature of these resources, they are classified as hazardous waste in which the toxicity is mainly due to the presence of different metals such as zinc and other metals [1].

During zinc extraction process, a large amount of residues is generated daily in zinc processing plants, which is generally called zinc plant residues (ZPR) which are classified as hazardous materials since in addition to zinc; they contain lead, cadmium, arsenic etc. Due to the composition of primary zinc resources and additive chemical compounds during zinc extraction process, these residues could be an valuable sources of zinc, lead, cadmium, copper, germanium, nickel, cobalt, manganese, silver and gold.

Based on Kul and Topkaya's studies on the recovery of germanium and other valuable metals from ZPR of Çinkur Zinc Plant, the best condition for extraction of germanium and other valuable metals was reported at a temperature between 333 to 358 Kelvin temperature (K) for a leaching duration of 1 h with sulfuric the recoveries of mentioned valuable metals were more than 90% [16]. In another study, acid concentration of 150 gr/L and using a solid-liquid ratio of 1/8 gr/cc. Under these conditions, Wang and Zhou recommended a hydrometallurgical process for the production of cobalt oxide after inspecting the recovery of cobalt from ZPR [17].

ZPR has been studied for the recovery of zinc using sulphuric acid [9,16-23] and other leaching agents [24,25]; however, there is no considerable information on its kinetic analysis. Zinc is leached into diluted sulfuric acid from ZPR according to the overall reaction stoichiometry represented in Equations (1) and (2).

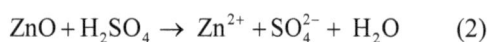

$$ZnSO_4 \xrightarrow{\text{At aqueous media}} Zn^{2+} + SO_4^{2-} \quad (1)$$

$$ZnO + H_2SO_4 \rightarrow Zn^{2+} + SO_4^{2-} + H_2O \quad (2)$$

In several investigations, summarized in **Table 1**, the dissolution of zinc from ZCM using hydrometallurgical method and the experimental conditions have been reported.

The aim of current study was to investigate factors affecting the selective leaching of zinc from ZPR with sulfuric acid. In this regard the effect of time, Solid-to-liquid-ratio(S/L), reaction temperature, agitation rate and pH were studied. The kinetics characterizations of the leaching process were analyzed in accordance with shrinking core model (SCM) and the order of reaction with respect to pH and Solid-to-liquid-ratios. Based on the experimental and calculated results, a semi-empirical rate equation was presented. The activation energy of the dissolution process was evaluated by applying the shrinking-core-model method in the temperature range of 298 - 353 K. In addition, the model-free method was employed in order to estimate the activation energy during the reaction and completion of kinetics investigation. The kinetics results by using model-free method could be useful and confirming for determined leaching mechanism by shrinking core model (SCM). The kinetics characterization of the selective leaching of zinc from ZPR is nessecery for future design of zinc recovery from this hazardous and high content zinc materials. Residue of zinc leaching from ZPR in diluet sulfuric acid may be collected for further cobalt, manganese and zinc recovery.

2. Theory of the Kinetic Modeling

2.1. Shrinking Core Model

In many practical cases, fluid-solid reactions can indeed be approximated as the first-order reactions for mathematical simplicity [28].

Leaching process of ZCM in acid [2-7,9,12] and ammonia [29,30] follows a kinetic model known as shrinking-core-model (SCM). This model should be employed for the first-order reactions and considers whether the rate of leaching process is controlled by the diffusion of reactant or the rate of the surface chemical reaction. The heterogeneous leaching reaction can be expressed as follows:

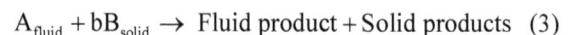

$$A_{fluid} + bB_{solid} \rightarrow \text{Fluid product} + \text{Solid products} \quad (3)$$

A review of the integrated forms of kinetic laws $(G(\alpha))$ for different control regimes known in ZCM leaching according to Levenspiel is detailed in **Table 2** [31].

Based on the Arrhenius law, the reaction rate constant is expressed by the Equation (7):

$$k = A \exp\left(\frac{-E}{RT}\right) \quad (7)$$

where, k is the kinetic constant, A is the pre-exponential factor, E is the activation energy, T is temperature and R is the gas constant.

The kinetic parameters E and A can be obtained using shrinking core model (SCM). When the integrated form of the kinetic law *i.e.* $G(\alpha) = kt$ is known, then one can

Table 1. Summary of the recent researches conducted by various workers on the leaching of Zn from ZCM.

ZCM	Leaching Media	Temp (K)	Description	Ref.
Tailings with high zinc content	H_2SO_4 (2 M)	333	Maximum zinc recovery = 98%, Duration = 7200 sec, Agitation rate = 480 rpm, Solid-Liquid-Ratio (S:L) = 1:4, Particle size = $-75 + 53$ μm	[9]
Pb bearing ZPR	$H_2O + H_2SO_4$	353	Maximum zinc recovery = 69.3%, Duration = 3600 sec, pH of solution = 2.5, Agitation rate = Good mixing, S:L ratio = 1:5, Particle size = $-180+30$ μm	[19]
ZPR	150 gr/L H_2SO_4	358	Maximum zinc recovery > 95%, Duration = 3600 sec, Agitation rate = Good mixing, S:L ratio = 1:4, About 84% of particle size < 147 μm	[16]
ZPR	H_2SO_4 (0.5 M)	348	Maximum zinc recovery >96%, Duration = 1800 sec, Agitation rate = Good mixing, S:L ratio = 1:10, Particle size = $-$	[17]
Roasted H_2SO_4/ZPR	H_2O	298	Step1: Roasting of equal weight ratio of H2SO4/ZPR at 473 K for 1800 sec Step2: Maximum zinc recovery = 86%, Duration = 3600 sec, Agitation rate = Good mixing, S:L ratio= 1:5, Particle size <74 μm	[26]
Blended leach residue	150 gr/L H_2SO_4	368	Maximum zinc recovery = 71.9%, Duration = 7200 sec, Agitation rate = 250 rpm, Solid:Liquid ratio = 1:5	[18]
Sphalerite concentrate	5% H_2SO_4 + 5%H_2O_2	333	Maximum zinc recovery = 80%, Duration = 14,400 sec, Agitation rate = 160 rpm, S:L ratio = 1:20, Particle size ≤ 38 μm	[27]
Sphalerite	5%(V/V) H_2SO_4 + 5%(W/V) ammonium persulphates	333	Maximum zinc recovery = 95%, Duration = 18000 sec, Agitation rate = Good mixing, S:L ratio = 1:10, Particle size ≤ 150 μm	[2]
	1M H_2SO_4 + 0.5 M [Fe^{3+}]	333	Maximum zinc recovery = 73%, Duration = 18000 sec, Agitation rate = 480 rpm, S:L ratio = 1:200, Particle size = $-75 + 53$ μm	[3]
Low-grade zinc silicate ore	2M H_2SO_4 + 1M H_2O_2	333	Maximum zinc recovery = 60%, Duration = 9000 sec, Agitation rate = $0 - 600$ rpm, S:L ratio = 1:500, Particle size = $-45 + 38$ μm	[4]
	2M H_2SO_4 + 0.2 M HNO_3	358	Maximum zinc recovery = 99.6%, Duration = 10800 sec, Agitation rate = Good mixing, PO_2 = 0.1 MPa, S:L ratio = 1:10, Particle size < 74μm, C_2Cl_4:Leaching solution = 1:20	[7]
	10% H_2SO_4	343	Maximum zinc recovery = 94%, Duration = 10800 sec, Agitation rate = 550 rpm, S:L ratio = 1:20, Particle size = $-200 + 270$ mesh	[12]
Zinc silicate calcine	0.4 M H_2SO_4	333	Maximum zinc recovery = 95%, Duration = 420 sec, Agitation rate = 480 rpm, S:L ratio = 1:100, Particle size = $75 - 53$ μm	[13]

Table2. Set of control regimes according to shrinking core model (SCM) [31].

Control regimes	$G(\alpha)$	k	Equation
Liquid film diffusion	$1-(1-\alpha)^{\frac{2}{3}}$	$\dfrac{2bDC_A}{\rho R^2}$	4
Solid product diffusion	$1-3(1-\alpha)^{\frac{2}{3}}+2(1-\alpha)$	$\dfrac{2bDC_A}{\rho R}$	5
Chemical reaction	$1-(1-\alpha)^{\frac{1}{3}}$	$\dfrac{2bDk_cC_A}{\rho R^2}$	6

$G(\alpha)$: Reaction model; k: Reaction rate constant; k_d: reaction rate constant when diffusion is rate controller; k_C: reaction rate constant when chemical reaction is rate controller; α: Fraction reacted; b: Stoichiometric coefficient of reaction; D: diffusion coefficient in the porous product layer; C_A: Concentration of the leachant in the solution; ρ: Density of the solid particle; R_o: Radius of the unreacted particle; K_o: Kinetic constant;

plot the values of $G(\alpha)$ against time for different iso-thermal experiments. Each of these plots should be linear, the slope being the value of k at that temperature. Since:

$$\ln k = \ln A - \frac{E}{RT} \qquad (8)$$

The plot of $\ln k$ versus reciprocal temperature would be a straight line. The slope of straight line gives $-E/R$ and intercepts $\ln A$. In this method, there is also an implicit assumption that the activation energy E does not change during the course of reaction.

2.2. Theory of Model-Free Method

If the $G(\alpha)$ is unknown then evaluation of activation energy E by the integral approach is useless. In addition, in the previous method it was assumed the the evaluated E does not change during the reaction. In model-free method the activation energy should be accepted as a variable parameter, therefore, E can be calculated at different level of α. In general, the integral form of the rate equation is written as:

$$G(\alpha) = kt \qquad (9)$$

where, $G(\alpha)$ is an appropriate function of α. By differentiating with regard to time:

$$G'(\alpha) \cdot \frac{d\alpha}{dt} = k \qquad (10)$$

$$\frac{d\alpha}{dt} = \frac{k}{G'(\alpha)} = k \cdot f(\alpha) \qquad (11)$$

So, $f(\alpha)$ in the differential form of rate equation equals $\frac{1}{G'(\alpha)}$. We have, from Equation (7) and (11):

$$\frac{d\alpha}{dt} = k \cdot f(\alpha) = A \cdot \exp\left(\frac{-E}{RT}\right) \cdot f(\alpha) \qquad (12)$$

Considering a fixed value of α, Equation (12) can be rewritten as follows:

$$\int_0^\alpha \frac{d\alpha}{f(\alpha)} = A \cdot \exp\left(\frac{-E}{RT}\right) \int_0^{t_\alpha} dt \qquad (13)$$

And/or:

$$\frac{1}{A \cdot \exp\left(\dfrac{-E}{RT}\right)} \int_0^\alpha G'(\alpha) d\alpha = \int_0^{t_\alpha} dt \qquad (14)$$

Then:

$$\frac{G(\alpha)}{A \cdot \exp\left(\dfrac{-E}{RT}\right)} = t_\alpha \qquad (15)$$

Hence:

$$-\ln t_{\alpha,i} = \ln\left[\frac{A_\alpha}{G(\alpha)}\right] - \frac{E_\alpha}{RT_i} \qquad (16)$$

In the Equation (16) subscripts, i refer to isothermal condition [32,33]. $t_{\alpha,i}$ is the time required to reach a certain conversion isothermal, and A_α and $G(\alpha)$ represent pre-exponential factor and integrated form of the reaction model.

For the fixed value of α, it follows that a plot of the left side of equation (16) against reciprocal temperature would be a straight line; the slope of which should yield the value of $-E/R$. So, E can be calculated at different levels of α.

3. Materials and Reagents

The zinc plant residue used in the experiments was obtained from the National Lead and Zinc Co. (NILZ); Zanjan, Iran which has a production capacity of 20000 t Zn/y. In this company, ZnO-rich calcine is first produced from the oxide-carbonate concentrates and then leached with hot sulfuric acid solution. After the separation of liquid and solid, the pregnant solution is purified by selective precipitation method and cementation. Finally, the purified solution is electrowon for metallic zinc production. The separated solids during each stage are called ZPR. A flow sheet which explains each purification stage in NILZ plant is reported in literature [34].

The ZPR used in this study initially contained ca. 20% moisture. Prior to the use, the sample was dried, ground, and homogenized using a riffle; and then was crushed using a ball mill. At this stage, due to the nature of ZPR, its fractionation by dry screen test was not possible. Therefore, the samples were directly used for the experiments without any particle size fractionation and some preliminary tests such as Scanning Electron Micrograph pictures taken (**Figure 1**), which showed that its particle size was generally less than 50 μm.

After homogenizing ZPR, a sample was taken for its kinetics study. The chemical analysis of the sample was carried out by atomic absorption spectrometer (Perkin-Elmer AA300 atomic). The details of the chemical analysis of ZPR sample was; Zn = 28.39%, Mn = 5.53%, Co = 0.66%, Ca = 9.77%. The XRF analytical results are shown in **Table 3**. The results indicated that the ZPR was mostly composed of zinc, calcium, manganese and cobalt.

In the previous researches, the mineralogical characterization of ZPR for zinc compounds have been identified as follows: $(Zn,Cu)_2(AsO_4)OH$, zinc oxysulphate $(Zn_3O(SO_4)_2)$ [16], hydrated zinc sulfate $(ZnSO_4 \cdot XH_2O)$ [16,18,19], hydrated zinc sulfite $(ZnSO_3 \cdot 2.5H_2O)$ [16,34], zinc oxide (ZnO) [34], zinc ferrite $(ZnFe_2O_4)$ [18]. Different compounds of zinc was detected from different

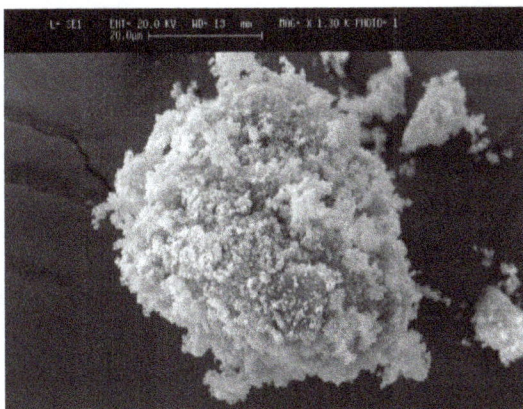

Figure 1. Scanning electron micrograph of ZPR particles.

Table 3. XRF analysis of used ZPR used in this study.

Chemical composition	Conc.%
Mn	4.97
Si	1.00
S	10.96
Ca	11.51
Fe	0.11
Co	0.71
Zn	25.96
Pb	0.17
Cd	0.10
Mg	0.22
Al	0.40
K	0.15

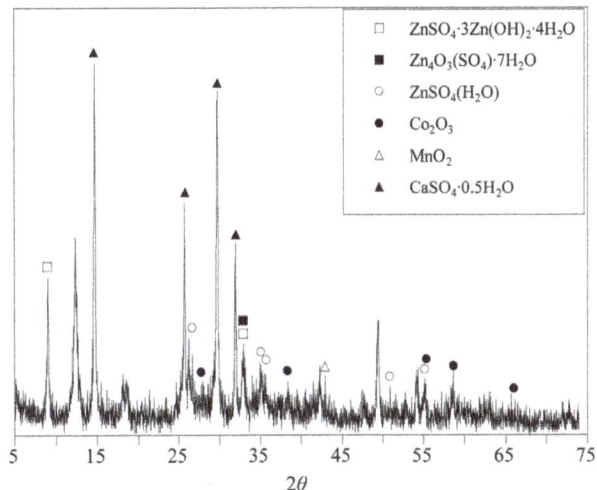

Figure 2. X-ray diffraction analysis of the ZPR used in this study.

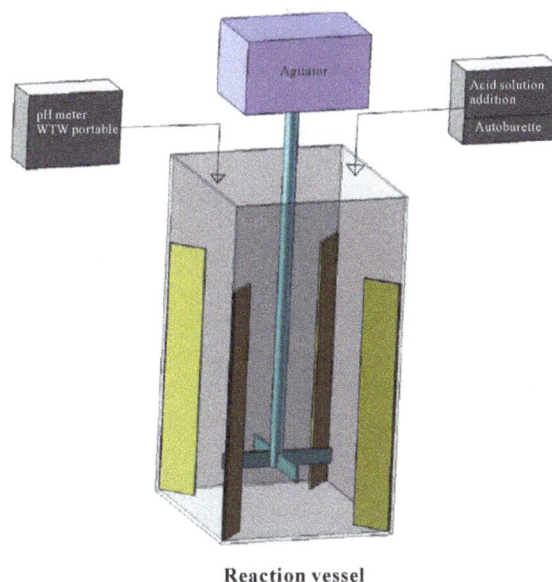

Figure 3. Schematic diagram of the reaction vessel during the kinetics study.

kind of produced ZPRs in zinc plants. Mineralogical analysis performed using Philips PW3710 model X-ray diffractometer, which indicated that $ZnSO_4 \cdot 3Zn(OH)_2 \cdot 4H_2O$, $Zn_4O_3(SO_4) \cdot 7H_2O$, $ZnSO_4 \cdot H_2O$, $CaSO_4 \cdot 0.5H_2O$, MnO_2 and Co_2O_3 were the main mineralogical composition in the used ZPR (**Figure 2**). Distilled water was used in the experiments which were performed under atmospheric pressure. Laboratory grade sulfuric acid was used to adjust the solution pH as required.

4. Experimental Method

For the kinetic study, experiments were carried out in 4.5 L stainless steel rectangular boxes ($14 \times 14 \times 23(L \times W \times H)$ cm) placed in a water bath equipped with a mechanical stirrer. The bath temperature was digitally controlled within ±273.5 K (**Figure 3**). There were four baffles (17×3 cm) in reaction vessel. The impeller was placed 3 cm above the bottom of the vessel. During the experiment, the agitation rate was adjusted in the range of 50 to 1000 rpm; temperature was varied in the range of 298 to 353 K; pH was adjusted at 1 to 5; solid to liquid ratio was changed in the range of 1/8 to 1/4 (%W/V); the applying maximum time of reaction is 7200 sec. When the solution temperature reached the desired value, the dried ZPR powder was added to the solution with an initial volume of 3 L. At certain intervals, samples of known volume were taken from the pulp. The samples were immediately vacuum filtered, diluted and analyzed for the zinc concentration, which was calculated with respect to the correction of volume [35]. The H_2SO_4 concentration in all cases was 238 gr/L. Any volume of sulfuric acid solution was used to adjust the pH of the solution as was required. The pH was measured using a WTW portable pH-meter equipped with a suitable electrode for aqueous solution and an automatic temperature compensation device.

5. Result and Discussion

5.1. Effect of Agitation

Agitation of pulps is usually necessary to maximize the kinetics and short reaction times in order to make it desirable for economic reasons [30]. In some leaching processes, maximization of reaction kinetics is being performed by increasing the agitation rate, so that the mineral particles could remain suspended in the liquor and induce a decrease in the thickness of the mass transfer boundary layer on the surface of the particles. By increasing the agitation rate, therefore, the diffusion of liquor to the surface of the particles increases [36]. The effect of agitation rate on the dissolution of zinc was investigated at various agitation rates (50, 250, 450, 700 and 1000 rpm) at 333 K and pH of 3 and with solid/liquid ratio of 1/6 %(W/V) after 3600 sec. As **Figure 4** indicates, dissolution of zinc was affected by changes in agitation rate. The results show that the leaching rate of zinc increases quickly below 450 rpm and remains almost constant beyond this speed to 1000 rpm. The maximum zinc recovery under this condition was more than 80% when 1000 rpm agitation rate was applied; however, this agitation rate may not be applicable in industrial operations due to the increase in the capital and opera compromised optimum operating agitation rate.

5.2. Effect of Temperature

Experiments were performed to study the temperature-dependency of the reaction to the amount of zinc extracted. **Figure 5** shows that, zinc extraction increased along with the leaching time. **Figure 5** also indicates that increasing the temperature from 298 to 353 K did not improve this leaching system. There was some improvement in the zinc extraction upon increasing the temperature up to 353 K, but in different reaction time. About 95% of zinc could be extracted at 333 K within 5400 sec.

Temperature dependency can be used to estimate the apparent activation energy, enthalpy of activation and entropy of activation [37]. It is widely accepted [31] that systems with an activation energy greater than 40 kJ/mol are controlled by a chemical reaction while those with an activation energy of about 10 kJ/mol or less are controlled by a transport process whether in the product layer or a boundary fluid film.

Based on shrinking core model (SCM) results, Equations (5) and (6) were applied for the results obtained from each temperature value. **Figure 6** represents the data plot according to chemical reaction control. The slope of the line is the rate constant k_c. **Figure 7** also shows the data plot according to diffusion control process. The slope of this line is considered as the apparent rate constant k_d. The rate constant values, k_d and k_c are calculated from Equations (5) and (6), respectively. The results obtained from **Figures 6** and **7** indicated that the dissolution rate of ZPR was controlled by the diffusion and not the surface chemical reactions. That was due to lower value of correlation coefficients (R^2) of chemical reaction model than the value of R^2 of ash diffusion

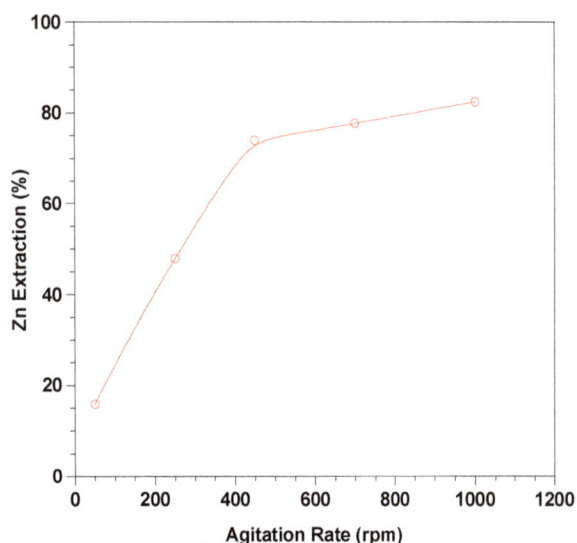

Figure 4. Effect of agitation rate on the zinc recovery from the ZPR, (pH: 3; temperature: 333 K; Solid-to-liquid-ratio: 1/6; time: 4200 sec).

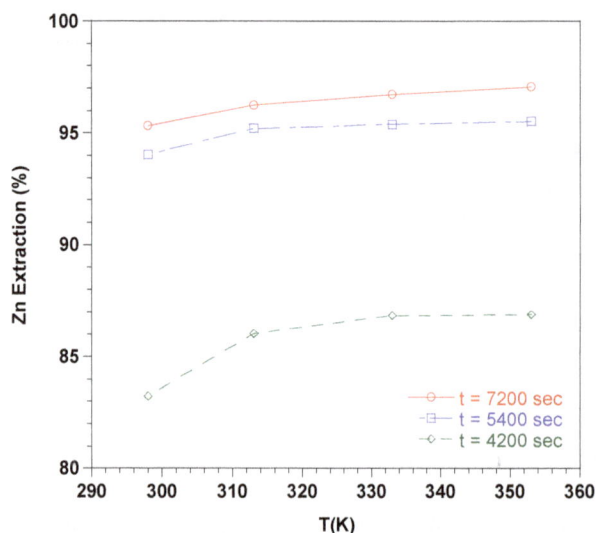

Figure 5. Effect of temperature on the zinc recovery from the ZPR in various times, (solid-liquid-ratio: 1/6; pH: 3; agitation rate: 700 rpm).

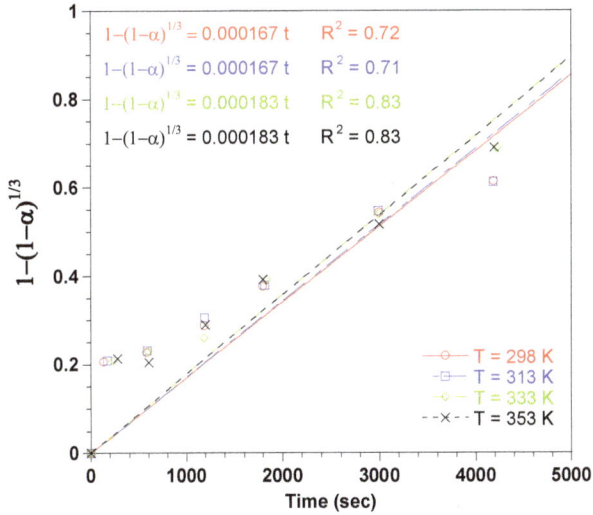

Figure 6. Plot of $1 - (1 - \alpha)^{1/3}$ vs time for different temperatures (Solid-to-liquid-ratio: 1/8; agitation rate: 700 rpm; pH: 3).

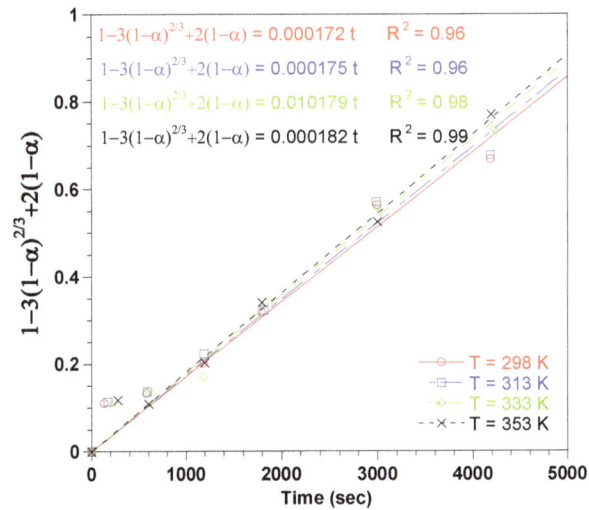

Figure 7. Plot of $1 - 3(1 - \alpha)^{2/3} + 2(1 - \alpha)$ vs time for different temperatures, (Solid-to-liquid-ratio: 1/8; agitation rate: 700 rpm; pH: 3).

model. The apparent rate constant for zinc dissolution increases slightly by increasing the temperature up to 353 K. Generally, a high value of activation energy indicates that the process is strongly influenced by temperature and therefore, the rate-controlling step could have a reaction at the solid surface. Conversely, a low value of activation energy indicates that the process is weakly influenced by the temperature and the rate-controlling step could be the mass transport of reagents or products through the reaction product.

The results that are given in **Table 4**, show that the dissolution of ZPR in the different Solid-to-liquid-ratios and temperature range of 298 - 353 K fitted the diffusion model given in Equation (5).

Arrhenius plot (**Figure 8**) presenting the apparent rate constants was obtained by applying Equation (5) to leaching experimental data (**Figure 8**). As seen in **Figure 8**, the calculated activation energy for each Solid to liquid ratio was about 1 kJ/mol which clearly suggests that diffusion controls the leach process [31]. Due to low value of activation energy, increasing the temperature would not affect its occurrence significantly.

Model-free method has been used by other investigators for solid-gas reactions to estimate the Arrhenius parameters, but the current method is also applied to investigate the kinetic study in solid-liquid reaction [38].

Table 4. zinc leaching experimental conditions and results in constant pH of 3.

S/L ratio (%wt/v)	Temp. (K)	Time (sec)	Zn % E	k_d (sec^{-1})	R^2
1/8	298	4200	94.230	0.000172	0.96
	313	4200	94.268	0.000175	0.96
	333	4200	96.970	0.000179	0.98
	353	4200	97.000	0.000182	0.99
1/6	298	5400	94.047	0.000125	0.97
	313	5400	95.204	0.000128	0.98
	333	5400	95.384	0.000131	0.98
	353	5400	95.534	0.000133	0.99
1/4	298	7200	87.104	6.83E−05	0.98
	313	7200	88.380	6.97E−05	0.97
	333	7200	91.547	7.14E−05	0.96
	353	7200	91.891	7.34E−05	0.93

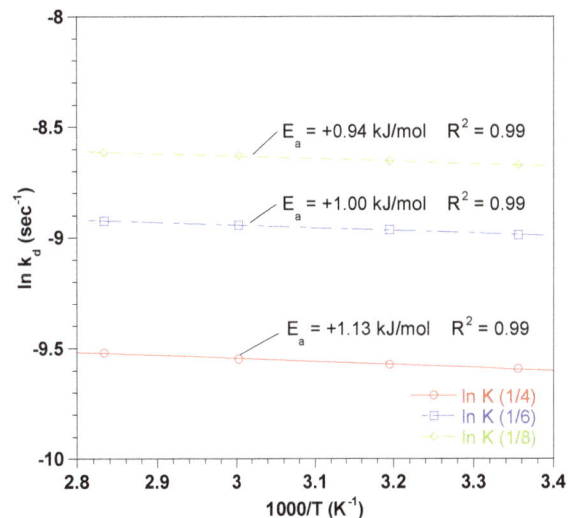

Figure 8. Arrhenius plot of reaction rate against reciprocal temperature in different Solid-to-liquid-ratios, (agitation rate: 700 rpm; pH: 3).

Vya zovkin and Wright compared the model-free and model fitting methods in solid reaction and the activation energy was estimated by the model-free method [32]. In this method, t_α, the time required for a given value of α was calculated first and then activation energy was obtained by applying Equation (16) to isothermal kinetic data within the range of 0.50 to 0.95. Values for E were achieved as a function of conversion with an average of 2.9 kJ/mol (**Figure 9**). This value is close to that determined previously, using shrinking core model (SCM).

It may be observed from **Figure 9** that dependency of activation energy on conversion is rather weak, decreasing at the range of 0.50 to 0.95. The changes in E with conversion may be described as follows. Nucleation of ash, nuclei growth and diffusion of the liquid reagent through the porous ash provide the effective parameters that will determine the activation energy. The results of BET surface area analyzing, show that by increasing the time of leaching the surface area of particles are risen. This could confirm the porosity in the product ash (**Figure 10**). As compared with the activation energy at range of 0.50 to 0.95, the high value of activation energy at $\alpha = 0.5$ may be explained by interaction of these phenomena. Decrease in activation energy could be illustrated by noting that at studied range; the product layer imposes a negligible resistance to the overall rate. However, as the result of nuclei growth during the reaction, the activation energy decreases.

5.3. Effect of pH

The effect of pH in the range of 1 - 5 was studied at 333 K with a stirring speed of 700 rpm for Solid-to-liquid-ratio of 1/6. Sulfuric acid solution was used to adjust the pH of the solution as was required. As seen in **Figure 11**, the

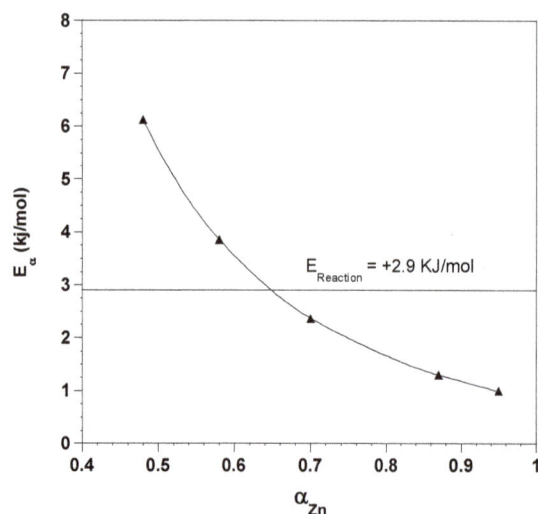

Figure 10. Variation of the surface area vs time, (Solid-to-liquid-ratio: 1/6; agitation rate: 700 rpm; pH: 3, temperature: 333 K).

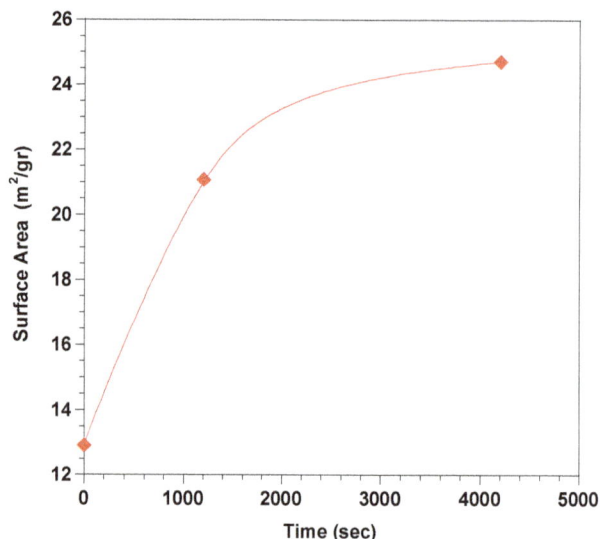

Figure 11. Effect of pH on the zinc recovery from the ZPR in various times (Solid-to-liquid-ratio: 1/6; temperature: 333 K; agitation rate: 700 rpm).

dissolution of zinc was improved by decreasing in pH until 3, but further decreasing of pH had no significant effect. It could be also seen in this figure that the zinc extraction efficiencies were developed along with the increasing leaching time at constant pH of 1, 2 and 3. For pH of 5, the maximum extraction of zinc is approximately 81%. On the other hand, the maximum zinc dissolution at pH of 3, 2 and 1 was about 95%, 98% and 98%, respectively. So, the lowest zinc recovery was at pH of 5. Based on the XRD analysis of ZPR, it could be concluded that most of the zinc in ZPR was as zinc sulfate and oxide. Zinc is leached from detected zinc minera-

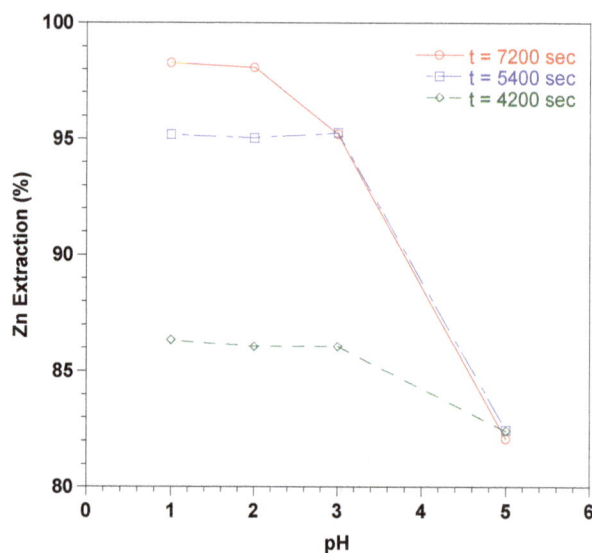

Figure 9. Variation of the activation energy with conversion derived from model-free method applying isothermal kinetic data.

logical composition of ZPR base on XRD analysis of ZPR in diluted sulfuric acid according to the following reaction stoichiometry:

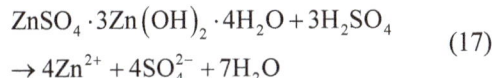

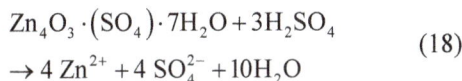

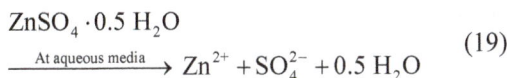

$$\begin{aligned}&ZnSO_4 \cdot 3Zn(OH)_2 \cdot 4H_2O + 3H_2SO_4 \\ &\rightarrow 4Zn^{2+} + 4SO_4^{2-} + 7H_2O\end{aligned} \tag{17}$$

$$\begin{aligned}&Zn_4O_3 \cdot (SO_4) \cdot 7H_2O + 3H_2SO_4 \\ &\rightarrow 4\,Zn^{2+} + 4\,SO_4^{2-} + 10H_2O\end{aligned} \tag{18}$$

$$ZnSO_4 \cdot 0.5\,H_2O \xrightarrow{\text{At aqueous media}} Zn^{2+} + SO_4^{2-} + 0.5\,H_2O \tag{19}$$

However, in all experiment less than 5% Co, Mn and Ca were leached from ZPR.

Due to the low acidity of the solution in pH of 5 most of the zinc was liberated from the zinc sulfate phase of the ZPR and smaller amount was released from zinc oxide phase. However, because of high acidity in pH of 1 and 2 during reaction, the remittance of the zinc was recovered from the zinc oxide and the maximum dissolution was improved further. Nevertheless, it was concluded that there was no benefit in decreasing the pH beyond 3 as the cost of consumption of acid to obtain pH of 1 and 2 is very high. Thus leaching at 333 K and pH of 3 after 5400 sec was selected to be practically optimum.

To further examine the effect of pH applied to this kinetic model and k_d values for each pH, a plot of log k_d versus pH was prepared. The order of the reaction obtained from the plot with respect to [H⁺] was proportional to a 0.016 power with a correlation coefficient of 0.99.

5.4. Effect of Solid-to-Liquid-Ratio

Figure 12 gives the zinc recovery as a function of Solid-to-liquid-ratio at 333 K in solutions with pH of 3 and agitation rate of 700 rpm in different constant times of leaching. According to experimental results presented in **Figure 12**, it was found that zinc recovery increased until the solid-to-liquid-ratio of 1/8, where about 98% of zinc recovery was reached. As seen in **Figure 12**, zinc extraction increases with leaching time at constant solid-to-liquid-ratio of 1/4 and 1/6. The maximum zinc recovery at solid-to-liquid-ratio of 1/6 after 5400 sec was about 96%. Therefore, this quantity could be considered an optimum value for the dissolution of zinc with respect to other ratios. A decrease in the solid-to-liquid-ratio increases the rate of leaching due to reduction in amount of ZPR in the solution. Depletion in the amount of ZPR in solution causes an easy diffusion of liquor to the surface of the ZPR particles.

From the effect of solid-to-liquid-ratio on zinc dissolution given in **Figure 12**, the apparent rate constant was determined. According to log k_d-log(S/L) curves, the order of reaction with respect to solid-to-liquid-ratio was found to be inversely proportional to a 1.34 power with a

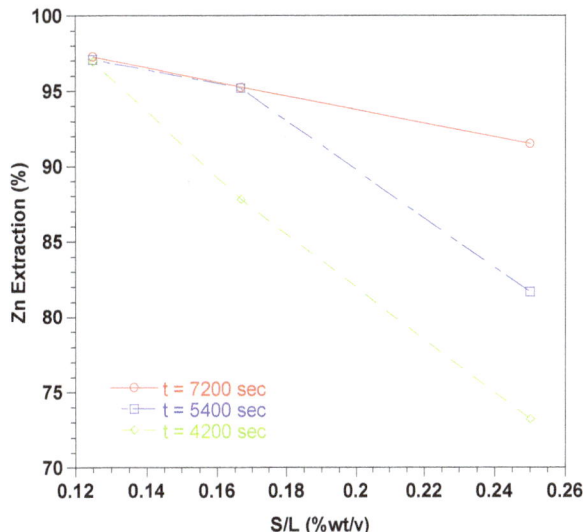

Figure 12. Effect of Solid-to-liquid-ratios on the zinc recovery from the ZPR in various times, (pH: 3; temperature: 333 K; agitation rate: 700 rpm).

correlation coefficient of 0.99.

5.5. Experimental Equation for Estimating Reaction Rate Constant

The detailed analysis of leaching kinetics show that activation energy and the order of reaction values with regard to acid concentration and solid-to-liquid-ratio substantiate the shrinking core model (SCM) results for a diffusion controlled process. Hence, the leaching of ZPR can be clearly presented by Equation (20).

$$\begin{aligned}&1 - 3(1-\alpha)^{\frac{2}{3}} + 2(1-\alpha) \\ &= 0.001187 \times [H^+]^{0.016} \times \left[\left(\frac{S}{L}\right)\right]^{-1.34} \times \exp\left(-\frac{1}{R\,T}\right) \times t\end{aligned} \tag{20}$$

As shown in **Figure 13**, Based on the experimental result, the left-hand side of Equation (20) is plotted against

$$[H^+]^{0.016} \times \left[\left(\frac{S}{L}\right)\right]^{-1.34} \times \exp\left(-\frac{1}{RT}\right) \times t$$

k_o value of 0.001187 with a regression coefficient of 0.97 was obtained.

To check the deviation of experimental values of conversion from calculated values from empirical equation, the plot of $G(\alpha)$-experimental against $G(\alpha)$-calculated was drawn. As seen in **Figure 14**, the agreement between experimental and calculated values was very substantial.

6. Conclusion

In the current work, dissolution kinetics of zinc from ZPR in dilute sulfuric acid solution was investigated. It was found that the reaction rate increases slightly along with an increase in pH, reaction temperature and also solid-to-liquid-ratio in the range studied. The shrinking core model (SCM) was applied in order to fit the experimental data. Concluded from the experimental work, a diffusion process in the porous product layer controls the leaching reactions. The activation energy of the dissolution process was found to be about 1 kJ/mol in the temperature range of 298 - 353 K. The low values of activetion energy confirm that zinc extraction from ZPR in dilute acid sulfuric solution is not sensitive to temperature. The evaluated activation energy using model-free method was found to be 2.9 kJ/mol. This value is close to that determined previously, using shrinking core model (SCM) and confirms that the transportation of ions has

been controlled by the rate in this leaching process. The dissolution rate which could be expressed by Equation (20) can be estimated by the reacted fraction. Furthermore, the agreement between the $G(\alpha)$-experimental and $G(\alpha)$-calculated is linear with a regression coefficient of 0.98. The order of the reaction with respect to $[H^+]$ and solid-to-liquid-ratio was found to be 0.016 and -1.34, respectively.

7. Acknowledgements

The authors would like to acknowledge the Iranian Zinc Mines Development Company, Zanjan, Iran, for financial support and the permission to publish this paper.

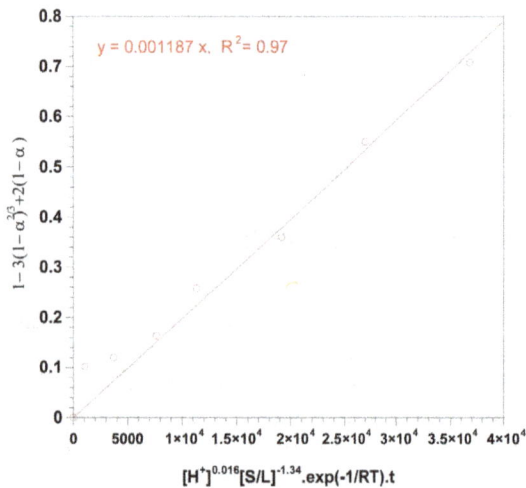

Figure 13. Plot of $1 - 3(1-a)^{2/3} + 2(1-a)$ **vs**

$$\left[H^+\right]^{0.016} \times \left[\left(\frac{S}{L}\right)\right]^{-1.34} \times \exp\left(-\frac{1}{RT}\right) \times t \text{ , (pH: 3; temperature:}$$

333 K; Solid-to-liquid-ratio: 1/6; agitation rate: 700 rpm).

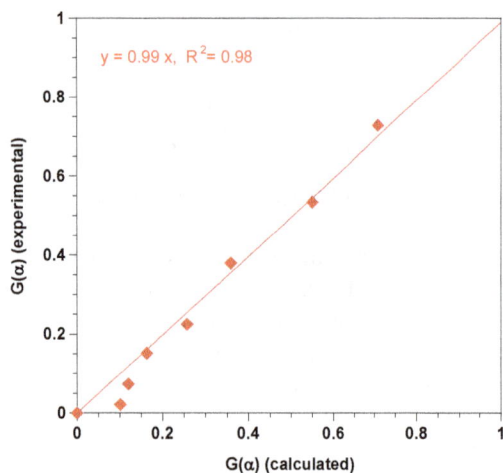

Figure 14. Comparison of experimental and calculated $G(\alpha)$ of zinc in sulfuric acid.

REFERENCES

[1] M. K. Jha, V. Kumar and R. J. Singh, "Review of Hydro-Metallurgical Recovery of Zinc from Industrial Wastes," *Resources, Conservation and Recycling*, Vol. 33, 2001, pp. 1-22. doi:10.1016/S0921-3449(00)00095-1

[2] M. N. Babu, K. K. Sahu and B. D. Pandey, "Zinc Recovery from Sphalerite Concentrate by Direct Oxidative Leaching with Ammonium, Sodium and Potassium Persulphates," *Hydrometallurgy*, Vol. 64, No. 2, 2002, pp. 119-129. doi:10.1016/S0304-386X(02)00030-0

[3] A. D. Souza, P. S. Pina, V. A. Leão, C. A. Silva and P. F. Siqueria, "The Leaching Kinetics of a Zinc Sulphide Concen-trate in Acid Ferric Sulphate," *Hydrometallurgy*, Vol. 89, No. 1-2, 2007, pp. 72-81.

[4] S. Aydogan, "Dissolution Kinetics of Sphalerite with Hydrogen Peroxide in Sulphuric Acid Medium," *Chemical Engineering Journal*, Vol. 123, 2006, pp. 65-70. doi:10.1016/j.cej.2006.07.001

[5] S. Aydogan, A. Aras and M. Canbazoglu, "Dissolution Ki-Netics of Sphalerite in Acidic Ferric Chloride Leaching," *Chemical Engineering Journal*, Vol. 114, No. 1-3, 2005, pp. 67-72.

[6] M. Al-Harahsheh and S. Kingman, "The Influence of Microwaves on the Leaching of Sphalerite in Ferric Chloride," *Chemical Engineering and Processing*, Vol. 46, No. 10, 2007, pp. 883-888. doi:10.1016/j.cep.2007.06.009

[7] P. Peng, H. Xie and L. Lu, "Leaching of a Sphalerite Concentrate with H_2SO_4-HNO_3 Solutions in the Presence of C_2C_{14}," *Hydrometallurgy*, Vol. 80, No. 4, 2005, pp. 265-271. doi:10.1016/j.hydromet.2005.08.004

[8] J. S. Niederkorn, "Kinetic Study on Catalytic Leaching of Sphalerite," *Journal of Marine Engineering and Technology*, Vol. 37, No. 7, 1985, pp. 53-56.

[9] S. Espiari, F. Rashchi and S. K. Sadrnezhaad, "Hydrometallurgical Treatment of Tailings with High Zinc Content," *Hydrometallurgy*, Vol. 82, No. 1-2, 2006, pp. 54-62.

[10] B. K. Thomas and D. J. Fray, "Leaching of Oxidic Zinc Materials with Chlorine and Chlorine Hydrate," *Metallurgical Transactions*, Vol. 12, No. 2, 1981, pp. 281-285. doi:10.1007/BF02654461

[11] J. Frenay, "Leaching of Oxidised Zinc Ores in Various Media," *Hydrometallurgy*, Vol. 15, 1985, pp. 243-253. doi:10.1016/0304-386X(85)90057-X

[12] E. A. Abdel-Aal, "Kinetics of Sulfuric Acid Leaching of Low-Grade Zinc Silicate Ore," *Hydrometallurgy*, Vol. 55, 2000, pp. 247-254. doi:10.1016/S0304-386X(00)00059-1

[13] A. D. Souza, P. S. Pina, E. V. O. Lima, C. A. Da Silva and V. A. Leão, "Kinetics of Sulphuric Acid Leaching of A Zinc Silicate Calcine," *Hydrometallurgy*, Vol. 89, No. 3, 2007, pp. 337-345. doi:10.1016/j.hydromet.2007.08.005

[14] M. G. Bodas, "Hydrometallurgical Treatment of Zinc Silicate Ore from Thailand," *Hydrometallurgy*, Vol. 40, No. 1-2, 1996, pp. 37-49.

[15] Y. Zhao and R. Stanforth, "Production of Zn Powder by Alkaline Treatment of Smithsonite Zn-Pb Ores," *Hydrometallurgy*, Vol. 56, No. 2, 2000, pp. 237-249. doi:10.1016/S0304-386X(00)00079-7

[16] M. Kul and Y. Topkaya, "Recovery of Germanium and Other Valuable Metals from Zinc Plant Residues," *Hydrometallurgy*, Vol. 92, No. 3-4, 2008, pp. 87-94.

[17] Y. Wang and Z. Chunshan, "Hydrometallurgical Process for Recovery of Cobalt from Zinc Plant Residue," *Hydrometallurgy*, Vol. 63, No. 3, 2002, pp. 225-234. doi:10.1016/S0304-386X(01)00213-4

[18] A. Ruşen, A. S. Sunkar and Y. A. Topkaya, "Zinc and Lead Extraction from Çinkur Leach Residues by Using Hydrometallurgical Method," *Hydrometallurgy*, Vol. 93, No. 1-2, 2008, pp. 45-50.

[19] F. Farahmand, D. Moradkhani, M. S. Safarzadeh and F. Rashchi, "Brine Leaching of Lead-Bearing Zinc Plant Residues: Process Optimization Using Orthogonal Array Design Methodology," *Hydrometallurgy*, Vol. 95, No. 3-4, 2009, pp. 316-324.

[20] S. Kikuchi, T. Goto and A. Nakayama, "Recovery of Zinc from Leaching Residue in Zinc Hydrometallurgy by TOPO," *Journal of the Mining and Metallurgical Institute of Japan*, Vol. 101, No. 1168, 1985, pp. 381-385.

[21] K. L. Bhat and K. A. Natrajan, "Recovery of Zinc from Leach Residues—Problems and Developments," *Transactions of the Indian Institute of Metals*, Vol. 40, No. 4, 1987, p. 361.

[22] A. V. Ropenack, W. Bohmer, G. Smykalla and V. Wiegand, "Method of Processing Residues from the Hydrometallurgical Production of Zinc," US Patent No. 4789446, 1987.

[23] F. J. J. B. Bodson, "Recovery of Zinc Values from Zinc Plant Residue," US Patent No. 3652264, 1972.

[24] Z. Youcai and R. Stanforth, "Extraction of Zinc from Zinc Ferrites by Fusion with Caustic Soda," *Mineral Engineering*, Vol. 13, No. 13, 2000, pp. 1417-1421.

[25] L. J. L. Blanco, V. F. M. Zapata and D. D. Garcia, "Statistical Analysis of Laboratory Results of Zn Wastes Leaching," *Hydrometallurgy*, Vol. 54, No. 1, 1999, pp. 41-48. doi:10.1016/S0304-386X(99)00057-2

[26] M. D. Turan, H. S. Altundoğan and F. Tumen, "Recovery of Zinc and Lead from Zinc Plant Residue," *Hydrometallurgy*, Vol. 75, No. 1-4, 2004, pp. 169-176. doi:10.1016/j.hydromet.2004.07.008

[27] T. Pecina, T. Franco, P. Castillo and E. Orrantia, "Leaching of a Zinc Concentrate in H_2SO_4 Solutions Containing H_2O_2 and Complexing Agents," *Minerals Engineering*, Vol. 21, No. 1, 2008, pp. 23-30. doi:10.1016/j.mineng.2007.07.006

[28] H. Y. Sohn and M. E. Wadsworth, "Rate Processes of Extractive Metallurgy," Plenum Press, New York, 1979. doi:10.1007/978-1-4684-9117-3

[29] M. K. Ghosh, R. P. Das and A. K. Biswas, "Oxidative Ammonia Leaching of Sphalerite—Part I: Noncatalytic Kinetics," *International Journal of Mineral Processing*, Vol. 66, No. 1, 2002, pp. 241-254. doi:10.1016/S0301-7516(02)00068-6

[30] M. K. Ghosh, R. P. Das and A. K. Biswas, "Oxidative Ammonia Leaching of Sphalerite: Part II: Cu(II)-Catalyzed Kinetics," *International Journal of Mineral Processing*, Vol. 70, No. 1-4, 2003, pp. 221-234.

[31] O. Levenspiel, "Chemical Reaction Engineering," 2nd Edition, Wiley, New York, 1999.

[32] S. Vyazovkin and W. A. Charles, "Model-Free and Model-Fitting Approaches to Kinetic Analysis of Isothermal and Nonisothermal Data," *Thermochimica Acta*, Vol. 340-341, 1999, pp. 53-68. doi:10.1016/S0040-6031(99)00253-1

[33] S. Vyazovkin, "Computational Aspects of Kinetic Analysis: Part C. The ICTAC Kinetics Project—the Light at the End of the Tunnel?" *Thermochimica Acta*, Vol. 355, No. 1, 2000, pp. 155-163. doi:10.1016/S0040-6031(00)00445-7

[34] M. S. Safarzadeh, D. Moradkhani and M. O. Ilkhchi, "Kinetics of Sulfuric Acid Leaching of Cadmium from Cd-Ni Zinc Plant Residues," *Journal of Hazardous Materials*, Vol. 163, No. 2-3, 2009, pp. 880-890.

[35] W. L. Choo, M. I. Jeffrey and S. G. Robertson, "Analysis of Leaching and Cementation Reaction Kinetics: Correcting for Volume Changes in Laboratory Studies," *Hydrometallurgy*, Vol. 82, No. 1-2, 2006, pp. 110-116.

[36] R. Vaghar, "Hydrometallurgy," Iranian Copper Industry Co., Iran, 1998.

[37] D. Dreisinger and N. Abed, "A Fundamental Study of the Reductive Leaching of Chalcopyrite Using Metallic Iron Part I: Kinetic Analysis," *Hydrometallurgy*, Vol. 66, No. 1, 2002, pp. 37-57. doi:10.1016/S0304-386X(02)00079-8

[38] M. K. Sarker, A. K. M. B. Rashid and A. S. W. Kurny, "Kinetics of Leaching of Oxidized and Reduced Ilmenite in Dilute Hydrochloric Acid Solutions," *International Journal of Mineral Processing*, Vol. 80, No. 2-4, 2006, pp. 223-228. doi:10.1016/j.minpro.2006.04.005

17

Effect of Stacking Fault Energy on the Mechanism of Texture Formation during Alternating Bending of FCC Metals and Alloys

Natalia Shkatulyak

South Ukrainian National Pedagogical University after K.D. Ushinskii, Odessa, Ukraine

Email: shkatulyak@mail.ru

ABSTRACT

Alternating bending shear stresses lead to the formation of twin orientations in the texture of FCC materials with middle and low stacking fault energy (SFE). Only in the stainless steel with a low SFE during alternating bending with different number of cycles components of shear texture {111}<hkl>; {hkl}<110>; {001}<110> were formed. Copper (middle SFE), along with orientations of twinning and cubic texture formed orientation of deformation {135}<211>. During alternating bending of aluminum (high SFE), a dynamic recovery occurred. The share of initial cubic texture increases with the increase of number of cycles of alternating bending and reaches its maximum after three cycles. Share of component of texture Goss increased slightly. The most significant change of the microstructure and texture occurred during the first 3 - 5 cycles

Keywords: Stacking Fault Energy; Texture; Alternating Bending

1. Introduction

Most of the modern technologies for processing sheet metal include mechanical and thermal treatment, which inevitably results in the generation of internal stresses in the material and warping of the resulting parts. Therefore, before using coil metal, the metallurgists subject it to straightening on a leveling machine. In the course of straightening, the material is subjected to alternating bending (AB), which ensures good planeness. This treatment minimizes the level of internal residual stresses in the metal and provides necessary planar characteristics, which significantly facilitates the subsequent metal processing and positively affects the quality of finished products. As a result of plastic deformation in the process of straightening, there, of course, occurs a change in the metal texture and microstructure. Mechanisms of metal texture and microstructure formation during the alternating bending, undoubtedly, are due to the stacking fault energy (SFE) value.

The purpose of this research is to establish the regularity of texture and microstructure formation during the alternating bending of aluminum, copper and stainless steel with face-centered cubic (FCC) lattice, which are characterized by high, medium and low SFE, corre-

spondingly.

2. Materials and Methods

2.1. Materials

As the material for the study, strips of aluminum (0.040 Fe, 0.25 Si, 0.07 Zn, 0.05 Ti, 0.05 Mg, 0.05 Mn, 99.5 Al mass. %), copper (0.004 Fe, 0.002 Ni, 0.003 Zn, 0.001 Sn, 0.002 Sb, 0.001 As, 0.003 Pb, 0.003 S, 0.04 O, 99.93 Cu + Ag mass. %) and stainless steel (0.02 C, 18 Cr, 10 Ni, mass. %) were used. The as-delivered aluminum and copper strips 1 mm thick were annealed in a vacuum at 350°C for 1 h. The as-delivered stainless steel strips 1 mm thick were annealed in a vacuum at 650°C for 1 h. The process of sheet straightening was simulated with AB using a specially constructed machine. The diameter of the bending roller was 50 mm. The speed of the metal during bending was ~150 mm/s. The samples for texture and structure studies were cut from the initial sheet and from sheets subjected to bending using 0.25, 0.5, 1, 3, and 5 cycles.

2.2. Methods

Before examining the texture, the samples were chemically polished to a depth of 0.1 mm to remove the distorted surface layer. The crystallographic texture was investigated by recording inverse pole figures (IPFs) in the normal direction (ND) and rolling directions (RD) using a DRON-3 m diffractometer in the filtered Mo Kα radiation on both surfaces of the samples after the above-mentioned numbers of the cycles and after 0.25 cycle. The sample without texture was produced from fine recrystallized filings of corresponding metal. In order to ensure a flat surface after 0.25 cycle bending for taking IPFs, a composite sample was manufactured from pieces 3 mm wide, which were cut across the metal strips. To take IPFs in the RD, we also produced composite samples.

The metallographic structure was studied in the reflection mode from the butt surfaces of the samples cut in the RD and transverse direction (TD) using microscope Axioplan 2 of the KARL ZEISS Company.

3. Results and Discussion

Figures 1-3 show some IPF ND of Al, Cu and stainless steel respectively. **Figures 4-6** show the relevant microstructure of Al, Cu and stainless steel.

3.1. The Initial Texture and Microstructure

The texture of initial Al sample is represented mainly of cube orientation {001}⟨100⟩ with the scattering and of

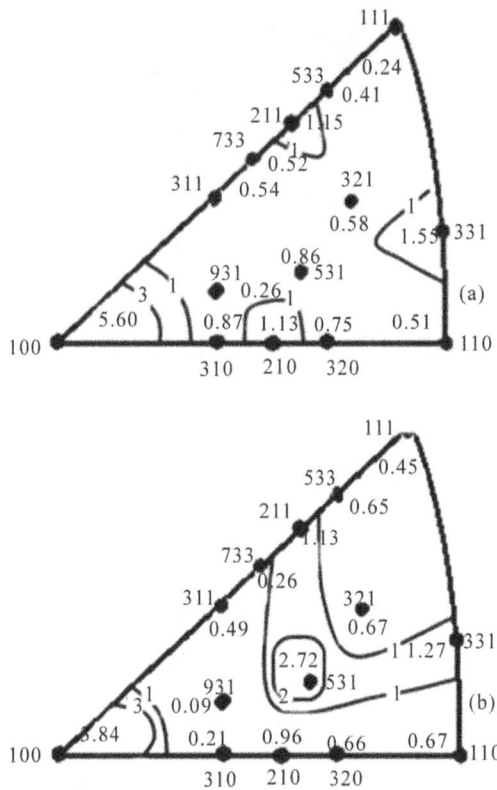

Figure 2. IPF ND of Cu samples in the initial state (a) and after 5 cycles of alternating bending (b).

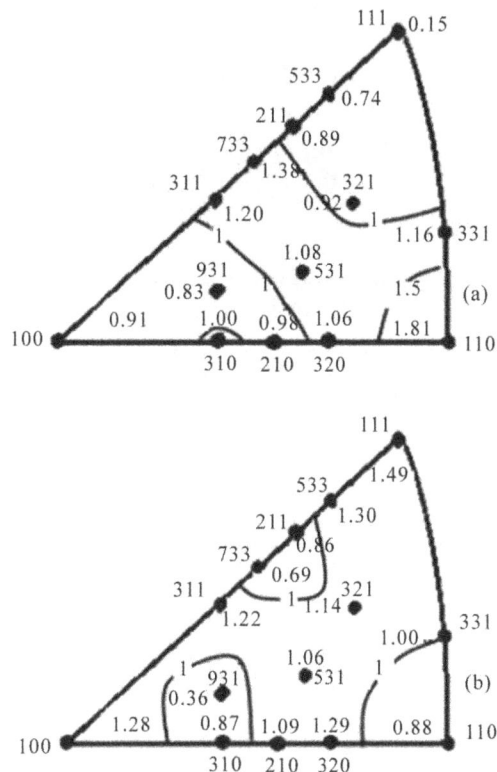

Figure 1. IPF ND of Al samples in the initial state (a) and after 5 cycles of alternating bending (b).

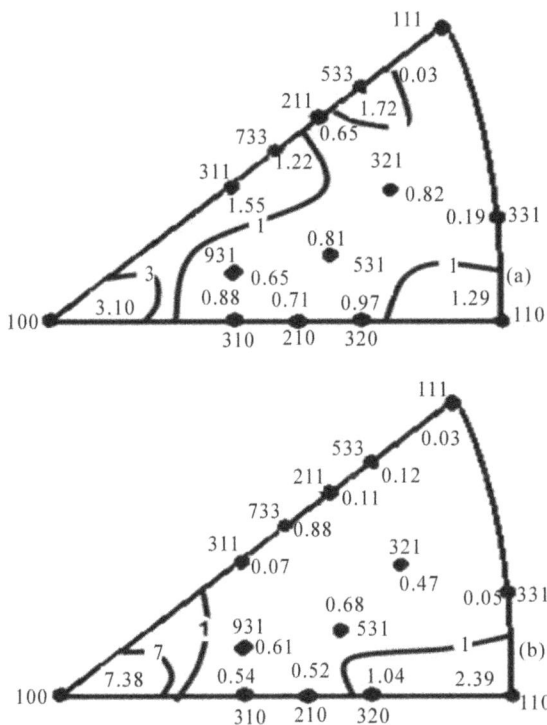

Figure 3. IPF ND of stainless steel samples in the initial state (a) and after 5 cycles of alternating bending (b).

(a)

(b)

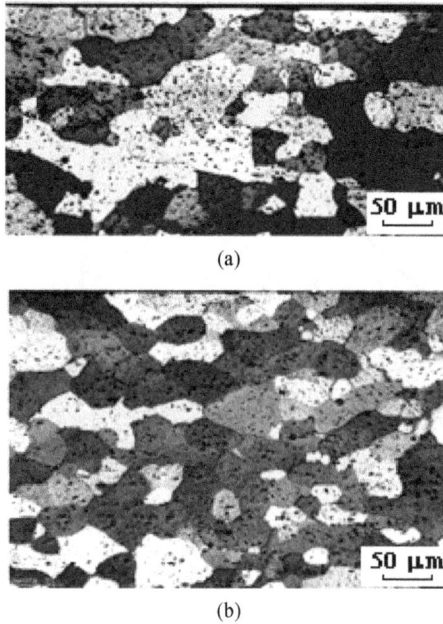

Figure 4. The microstructure of the cross section perpendicular to the ND of Al samples in the initial state (a) and after 5 cycles of alternating bending (b).

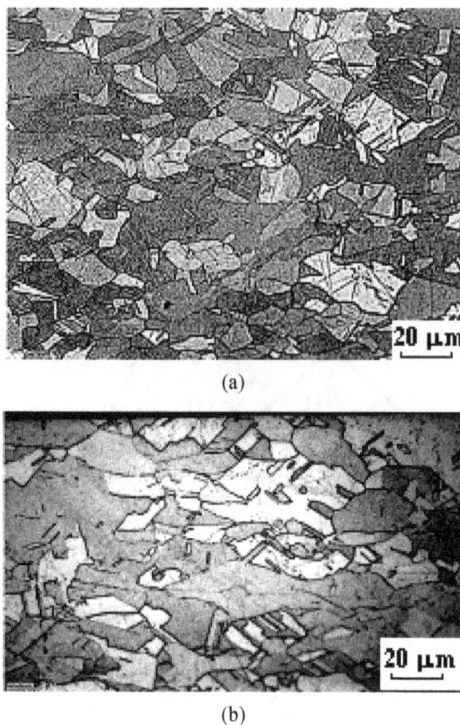

(a)

(b)

Figure 5. The microstructure of the cross section perpendicular to the ND of Cu samples in the initial state (a) and after 5 cycles of alternating bending (b).

(a)

(b)

Figure 6. The microstructure of the cross section perpendicular to the ND of stainless steel samples in the initial state (a) and after 5 cycles of alternating bending (b).

Goss orientation $\{110\}\langle100\rangle$. The cube texture of the most frequently found during recrystallization of rolled aluminum. Goss orientation formed in aluminum rolled with the small (~30%) and medium (~50% - 70%) de-

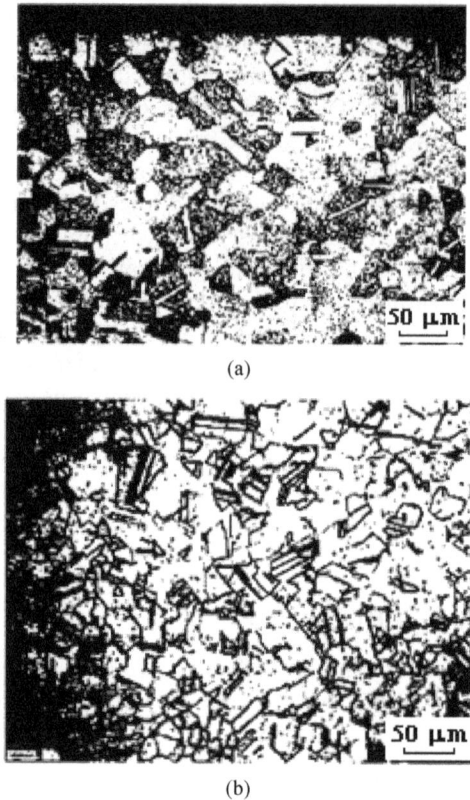

formation ratio [1]. Saving Goss texture after recrystallization annealing of aluminum is usually associated with the recovery process. This process is also called recrystallization "*in situ*", since the texture does not change by annealing [2]. The microstructure of initial sample (**Figure 4(a)**), consisting of relatively large grains (>50 micrometers), is typical of recrystallization [2]. Thus, the original sample is recrystallized after annealing at 350°C, 1 h [3].

It is seen (**Figure 2(a)**), the texture of initial sample Cu is represented mainly by cube $\{001\}\langle100\rangle$ orientation and by the remains of the rolling texture $\{112\}\langle111\rangle$. Also present orientations $\{012\}\langle120\rangle$; $\{012\}\langle121\rangle$; $\{012\}\langle100\rangle$; $\{012\}\langle123\rangle$; $\{331\}\langle110\rangle$; $\{331\}\langle013\rangle$; $\{331\}\langle123\rangle$. In the corresponding micrographs (**Figure 5(a)**), annealing twins are observed. The presence of a cube orientation in the annealing texture and of annealing twins in corresponding micrographs indicates that the initial material was recrystallized [4,5]. The formation of the above cube orientations and $\{012\}\langle uvw\rangle$ texture components during copper recristallization, as was described in the earlier work [6], can be activated by untwinning of the rolling texture component $\{112\}\langle111\rangle$ through the inverse Rowland transformation. Later, this has also been demonstrated upon the recrystallization annealing of preliminarily rolled copper bicrystals with the initial twin orientations $(112)[11\bar{1}]/(112)[1\bar{1}\bar{1}]$ and $(1\bar{1}0)[112]/(\bar{1}10)[112]$ in [7]. The formation of the $\{331\}\langle uvw\rangle$ orientations was described in the study of texture in copper rolled to 90% (in thickness) and annealed for half-hour at 250°C [8]. Together with the cube component, the recrystalliization texture of copper contains $\{122\}\langle212\rangle$ orientations, which are twinned with respect to the cube orientation [9,10]. As was said above, twins in the copper microstructure were found in our study as well (**Figure 5(a)**). On the one hand, it is impossible to quantitatively analyze $\langle221\rangle$ poles in the IPFs of an FCC metal because of the ambiguity of their indexing (superposition of 511 and 333 reflections, and 600 and 442 reflections). On the other hand, the $\langle221\rangle$ directions are deviated from $\langle331\rangle$ directions shown in the IPFs by an angle of $\approx6.3°$. In [11], the authors performed a thorough analysis of twin maxima in direct experimental pole figures in a coarse-grained rolled copper with grain sizes of 50 and 500 μm and in a fully annealed copper. The analysis has shown that the exact location of the twin maxima in the pole figures deviates from the ideal positions of $\{122\}\langle212\rangle$ twins by 2° - 3° to 9°, depending on the degree of preliminary rolling. This makes it possible to suppose that our original micrographs of the initial recrystallized copper sample depict annealing twins with orientations close to $\{122\}\langle212\rangle$, which in the IPFs corresponding to the normal direction (ND) lie in the region of spread of the 331 pole.

Texture analysis of initial sample stainless steel (**Figure 3(a)**) showed that the texture consists of two limited axial components. The first component with the axis $\langle110\rangle$ parallel to the ND runs from $\{011\}\langle100\rangle$ to $\{011\}\langle112\rangle$. The second component with the axis $\langle110\rangle$ that is inclined at about 60° to ND extends from ~ $\{112\}\langle111\rangle$ through $\{135\}\langle211\rangle$ to $\{011\}\langle112\rangle$. The development of these two limited axial components corresponds to Taylor prediction model based on normal octahedral sliding [12,13]. In addition there are the twinning orientation $\{113\}\langle211\rangle$, formed probably by annealing [4]. The microstructure of steel consists of equiaxed grains of size of ~50 microns. The grains contain twins of recrystallization after annealing at 600°C (**Figure 4(a)**).

3.2. Effect of Alternating Bending

Texture is changed after deformation by alternating bending with different number of cycles. For interpretation of texture changes is suggested the following deformation model. Grains in the layers of metal on the convex side of the sample are exposed to tensile stresses when bending to one side (1/4 cycles). At the same time, on the concave side of the sample appear the compressive stresses. As a result of the opposite sign stresses in

the strip can occurs shear strain. Thus, the shear strains arised at alternating bending of metal strip can lead to the formation in FCC metals and alloys of shear components texture such as [14]: A—$\{111\}\langle hkl\rangle$; B—$\{hkl\}\langle110\rangle$; C—$\{001\}\langle110\rangle$. Designations $\{hkl\}\langle uvw\rangle$, listed here, indicate that the plane $\{hkl\}$ coincide with the shear plane, and the directions $\langle uvw\rangle$ coincide with the shear direction. In addition, the twinning should play a significant role in the presence of alternating sign shear stress components. Therefore, should expect of the development of twin orientations during the alternating bending, all the more that as is knows the role of twinning is grows in materials with lowering their SFE [15].

3.3. Effect of Stacking Fault Energy

It was found that the textures of different investigated metals after deformation by alternating bending are significantly different (**Figures 1-3** and **Table 1**).

During the alternating bending of aluminum, which has high stacking fault energy, occurs a dynamic recovery. The type of the initial crystallographic texture

Table 1. Texture acquired during the bending of the metal and alloys with differ SFE.

Texture	SFE of material, mJ/m²		
	Al (~200) [16]	Cu (~50) [17,18]	Stainless Steel (~23) [19]
Initial	$\{001\}\langle100\rangle$+ $\{110\}\langle100\rangle$	$\{001\}\langle100\rangle$+ $\{112\}\langle111\rangle$; $\{012\}\langle uvw\rangle$+ $\{331\}\langle uvw\rangle$+ twins close to $\{122\}\langle212\rangle$	From $\{011\}\langle100\rangle$ to $\{011\}\langle112\rangle$; from $\{112\}\langle111\rangle$ trough $\{135\}\langle211\rangle$ to $\{011\}\langle112\rangle$; twins of $\{113\}\langle211\rangle$
After of 1/2 cycle AB	$\{001\}\langle100\rangle$+ $\{110\}\langle100\rangle$	$\{001\}\langle100\rangle$+ twins close to $\{331\}\langle uvw\rangle$, $\{012\}\langle uvw\rangle$	$\{hkl\}\langle110\rangle$+ twins of $\{110\}\langle112\rangle$+ $\{110\}\langle001\rangle$
After of 1 cycle AB	$\{001\}\langle100\rangle$+ $\{110\}\langle100\rangle$	$\{001\}\langle100\rangle$+ twins close to $\{331\}\langle uvw\rangle$, $\{012\}\langle uvw\rangle$	$\{001\}\langle110\rangle$+ twins of $\{110\}\langle112\rangle$+ $\{110\}\langle001\rangle$
After of 3 cycle AB	$\{001\}\langle100\rangle$+ $\{110\}\langle100\rangle$	$\{001\}\langle100\rangle$+ twins close to $\{331\}\langle uvw\rangle$, $\{012\}\langle uvw\rangle$	$\{001\}\langle110\rangle$+ twins of $\{110\}\langle112\rangle$+ $\{110\}\langle001\rangle$
5 cycle of AB	$\{001\}\langle100\rangle$+ $\{110\}\langle100\rangle$	$\{001\}\langle100\rangle$+ $\{112\}\langle111\rangle$+ twins close to $\{331\}\langle uvw\rangle$	$\{111\}\langle hkl\rangle$ + $\{001\}\langle110\rangle$ twins of $\{110\langle112\rangle$+ $\{110\}\langle001\rangle$

{001}⟨100⟩ + {110}⟨100⟩ does not change (**Figures 1(a)** and **(b)**). The share of cube texture with the increases of cycle's number of alternating bending increases and reaches its maximum after three cycles. Share of component of the texture Goss increased slightly. This is due to the fact that high stacking fault energy (aluminum and bcc metals) contributes to a very fast kinetics of static and dynamic recovery. In these metals climb and cross slip of dislocation rarely happens. The recovery (static or dynamic) reduces the density of crystalline defects and driving force of recrystallization. In the initial phase, for small strains, dislocations interact and multiply. With an increase of dislocation density during the increasing of strain will grow the driving force and, therefore, the share of recovered material with developed microstructure of low-angle boundaries and sub-boundaries. For strains higher than the critical strain $(\varepsilon > \varepsilon_c)$, the degree of strain hardening and recovery achieved of dynamic equilibrium, at where remains unchanged the disorientation and average size of equiaxed subgrains (steady state), which are a function of the level of strain and temperature [20]. This is confirmed by corresponding images of microstructure that is representing oneself equiaxed grains, as in the original sample. Deformation of aluminum and aluminum alloys at the room temperature is called often the cold deformation. However, since aluminum melting temperature (T_m) is quite low, the deformation at room temperature (0.33 T_m) is close to the warm deformation (0.4 T_m - 0.6 T_m). Alternating bending in our study occurs at a rate of deformation 0.2 s^{-1}. Heat, which released during deformation at the room temperature, warms the material. This brings closer the conditions of deformation to the warm mode.

Some overall reduction in the pole density in the IPFs of Cu is observed after bending of 0.25 cycles, which indicates about increased spread of texture. With an increase in the number of AB cycles to three, the number of twins tends to increase in the corresponding micrographs. This is accompanied by an increase in the pole density at the 210 and 331 points in the IPFs taken in the ND. The bending deformation can be simulated as a combination of tension of the convex side and compression on the concave side of the strip. The authors of [4] have analyzed the possibility of deformation induced twinning in FCC metals under tension and compression. It was shown that, since the twinning in the FCC structure is always preceded by slip [4], the twinning plane is $(\bar{1}11)$, and that Shockley and Frank partial dislocations are formed [4] in the process of twinning. These partials act as twinning dislocations under compression for the initial orientations that in the unit triangle of the standard stereographic projection are adjacent to the 011 pole, and, under tension, for the orientations adjacent to 001. Thus, under compression, we should expect the formation of

twins with orientations that lie near the 331 pole; and under tension, in the region bounded by the 210, 931, 311, and 211 poles [4]. These orientations are present in our IPFs, as was mentioned above. Thus, in the AB process there occurs deformation induced twinning, which becomes somewhat weaker as the number of AB cycles increases to 5 (**Figure 5(b)**). Cubic texture and pole density of twins orientation are weakens after increasing the number of cycles of alternating bending to 5, and formed component of deformation texture of type {135}⟨211⟩ (**Figure 2(b)**).

It was found that only the stainless steel, which has a low SFE, results in formation of various combinations of initial rolling texture, shear components at a different (small) number of cycles of alternating bending. These are: A—{111}⟨hkl⟩; B—{hkl}⟨110⟩; C—{001}⟨110⟩ and twins orientations (**Figure 6(b)**). The shear texture component A is formed after bending on 1/4 part of cycle on both sides of stainless steel specimen. After bending on 1/2 part of cycle on the one side of stainless steel sample is formed component B of shear texture. There are also orientations of twinning, which form the orientations of texture {110}⟨112⟩ and {110}⟨001⟩ in metals and alloys with low SFE [11,12]. In the sample after 1 cycle of alternating bending is formed component C of shear texture. At the same time, on the opposite side of the same sample is observed orientation of initial texture. The twinning orientations, as in the previous case also take place. After three cycles of alternating bending on the one side of specimen is formed texture that is similar to the initial texture. Texture on the opposite side the same specimen is represented fairly intense shear texture component C. The shear components A and C characterize the texture on the one side of stainless steel specimen after five cycles of alternating bending. Region of high pole density on IPF significantly increased in comparison with the other samples that probably due to the formation of twin orientations [4]. Texture the same specimen on the opposite side is characterized by the formation of shear component of the texture and twinning orientations. In general, the scattering texture increased compared to initial state if you take into account the both surface of stainless steel strip after 5 cycles of bending.

A common feature of the studied materials is that the most significant change in texture and structure occur within the first three to five cycles of alternating bending.

4. Conclusions

Thus, it is shown that in the presence of shear stress components of opposite sign during the process of alternating bending the mechanism of the texture formation of FCC metals depends on the stacking fault energy of material.

In aluminum that is characterized of high stacking fault energy and a low melting point, the process of recovery occurs. The share of cubic component in the texture with the number of cycles of alternating bending increases and reaches a maximum after three cycles. The share component of the texture Goss increased slightly.

In copper, which is material with an average of stacking faults energy, during the alternating bending the deformation twinning process occurs, which is developed most rapidly during the first three cycles of alternating bending. Increasing the number of cycles to 5 leads to a weakening of the texture of the cube and the pole density of twinning orientations and development of deformation texture of type $\{135\}\langle211\rangle$.

Only stainless steel strips, which have a low stacking fault energy, result in formation of various combinations of initial texture, components of shear texture at a different (small) number of cycles; these are $\{111\}\langle hkl\rangle$, $\{hkl\}\langle110\rangle$, $\{001\}\langle110\rangle$ and twinning orientations.

The most significant change in texture and structure occur within the first three to five cycles of alternating bending.

REFERENCES

[1] W. M. Mao, "Influence of Rolling Reduction on Recrystallization Texture in Commercially Pure Al," *Journal of Materials Science & Technology*, Vol. 6, No. 4. 1990, pp. 257-262.
http://www.jmst.org/EN/abstract/abstract15490.shtml

[2] F. J. Humphreys and M. Hatherly, "Recrystallisation and Related Annealing Phenomena," Elsevier, Oxford, 2004, 628 p.

[3] N. M. Shkatulyak and N. P. Pravedna, "Effect of Alternating Bending at the Texture, Structure and Mechanical Properties of Aluminum Sheets," *Metallovedenie i Termicheskaja Obrabotka Metallov*, Vol. 34, No. 2, 2012, p. 209 (in Russian).

[4] Y. D. Vishnyakov, A. A. Babareko, S. A. Vladimirov and I. V. Egiz, "Teoriya Obrazovaniya Tekstur v Metallakh i Splavakh (Texture Formation Theory in Metals and Alloys)," Nauka, Moscow, 1979, 343 p (in Russian).

[5] N. M. Shkatulyak, A. A. Bryukhanov, M. Rodman, V. V. Usov, M. Schaper, G. Haferkamp and V. A. Nastasyuk, "Reverse Bending Effect on the Texture, Structure, and Mechanical Properties of Sheet Copper," *The Physics of Metals and Metallography*, Vol. 113, No. 8, 2012, pp. 810-816. doi:10.1134/S0031918X1208011X

[6] C. A. Verbraak, "The Formation of Cube Recrystallization Textures by 112 Slip," *Acta Metallurgica*, Vol. 6, No. 9, 1958, pp. 580-597. doi:10.1016/0001-6160(58)90100-7

[7] W. F. Hellerph, C. A. Verbraaks and B. H. Kolster, "Recrystallization at Grain Boundaries in Deformed Copper Bicrystals," *Acta Metallurgica*, Vol. 32, No. 9, 1984, pp. 1395-1406. doi:10.1016/0001-6160(84)90085-3

[8] T. Kamijo "Study on the Inverse Rowland Mechanism for the Nucleation of a Cube Recrystallization Texture," *Journal of the Japan Institute of Metals*, Vol. 31, No. 6, 1967, pp. 741-746.
http://www.jim.or.jp/journal/j/31/06/741-746.html

[9] W. Mao, "Formation of Recrystallization Cube Texture in High Purity Face-Centered Cubic Metal Sheets," *Journal of Materials Engineering and Performance*, Vol. 8, No. 5, 1999, pp. 556-560. doi:10.1007/s11665-999-0009-3

[10] K. Sztwiertnia, "Orientation Aspects of the Recrystallization Nucleation in Highly Deformed Polycrystalline Copper," *Materials Science Forum*, Vol. 467-470, 2004, pp. 99-106.
doi:10.4028/www.scientific.net/MSF.467-470.99

[11] M. Sindel, G. D. Kohlhoff, K. Lücke and B. J. Duggan, "Development of Cube Texture in Coarse Grained Copper," *Textures and Microstructures*, Vol. 12, No. 1-3, 1990, pp. 37-46. doi:10.1155/TSM.12.37

[12] C. D. Singh, V. Ramaswamy and C. Suryanarayana, "Development of Rolling Textures in an Austenitic Stainless Steel," *Textures and Microstructures*, Vol. 19, No. 1-2, 1992, pp. 101-121. doi:10.1155/TSM.19.101

[13] C. D. Singh, "On the Development of the Brass-Type Texture in Austenitic Stainless Steel," *Textures and Microstructures*, Vol. 22, No. 1, 1993, pp. 59-72. doi:10.1155/TSM.22.59

[14] G. R. Canova, U. F. Kocks and J. J. Jonas, "Theory of Torsion Texture Development," *Acta metal.*, Vol. 32, No. 2, 1984, pp. 211-226. doi:10.1016/0001-6160(84)90050-6

[15] E. B. Tadmor and N. Bernstein, "A First-Principles Measure for the Twinnability of FCC Metals," *Journal of the Mechanics and Physics of Solids*, Vol. 52, No. 11, 2004, pp. 2507-2519. doi:10.1016/j.jmps.2004.05.002

[16] L. E. Murr, "Interfacial Phenomena in Metals and Alloys". Addison-Wesley Pub. Co., 1975. 376 p.

[17] J. A. Venables, "The Electron Microscopy of Deformation Twinning," *Journal of Physics and Chemistry Solids*, Vol. 25, No. 7, 1964, pp. 685-690. doi:10.1016/0022-3697(64)90177-5

[18] Y. H. Zhao, X. Z. Liao, Y. T. Zhu, Z. Horita and T. G. Langdon, "Influence of Stacking Fault Energy on Nanostructure Formation under High Pressure Torsion," *Materials Science and Engineering: A*, Vol. 410-411, 2005, pp. 188-193. doi:10.1016/j.msea.2005.08.074

[19] G. R. E. Schramm and R. P. Reed, "Stacking Fault Energies of Austenitic Stainless Steels," *Metallurgical Transactions A*, Vol. 6, No. 7, 1974, pp. 1345-1351. doi:10.1007/BF02641927

[20] E. Totten and D. S. Mackenzie, "Handbook of Aluminum: Alloy Production and Materials Manufacturing," Marcel Dekker Inc., New York, Basel, 2003.

18

Sono-Photo Fenton Treatment of Liquid Waste Containing Ethylenediaminetetraacetic Acid (EDTA)

S. Chitra[1*], K. Paramasivan[1], P. K. Sinha[2]

[1]Centralised Waste Management Facility, Bhabha Atomic Research Centre Facilities, Kalpakkam, India
[2]Waste Management Division, Bhabha Atomic Research Centre (BARC), Mumbai, India
Email: [*]schitra@igcar.gov.in

ABSTRACT

Ethylenediaminetetraacetic acid (EDTA) is a chelating agent that has been used for decontamination purposes in nuclear industry. The presence of EDTA in decontamination wastes can cause complexation of the cations resulting into interferences in their removal by various treatment processes such as chemical precipitation, ion exchange etc. Further, it might also impart elevated leachability of cationic contaminants from the conditioned wastes immobilized in cement or other matrices and can negatively influence the quality of the final form of waste. In the present study, kinetics of degradation of EDTA (20,000 mg/l) by employing either Photo-Fenton process using UV (15 W λ_{max} = 253.7 nm) or Sono-Fenton process using ultrasound at 130 KHz or simultaneous Sono-Photo Fenton process has been investigated. EDTA is effectively degraded by the synergistic effect of both Photo Fenton and Sono Fenton process. All the above mentioned processes were found to follow a first order kinetics reaction. From the observed pH changes during the oxidation processes, it can be concluded that there is a loss of chelating ability of EDTA. Formation of amides was confirmed during the degradation processes.

Keywords: Advanced Oxidation Process (AOP); EDTA; Fenton's Reagent; Sono-Photofenton

1. Introduction

Ethyelenediaminetetraacetic acid (EDTA) has numerous applications, based on its ability to control the action of different metal ions through complexation. EDTA is one of the most important components found in bath used for electroless plating, metal etching, surface cleaning, selective removal of precious metals from waste water emerging from non ferrous metallurgical extraction processes [1-3]. The influence of EDTA as crystal habit modifier an important stage of the Bayer process for production of alumina from bauxite ore is worth mentioning [4]. EDTA is used as an important decontaminating agent in nuclear industry. The presence of EDTA in decontamination wastes can cause complexation of the cations resulting into interferences in their removal by various treatment processes such as chemical precipitation, ion exchange etc. [5]. Further, it might also impart elevated leachability and higher mobility of cationic contaminants from the conditioned wastes *i.e.* waste immobilized in cement or other matrices and can negatively influence the quality of the final form of waste [6]. EDTA is not

easily biodegradable [7], scarcely degradable by chlorine [8] and hardly retained by activated carbon filters [9].

As a consequence, techniques suitable for destruction of EDTA are needed to render the liquid waste amenable to treatment and thus protect the environment. In light of the increasing concern over the contamination of the environment by hazardous chemicals, there is a great need to develop innovative technologies for the safe destruction of toxic pollutants. The processes must be cost effective, easy to operate, and capable of achieving a total or near-total mineralization.

Advanced Oxidation Processes (AOP) involving hydroxyl radicals, which are one of the strongest inorganic oxidants next to elemental fluorine, have been extremely effective in the destruction of organic pollutants. These advanced chemical oxidation process (AOP) generally use a combination of oxidation agents (such as H_2O_2 or O_3), irradiation (such as UV or ultrasound), and catalysts (such as metal ions or photo catalysts), as a means to generate hydroxyl radicals [10]. The reason that H_2O_2 can be used for such diverse applications is the different ways in which its selectivity can function. H_2O_2 has none of the problems of gaseous release or chemical residues

---

[*]Corresponding author.

that are associated with other chemical oxidants. By simply adjusting the conditions of the reaction (e.g., pH, temperature, dose, reaction time, and/or catalyst addition), H_2O_2 can often be made to oxidize one pollutant over another, or even to favor different oxidation products from the same pollutant.

The degradation of EDTA has been attempted by ozonation [11], UV + H_2O_2 [12], phototcatalysis [5], UV + oxidants [13], radiolysis [14], radio-photocatalysis [15] and combined techniques [16] with variable results. Degradation of EDTA using H_2O_2 alone at alkaline pH has also been reported [17].

The treatment of liquid waste generated during the chemical decontamination of heat exchangers of the boilers of the Pressurised Heavy Water Reactor (PHWR) was planned to be taken up at Centralised Waste Management Facility (CWMF), Kalpakkam, using Advanced Oxidation Processes. The decontamination wastes contain EDTA (2%), Ethylenediamine (EDA) (0.6%), Ammonia (0.6%) and hydrazine (0.1%).

For the optimization of the waste treatment conditions, information on the influence of the experimental conditions on the degradation is needed. Since the concentration of EDTA is the maximum (20,000 mg/L), a systematic study of the degradation of 20,000 mg/L of EDTA alone using advanced oxidation processes was attempted initially. In the present study, without taking into consideration the formation of degradation intermediates, a comparison of kinetics of degradation of 20,000 mg/L of EDTA was carried out using stoichiometric quantities of H_2O_2 (30% w/v) in all the experiments viz., using UV (15 W) + Fenton's reagent (FeII + H_2O_2), US (130 KHz) + FeII + H_2O_2, and UV (15 W) + US (130 KHz) + FeII + H_2O_2 was carried out at pH 3.0.

2. Experimental

2.1. Experimental Setup and Materials

The photoreactor was a glass trough (Borosil) with inside dimensions of 210 × 210 × 50 mm deep. A UV lamp (8 W with λ_{max} = 253.7 nm) was positioned horizontally over the reactor. The distance of the UV lamp from the surface of the sample was maintained at 70 mm. The set-up was housed inside a fume hood. For the experiments carried out using UV lamp (8 W), an aluminum cover was used to house the glass trough to control the UV radiation. The light intensity inside the reacting medium as determined by potassium ferrioxalate actinometry was 3.5E−3 einsteins/cm^2. The schematic diagram of the experimental set-up used in the study is shown in **Figure 1**.

Sonication experiments were performed in an ultrasonic cleaning bath of frequency 130 KHz ELMA Transonic industrial table top model T1-H-20-MF of power

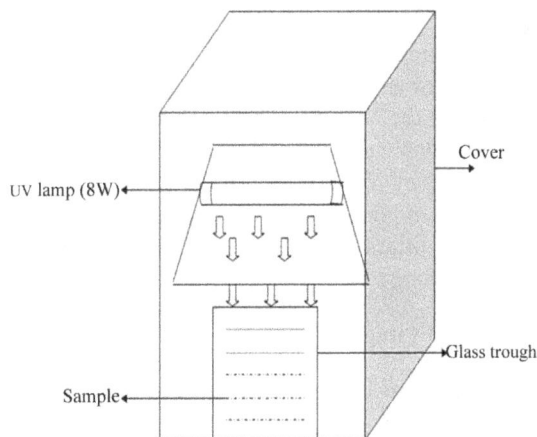

Figure 1. Experimental setup.

300 W. The reactions were carried out using a similar glass trough (Borosil) as used for Photo Fenton experiment with inside dimensions of 210 × 210 × 50 mm deep which was closed during ultrasonic irradiation. The glass trough was placed inside the ultrasonic bath. In addition to the above-mentioned instrumentation, a standard pH meter was used. The pH value of the solution was adjusted using H_2SO_4 solution.

For Sono-Photo Fenton experiments the photoreactor was placed inside the ultrasonic cleaning bath of frequency 130 KHz ELMA Transonic industrial table top model T1-H-20-MF of power 300 W.

All the chemicals EDTA disodium salt, hydrogen peroxide (30%), $FeSO_4$, $MgSO_4 \cdot 7H_2O$ was of analytical grade, 99.9% purity.

2.2. Degradation Procedures

By stoichiometric calculations, 17 moles of H_2O_2 are required to completely oxidize 1 mol of EDTA to CO_2, NH_3 and H_2O. The balanced stoichiometric Equation (1) is as follows:

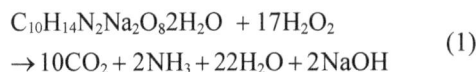

$$C_{10}H_{14}N_2Na_2O_8 2H_2O + 17H_2O_2 \qquad (1)$$
$$\rightarrow 10CO_2 + 2NH_3 + 22H_2O + 2NaOH$$

To 898 ml of 22,500 mg/L EDTA solution taken in the photoreactor, 0.002% (w/v) of Fe^{2+} was added and mixed thoroughly before the addition of 102 ml H_2O_2 (30% w/v). The amount of Fe^{2+} added was arrived at by considering the maximum concentration of interference of Fe^{2+} that may be present in the sample and still permit the titration with EDTA [18]. The time at which the UV lamp/US system was turned on was considered time zero or the beginning of the experiment.

2.3. Analytical Methods

The pH of the solution was measured using a calibrated pocket pH meter (HACH) at appropriate time intervals and during those time intervals 5 ml of the sample was pipetted into a conical flask. One drop of hydrazine was also added to prevent H_2O_2 from reacting with organic substrates during the analysis. The residual EDTA present in the samples was analysed titrimetrically against standard Mg^{2+} using Eriochrome Black T as indicator [19]. The hydrogen peroxide was measured by the standard iodometric titration [20]. At least triplicate runs were carried out for each condition averaging the results.

3. Results and Discussion

3.1. Comparison of UV/Fenton, US/Fenton and US-UV/Fenton

Figure 2 illustrates the comparison of percent degradation of EDTA (20,000 mg/L) as a function of time using $UV + Fe(II) + H_2O_2$, $US + Fe(II) + H_2O_2$ and $US + UV + Fe(II) + H_2O_2$. From **Figure 2** it can be observed that there was complete degradation of EDTA at 20 hours, 10 hours and 5 hours for samples treated with US (130 KHz) + Fe(II) + H_2O_2, UV (15 W) + Fe(II) + H_2O_2, US + UV + Fe(II) + H_2O_2 respectively.

The rate of degradation by UV (8 W) + FeII + H_2O_2 was 2.3 times faster than that in the system using US (130 KHz) + FeII + H_2O_2. The formation of hydroxyl radicals by using the photo-Fenton process occurs according to the following Equation (2) [10]

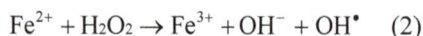

$$Fe^{2+} + H_2O_2 \rightarrow Fe^{3+} + OH^- + OH^{\bullet} \quad (2)$$

Equation (2), known as the Fenton reaction, possesses a high oxidation potential, but its application to wastewater treatment began only recently [21].

In combination of thermal process and UV irradiation, the oxidation power of Fenton's reagent was significantly increased mainly due to photo-reduction of Fe^{3+} to Fe^{2+}, which could react with H_2O_2 establishing a cycle mechanism of generating additional hydroxyl radicals which is

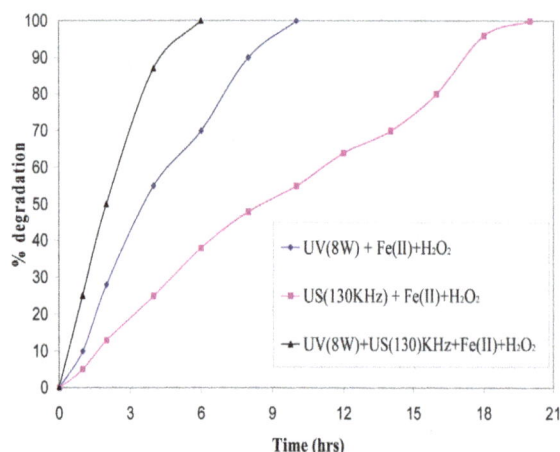

Figure 2. Comparison of advanced oxidation processes for degradation of EDTA (2%).

shown by Equation (3) [22].

$$Fe^{3+} + H_2O_2 + h\nu \rightarrow {}^*OH + Fe^{2+} + H^+ \quad (3)$$

The temperature of the reaction process was found to vary from 28°C - 32°C. Furthermore, the effect of UV light also attributed to the direct hydroxyl radical formation and regeneration of Fe^{2+} from the photolysis of the complex $Fe(OH)^{2+}$ in solution Equation (4).

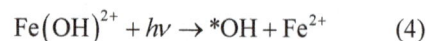

$$Fe(OH)^{2+} + h\nu \rightarrow {}^*OH + Fe^{2+} \quad (4)$$

From the simultaneous application of UV and Fenton's reagent there is considerable synergistic effect on the rate of degradation of EDTA.

Irradiation by high power ultrasound in a liquid leads to the acoustic cavitations phenomenon, such as the formation, growth and collapse of bubbles, accompanied by the generation of local high temperature, pressure and reactive radical species. This conditions lead to some chemical reactions in three phases: internal cavity, interface boundary layer and the bulk liquid Thermal decomposition may take place in the internal parts of the cavities. The reactions with radicals occur in the interface boundary layer. In wastewater treatment a bubble of cavitation may function as a micro-reactor inside which, the volatile compounds are destroyed. During sonolysis of water and in the presence of oxygen, the cavity may also be considered as source of $H^{\bullet}$, ${}^{\bullet}OH$, $HOO^{\bullet}$ radicals, which have been extremely effective in the destruction of organic pollutants [22].

To enhance the efficiency of degradation, a more effective utilization of OH radicals is desirable. It is expected that addition of Fe(II) + H_2O_2 will regenerate ${}^{\bullet}OH$, thus accelerating the rate of degradation and thereby increasing the efficiency of ultrasonic degradation [23].

In spite of the fact that irradiation of a treatment medium using ultrasound in combination with Fenton's reagent also lead to formation of free radicals similar to a Photo Fenton process the kinetics of degradation of EDTA in the Sono Fenton process is slow compared to Photo Fenton process. In the case of photochemical degradation there was direct exposure of UV radiation onto the reaction medium whereas in the case of sonodegradation there could have been attenuation of the sound wave by the glass medium due to the placement of the glass trough inside the ultrasonic bath. Due to the above, the number of cavitation bubbles produced in the reaction vessel may be less. The decomposition rate is dependent on the rate of formation of ${}^{\bullet}OH$ that is the rate of cavitation occurrence, which is a function of cavitation period and number of transient cavitation bubbles and also on the concentration of the pollutant [24]. The factors viz., the operating conditions (frequency and power dissipation per unit volume) and the equipment configuration decide the rate of generation of free radicals due to the ultrasonic action. Sono-Fenton degradation process is very much dependent on the utilization of free radicals by the pollutant molecules, which in turn is dependent on the efficiency of contact of the generated free radicals with the pollutant over a specified time period [25]. The kinetics of the Sono-Fenton process for degradation of EDTA (20,000 mg/l) can be accelerated by the use of a higher frequency and higher power of ultrasound or by usage of a horn type sonicator for direct production and interaction of hydroxyl free radicals with the pollutant. The above observation is also in conformity with literature where it has been cited that multiple frequency reactors have been found to generate more intense and spatially uniform cavitation as compared to reactors with single frequency and/or single transducer operation [26] and hence these give better destruction efficiency [27].

The time taken for EDTA degradation due to the integral effect of both US (130 KHz) + UV (8 W) + Fe(II) + H_2O_2 was found to be at 5 hours which is twice and four times lesser than application of UV/Fenton and US/ Fenton respectively. The expected synergism between different hybrid methods discussed in the present work is mainly due to an identical controlling reaction mechanism, *i.e.* the free radical attack. Generally, combination of two or more advanced oxidation processes, sono-photochemical/sono-photocatalytic oxidation etc. leads to an enhanced generation of the hydroxyl radicals, which eventually results in higher oxidation rates. The efficacy of the process and the extent of synergism depend not only on the enhancement in the number of free radicals but also on the alteration of the reactor conditions or configuration leading to a better contact of the generated free radicals with the pollutant molecules and also better utilization of the oxidants and catalytic activity [26]. The synergistic process may lead to complete mineralization by oxidizing the organic intermediates.

In all the UV/US/US + UV Fenton processes there was a change in pH from 3.0 to 8.5 as the reaction proceeded. From the above, it can be inferred that there was loss of acidity and hence the loss of chelating ability of EDTA [13]. This was confirmed titrimetrically. Since there was a change in pH from acidic range to alkaline range as the reaction proceeded, formation of acidic intermediates is ruled out and formation of either amines or amides was suspected. The qualitative hydroxamic acid test was positive wherein a deep red colored solution was obtained on treatment with hydroxylamine and ferric chloride and thus formation of amides was confirmed. The above observation is in agreement with the results obtained for the mechanistic formation of intermediates in UV/US + Fenton processes [21].

3.2. Kinetics of Degradation of EDTA by UV/US/UV + US-Fenton Processes

The rate of degradation of EDTA was investigated for the systems UV + US+ FeII + H_2O_2, US (130 KHz) +

FeII + H_2O_2 and UV (8 W) + FeII + H_2O_2. The degradation of EDTA was observed as a function of time and the data were fitted to a first-order rate model as shown in Equation (5)

$$\ln C_1/C_0 = -k_0 t \qquad (5)$$

where C_0 and C_t are the concentration of EDTA at times 0, and t, k_0 is a first order rate constant (hr^{-1}) and t is the time (hrs). The rate constants were determined using a first order rate model [Equation (5)]. The results are listed in **Table 1**. From the data in **Table 1**, it can be observed that the rate of EDTA degradation due to the integral effect of both US (130 KHz) + UV (8 W) + Fe(II) + H_2O_2 is found to be 1.7 and 3.8 times greater than application of UV/Fenton and US/Fenton respectively. Hence, rate of degradation follows the order UV + US + FeII + H_2O_2 > UV (8 W) + FeII + H_2O_2 > US (130 KHz) + FeII + H_2O_2.

The ratio of the Sono-Photo-Fenton rate constant to the sum of the rate constants of the individual processes was used to evaluate the synergistic effect of the combined system as shown in Equation (6)

$$\text{Synergistic Index} = k_{\text{sonophoto}} / \left[k_{\text{sono}} + k_{\text{photo}} \right] \qquad (6)$$

where $k_{\text{sonophoto}}$, k_{sono} and k_{photo} are the rate constants for Sono-Photo Fenton, Sono-Fenton and Photo-Fenton process respectively [28]. The synergistic index was found to be 1.16 for the Sono-Photo-Fenton degradation of EDTA (20,000 mg/l). The rate constant for the Sono-Photo Fenton process was found to be additive of the rate constants of the Sono-Fenton and photo-Fenton process as shown in **Table 1**.

4. Conclusion

The results from the comparative study on the kinetics of EDTA degradation revealed that the reaction using UV + US + FeII + H_2O_2 was faster than that obtained in any of the other processes viz., UV (8 W) + FeII + H_2O_2, US (130 KHz) + FeII + H_2O_2 respectively. Sono-Photo Fenton process effectively degrades the EDTA. The synergistic effect is attributable to effective enhancement of Photo-Fenton oxidation by Sono-Fenton oxidation. From the observed pH changes during either Photo-Fenton or

Table 1. Reaction rate constants for the degradation of EDTA (2%) using.

Types of oxidation process	K_0 (hr^{-1})	R^2
UV (8 W) + US (130 kHz) + Fe(II) + H_2O_2	0.5 ± 0.02	0.97
UV (8 W) + Fe(II) + H_2O_2	0.3 ± 0.01	0.96
US (130 kHz) + Fe(II) + H_2O_2	0.13 ± 0.03	0.99

Sono-Fenton or Sono-Photo Fenton processes, it can be concluded that there is a loss of chelating ability of EDTA and formation of amides was confirmed.

REFERENCES

[1] Q.-H. Tian and X.-Y. Guo, "Electroless Copper Plating on Microcellular Polyurethane Foam," *Transactions of Nonferrous Metals Society of China*, Vol. 20, Suppl. 1, 2010, pp. s283-s287. doi:10.1016/S1003-6326(10)60057-X

[2] P. Pitter and V. Sykora, "Biodegradability of Ethylene Diamine-Based Complexing Agents and Related Compounds," *Chemosphere*, Vol. 44, No. 4, 2001, pp. 823-826. doi:10.1016/S0045-6535(00)00512-9

[3] D. Li, X. Wu, D. Wang and J. A. Finch, "Selective Removal of Nickel from Iron Substrate by Non-Cyanide Stripper," *Transactions of Nonferrous Metals Society of China*, Vol. 14, No. 3, pp. 599-602.

[4] B.-L. Lu, Q.-Y. Chen, Z.-L. Yin and H.-P. Hu, "Effects of Na4EDTA and EDTA on Seeded Precipitation of Sodium Aluminate Solution," *Transactions of Nonferrous Metals Society of China*, Vol. 20, No. z1, 2010, pp. s37-s41.

[5] S. Chitra, S. Chandran, P. Sasidhar, K. B. Lal and R. V. Amalraj, "Biodegradation of Surfactant Bearing Wastes," *Indian Journal of Environmental Protection*, Vol. 11, No. 9,1991, pp. 689-692.

[6] K. Rosikova, J. John, E. Danacikova-Popelova, F. Sebesta and E. W. Hooper, "Study of EDTA Photodegradation," *Proceedings of 4th Institute for International Cooperative Environmental Research*, Florida State University, Tallahassee, 1998, pp. 379-385.

[7] M. L. Hinck, J. Ferguson and J. Puhaakka, "Resistance of EDTA and DTPA to Aerobic Biodegradation," *Water Science and Technology*, Vol. 35, No. 2-3, 1997, pp. 25-31. doi:10.1016/S0273-1223(96)00911-0

[8] H. J. Brauch and S. V. Schullerer, "EDTA and NTA Beider Trinkwasseraufbereitung," V*om Wasser*, Vol. 69, 1987, pp. 155-164.

[9] S. V. Schullerer and H. J. Brauch, "Oxidative and Adsorptive Behandlung EDTA and NTA Haltiger Wasser," *Vom Wasser*, Vol. 72, 1989, pp. 23-29.

[10] C. P. Huang, D. Cheng and Z. H. Tang, "Advanced Chemical Oxidation: Its Present Role and Potential Future in Hazardous Waste Treatment," *Waste Management*, Vol. 13, No. 5-7, 1993, pp. 361-377. doi:10.1016/0956-053X(93)90070-D

[11] E. Gilbert and S. Hoffmann-Glewe, "Ozonation EDTA in Aqueous Solution, Influence of pH Value and Metal Ions," *Water Research*, Vol. 24, No. 1, 1990, pp. 39-44. doi:10.1016/0043-1354(90)90062-B

[12] M. Sorensen and F. H. Frimmel, "Photodegradation of EDTA and NTA in the UV/H_2O_2 Process," *Zeitschrift für Naturforschung*, Vol. 50, No. 12, 1995, pp. 1845-1853.

[13] M. D. Tucker, L. L. Barton, B. M. Thomson, B. M. Wagener and A. Aragon, "Treatment of Waste Containing EDTA by Chemical Oxidation," *Waste Management*, Vol. 19, No. 7-8, 1999, pp. 477-482.

[14] K. Krapfenbauer and N. Getoff, "Comparative Studies of Photo-and-Radiation-Induced Degradation of Aqueous EDTA. Synergistic Effects of Oxygen, Ozone and TiO_2," *Radiation Physics and Chemistry*, Vol. 55, No. 4, 1999, pp. 385-393. doi:10.1016/S0969-806X(99)00205-4

[15] Y. Su, Y. Wang, J. L. Daschbach, T. B. Fryberger, M. A. Henderson, J. Janata and C. H. F. Peden, "Gamma Ray Destruction of EDTA Catalyzed by Titania," *Journal of Advanced Oxidation Technology*, Vol. 3, 1998, pp. 63-69.

[16] S. Kagaya, Y, Bitoh and K. Hasegawa, "Photocatalyzed Degradation of Metal-EDTA Complexes in TiO_2 Aqueous Suspensions and Simultaneous Metal Removal," *Chemical Letters*, Vol. 26, No. 2, 1997, pp. 155-156. doi:10.1246/cl.1997.155

[17] J. Ramo and M. Sillanpaa, "Degradation of EDTA by Hydrogen Peroxide in Alkaline Conditions," *Journal of Cleaner Production*, Vol. 9, No. 3, 2001, pp. 191-195. doi:10.1016/S0959-6526(00)00049-4

[18] American Public Health Association, "Standard Methods for the Examination of Water and Wastewater," New York, 1975.

[19] A. I. Vogel, "Quantitatative Inorganic Analysis," 3rd Edition, Longman Publishing, London, 1939.

[20] M. Y. Ghaly, H. Georg, M. Roland and R. Haseneder, "Photochemical Oxidation of *p*-Chlorophenol by UV/H_2O_2 and Photo-Fenton Process. A Comparative Study," *Waste Management*, Vol. 21, No. 1, 2001, pp. 41-47. doi:10.1016/S0956-053X(00)00070-2

[21] P. A. Babay, D. A. Batistoni, D. A. Ferreyra, E. A. Gautier, R. T. Gettar and M. I. Litter, "Kinetics and Mechanisms of EDTA Photocatalytic Degradation with TiO_2," *Water Science and Technology*, Vol. 44, No. 5, 2001, pp. 179-185.

[22] Y. G. Adewuyi, "Sonochemistry: Environmental Science and Engineering Applications," *Industrial & Engineering Chemistry Research*, Vol. 40, No. 22, 2001, pp. 4681-4715. doi:10.1021/ie0100961

[23] B. Yim, H. Okuno, N. Nagata and Y. Maeda, "Sonolysis of Surfactants in Aqueous Solutions: An Accumulation of Solute in the Interfacial Region of the Cavitation Bubbles," *Ultrasonics Sonochemistry*, Vol. 9, No. 4, 2002, pp. 209-213. doi:10.1016/S1350-4177(01)00123-7

[24] Y. Suzuki, Warsito, A. Maezawa and S. Uchida, "Effects of Frequency and Aeration Rate on Ultrasonic Oxidation of a Surfactant," *Chemical. Engineering Technology*, Vol. 22, No. 6, 1999, pp. 507-510. doi:10.1002/(SICI)1521-4125(199906)22:6<507::AID-CEAT507>3.0.CO;2-D

[25] R. Parag Gogate and B. Aniruddha Pandit, "A Review of Imperative Technologies for Wastewater Treatment II: Hybrid Methods," *Advances in Environmental Research*, Vol. 8, No. 3-4, 2004, pp. 553-597. doi:10.1016/S1093-0191(03)00031-5

[26] P. A. Tatake and A. B. Pandit, "Modelling and Experimental Investigation into Cavity Dynamics and Cavitational Yield: Influence of Multiple Frequency Ultrasound Sources," *Chemical Engineering Science*, Vol. 57, No. 22, 2002, pp. 4987-4995.

doi:10.1016/S0009-2509(02)00271-3

[27] M. Sivakumar, P. A. Tatake and A. B. Pandit, "Kinetics of p-Nitrophenol Degradation: Effect of Reaction Conditions and Cavitational Parameters for a Multiple Frequency System," *Chemical Engineering Journal*, Vol. 85, No. 2-3, 2002, pp. 327-338. doi:10.1016/S1385-8947(01)00179-6

[28] Y. He, F. Grieser and M. Ashokkumar, "The Mechanism of Sonophotocatalytic Degradation of Methyl Orange and Its Products in Aqueous Solutions," *Ultrasonics Sonochemistry*, Vol. 18, No. 5, 2011, pp. 974-990. doi:10.1016/j.ultsonch.2011.03.017

19

Thermodynamic Assessment of the Pt-Sb System

Jinming Liu[1]*, Yinghui Zhang[1], Cuiping Guo[2]

[1]School of Material Science and Engineering, Jiangxi University of Science and Technology, Ganzhou, China
[2]Department of Materials Science and Engineering, University of Science and Technology Beijing, Beijing, China
Email: *liujm2011@sina.com

ABSTRACT

The Pt-Sb system was critically assessed by means of CALPHAD technique. Based on the experimental data in the literature, the excess Gibbs energies of the solution phases (liquid, rhombohedral, fcc) were modeled with the Redlich-Kister equation. The five intermetallic compounds, Pt_7Sb, Pt_3Sb, Pt_3Sb_2, $PtSb$, and $PtSb_2$ were treated as stochiometric compounds and expressed as the formula $(Pt)_m(Sb)_n$. The intermetallic compound, Pt_5Sb with a homogenerity ranges 0.155 - 0.189 Sb, were treated as the formula $(Pt,Sb)_m(Pt,Sb)_n$. A set of self-consistent thermodynamic parameters of the Pt-Sb system was obtained.

Keywords: Pt-Sb Phase Diagram; Thermodynamic Assessment; Thermodynamic Properties; CALPHAD Technique

1. Introduction

The precious metals, Pt, Sb, Au, Ag and their alloys, have been widely applied as catalysts for their high thermal stability and high activity, such as decompounding hydrazine, purifying automobile gas, oxygenating selectively and so on [1-5]. Understanding the phase equilibria as well as the thermochemical behaviour of these systems is helpful for the development of the related materials. This paper intends to assess the Pt-Sb system thermodynamically and provides a set of self-consistent parameters for calculation of the phase equilibria and thermochemical properties of the system.

2. Literature Review

The Pt-Sb phase diagram consists of the liquid, the face-centered cubic (fcc), terminal solid solution (Pt); the terminal solid solution, rhombohedral (Sb) with a negligible solubility of Pt. The gas phase was estimated by Itkin and Alcock [6], which is not considered in the present work.

The liquidus was determined by thermal analysis in Refs. [7-10]. The liquidus in the Pt-Sb system was based mainly on the results made by Durussel and Feschotte [11] using differential thermal analysis (DTA), which was supported by the majority of the data [7-10].

In the Pt-Sb system, there are six intermediate phases,

Pt_7Sb, Pt_5Sb, Pt_3Sb, Pt_3Sb_2, $PtSb$, and $PtSb_2$. The Pt_7Sb compound was discovered by [11], using DTA, microprobe analysis, and X-ray powder (XRD). Pt_7Sb, a cubic structure, is formed by a peritectoid reaction. Different designations for the phase Pt_5Sb are "Pt_4Sb" [8], "$Pt_{34}Sb_7$" [10], "$Pt_{4+}Sb$" [12], and "$Pt_{82}Sb_{18}$" [13]. A compound, "Pt_5Sb_2" was reported in Ref. [7] and not confirmed in other publications [8,10,12,13]. The compound as Pt_5Sb has a cubic structure wth a homogeneity range of 0.155 to 0.189 Sb [11]. The boundaries of the phase were determined using a temperature dependence of the lattice parameter. This phase is formed at 748°C ± 4°C by peritectic reaction, which was consistent with Refs. [7-9]. It decomposed at 560°C by eutectoid reaction.

The compound Pt_3Sb was formed by peritectic reaction [9,11]. The compound was investigated using thermal arrest by [8] and explained by the phase transformation in "Pt_4Sb". The compound Pt_3Sb was examined and observed as a single phase with a composition of 0.275 Sb and as a second phase at 0.25 and 0.30 Sb by Srivastava et al. [14]. However, Pt_3Sb was confirmed as a stoichiometric compound in subsequent research [10].

The compound Pt_3Sb_2 was determined by Bhan et al. [9] and confirmed by Kim [10]. It is formed by a peritectic reaction at 732°C [9] or 739°C ± 4°C [11]. The compound PtSb was determined by several researchers and the formed temperature was different, 1045°C [7], 1040°C [8], 1043°C [9], and 1046°C ± 3°C [11]. The composition of PtSb was determined by Refs. [10,11,15,16]. The results of made by Durussel and Feschotte [11], Kim [10], and Kjekshus [17] showed that the composition of PtSb can be adopted as only approximately stoichiometric compound.

The compound $PtSb_2$ was first found by chemical analysis of crystals precipitating from Pt-Sb melts containing an excess of Sb [18]. The temperature of congruent melting of $PtSb_2$ was measured as 1226°C [7], 1210°C [8], and 1225°C [11]. The enthalpy increments of $PtSb_2$ measured by drop calorimetry in the temperature interval 196°C to 620°C [19].

From the above literature [6-19], the present assessed Pt-Sb phase diagram consists of the liquid, the face-centered cubic (fcc) terminal solid solution (Pt), the terminal solid solution (Sb) with a negligible solubility of Pt and six intermediate phases, Pt_7Sb, Pt_5Sb, Pt_3Sb, Pt_3Sb_2, PtSb and $PtSb_2$.

3. Thermodynamic Models

3.1. Unary Phases

The Gibbs energy function $G_i^\phi(T) = G_i^\phi(T) - H_i^{SER}$ (298.15 K) for the element i (i = Pt, Sb) in the phase ϕ (ϕ = liquid, face-centered cubic(fcc) and rhombohedral) is described by an equation of the following form:

$$G_i^\phi(T) = a + bT + cT\ln(T) + dT^2 + eT^3 + fT^{-1} + gT^7 + hT^{-9} \tag{1}$$

where H_i^{SER} (298.15 K) is the molar enthalpy of the element i at 298.15 K in its standard element reference (SER) state, fcc for Pt and rhombohedral for Sb. The Gibbs energy of the element i, $G_i^\phi(T)$, in its SER state, is denoted by GHSER$_i$, i.e.,

$$\text{GHSER}_{Pt} = {}^0G_{Pt}^{fcc}(T) - H_{Pt}^{SER}(298.15\,\text{K}) \tag{2}$$

$$\text{GHSER}_{Sb} = {}^0G_{Sb}^{rhomb}(T) - H_{Sb}^{SER}(298.15\,\text{K}) \tag{3}$$

In the present work, the Gibbs energy functions are taken from the SGTE (Scientific Group Thermodata Europe) pure elements database compiled by Dinsdale [20].

3.2. Solution Phases

In the Pt-Sb system, there are three solution phases: liquid, fcc and rhombohedral. The gas phase was not considerated in the Pt–Sb system, similar to the results reported by Durussel and Feschotte [11]. Their Gibbs energies are described by the following expression:

$$G_m^\phi = x_{Pt}G_{Pt}^\phi(T) + x_{Sb}G_{Sb}^\phi(T) + RT(x_{Pt}\ln x_{Pt} + x_{Sb}\ln x_{Sb}) + {}^EG_m^\phi \tag{4}$$

where R is the gas constant, x_{Pt} and x_{Sb} are the mole fraction of Pt and Sb, respectively, and ${}^EG_m^\phi$ is the excess Gibbs energy, expressed by the Redlich–Kister polynomial [21].

$$^EG_m^\phi = x_{Pt}x_{Sb}\sum_j {}^jL^\phi(x_{Pt} - x_{Sb})^j \tag{5}$$

where ${}^jL^\phi$ is the interaction parameter between element Pt and Sb, which is to be evaluated in the present work. Its general form is

$$L^\phi = a + bT + cT\ln(T) + dT^2 + eT^3 + fT^{-1} \tag{6}$$

In most cases, only the first one or two terms are used according to the temperature dependence on the experimental data.

3.3. Intermetallic Compounds

There were six intermediate phases, Pt_7Sb, Pt_5Sb, Pt_3Sb, Pt_3Sb_2, PtSb, and $PtSb_2$ in the Pt-Sb system. The phases crystal structure data was shown in **Table 1**. According to the composition and crystal structure data , the Sb in the five compounds, Pt_7Sb, Pt_3Sb, Pt_3Sb_2, PtSb and $PtSb_2$ of the system with no or little solid solubility will be treated as the stoichiometric compounds. While the compound Pt_5Sb with more homogenertiy ranges 0.155 - 0.189 Sb will treated as solid solution compound.

So the Pt_5Sb phase had a homogenertiy ranges, 0.155 - 0.189 Sb. The two-sublattice model, $(Pt,Sb)_{0.833}$ $(Pt,Sb)_{0.167}$ is used to describe this phase in the present work.

The parameters y_i' and y_i'' are the site fractions of Pt or Sb on the first and second sublattices, respectively; the parameter $G_{*:*}^{Pt_5Sb}$ represents the Gibbs energies of the compound Pt_5Sb when the first and second subletices are occupied by only one element Pt or Sb, respectively, which are relative to the enthalpies of pure fcc for Pt and rhombohedral for Sb in their SER state; ${}^jL_{Pt,Sb:*}^{Pt_5Sb}$ and ${}^jL_{*:Pt,Sb}^{Pt_5Sb}$ represent the jth interaction parameters (j = 0) between the element Pt and Sb on the first and second sublattice, respectively.

$$G_m^{Pt_5Sb} = y_{Pt}'y_{Pt}''G_{Pt:Pt}^{Pt_5Sb} + y_{Pt}'y_{Sb}''G_{Pt:Sb}^{Pt_5Sb} + y_{Sb}'y_{Pt}''G_{Sb:Pt}^{Pt_5Sb} + y_{Sb}'y_{Sb}''G_{Sb:Sb}^{Pt_5Sb} + 0.833RT(y_{Pt}'\ln y_{Pt}' + y_{Sb}'\ln y_{Sb}')$$

$$+ 0.167RT(y_{Pt}''\ln y_{Pt}'' + y_{Sb}''\ln y_{Sb}'') + y_{Pt}'y_{Sb}'\left(y_{Pt}''\sum_j {}^jL_{Pt,Sb:Pt}^{Pt_5Sb}(y_{Pt}' - y_{Sb}')^j + y_{Sb}''\sum_j {}^jL_{Pt,Sb:Sb}^{Pt_5Sb}(y_{Pt}' - y_{Sb}')^j\right) \tag{7}$$

$$+ y_{Pt}''y_{Sb}''\left(y_{Pt}'\sum_j {}^jL_{Pt:Pt,Sb}^{Pt_5Sb}(y_{Pt}'' - y_{Sb}'')^j + y_{Sb}'\sum_j {}^jL_{Sb:Pt,Sb}^{Pt_5Sb}(y_{Pt}'' - y_{Sb}'')^j\right)$$

Table 1. Pt-Sb Crystal Structure Data.

Phase	Composition, at% Sb	Pearson symbol	Space group	Strukturbericht designation	Prototype	Refs.
Fcc (Pt)	0 to 10.4	$cF4$	$Fm\bar{3}m$	$A1$	Cu	[22]
Pt_7Sb	12.5	$cF32$	$Fm\bar{3}m$	...	Ca_7Ge	[11]
Pt_5Sb	15.5 to 18.9	$cP4$	$Fm\bar{3}m$	$L1_2$	$AuCu_3$	[13]
Pt_3Sb	25.0	$tI16$	$I4/mmm$	$D0_{23}$	Al_3Zr	[13]
Pt_3Sb_2	40.0	$oI20$	$Ibam$	...	...	[9]
$PtSb$	50.0	$hP4$	$P6_3/mmc$	$B8_1$	NiAs	[23]
$PtSb_2$	66.7	$cP12$	$Pa\bar{3}$	$C2$	FeS_2(pyrite)	[24]
Rhombohedral (Sb)	100.0	$hR2$	$R\bar{3}m$	$A7$	αAs	[22]

The other intermediate phases, Pt_7Sb, Pt_3Sb, Pt_3Sb_2, $PtSb$, and $PtSb_2$ in the Pt-Sb system were treated as stochiometric compounds. The two-sublattice model, $(Pt)_m(Sb)_n$, is used to describe these phases in the present work. The Gibbs energy per mole of formula unit $(Pt)_m(Sb)_n$, was expressed as the following:

$$G_m^{Pt_mSb_n} = m\text{GHSER}_{Pt} + n\text{GHSER}_{Sb} + \Delta G_f^{Pt_mSb_n} \quad (8)$$

where $\Delta G_f^{Pt_mSb_n}$ is the Gibbs energy of formation per mole of formula unit Pt_mSb_n. And $\Delta G_f^{Pt_mSb_n}$ can be given by the following expression:

$$\Delta G_f^{Pt_mSb_n} = a + bT \quad (9)$$

where the parameters a and b were to be evaluated in the present work.

4. Optimization

Most of the above experimental information was selected for the evaluation of the thermodynamic model parameters. In the present work, the phase relation and transformation temperatures based on the phase diagram of the Pt-Sb system compiled by Itkin and Alcock [6], determinded by Bhan et al. [9], and Durussel and Feschotte [11].

The optimization was carried out by means of the Thermo-Calc software [25], which can handle various kinds of experimental data. The program works by minimizing an error sum where each of the selected data values is given a certain weight. The weight is chosen by personal judgment and changed by trial and error during the work until most of the selected experimental information is reproduced within the expected uncertainty limits.

The ways to get the phase diagram summarized as several steps using the Thermo-Calc software. Firstly, it is the Unary phases to get the GES file, which is the basis of the assessment. Secondly, it is the thermodynamic models of the solution phases and the compounds to form the TCM file. The expression of the thermodynamic model was shown in Section 3.

Last is the optimization. The optimization was carried out by means of the Thermo-Calc software based on the POP file to get the PAR file. The POP file was the phases equilibria, while the PAR file was the results of the optimization. When the phase diagram is successfully optimized, all the phases in the system will get a set of consistent thermodynamic parameters to express Gibbs energies. Using the set of consistent thermodynamic parameters to reproduce the phase diagram and build the database such as the TDB file.

5. Results and Discussions

A thermodynamic description of the Pt-Sb system obtained in the present work is shown in **Table 2**. The Pt-Sb phase diagram calculated by means of the present thermodynamic parameters is presented in **Figure 1**, and nearly identical to the one determined by Bhan et al. [9], Durussel and Feschotte [11]. Because of the symmetry of liquid at both sides of the compound $PtSb_2$, there is great different about the liquid at 0.20 - 0.40 Sb between the calculated results and the experimental data [7-9].

The invariant equilibria of the Pt-Sb system are listed in **Table 3**. In the **Table 3**, some calculating data is nearly to experiment data but some is very different to experiment data. There is some reasons about these. The one is that the liquid at 0.20 - 0.40 Sb is deeper than others. Other is that the liquid of the sides of $PtSb_2$ is treated the symmetry during the assessed procedure. So it is very difficult to treat and optimize to get the same as the experimental data. In order not to change the types of reactions, the temperatures of reactions were revised in this work. As shown in the table, most satisfactory agreement is obtained between the calculations and experiments [7-9,11], where there is uncertainty in the invariant reaction temperature at 0.20 - 0.40 Sb.

Table 2. Thermodynamic parameters of the Pt-Sb system[a].

Phase	Models	Thermodynamic parameters
liquid	$(Pt, Sb)_1$	$^0L^{liq} = -33358.7 - 29.6508T$
		$^1L^{liq} = +4402.1 - 17.9740T$
		$^2L^{liq} = -11140.9$
	rhombohedral $(Pt, Sb)_1$	$G(\text{rhombohedral, Pt}) = + \text{GHSER}_{Pt} + 5000.0$
		$G(\text{rhombohedral, Sb}) = + \text{GHSER}_{Sb}$
fcc	$(Pt, Sb)_1$	$^0L^{fcc} = -9650.4 - 26.3403T$
		$^1L^{fcc} = -10650.4$
Pt_7Sb	$(Pt)_{0.865}(Sb)_{0.125}$	$G_{Pt:Sb}^{Pt_7Sb} = +0.865\text{GHSER}_{Pt} + 0.125\text{GHSER}_{Sb} - 7499.0 + 0.0120T$
Pt_3Sb	$(Pt)_{0.75}(Sb)_{0.25}$	$G_{Pt:Sb}^{Pt_3Sb} = +0.75\text{GHSER}_{Pt} + 0.25\text{GHSER}_{Sb} - 14002.3$
Pt_3Sb_2	$(Pt)_{0.6}(Sb)_{0.4}$	$G_{Pt:Sb}^{Pt_3Sb_2} = +0.6\text{GHSER}_{Pt} + 0.4\text{GHSER}_{Sb} - 20702.3$
$PtSb$	$(Pt)_{0.5}(Sb)_{0.5}$	$G_{Pt:Sb}^{PtSb} = +0.5\text{GHSER}_{Pt} + 0.5\text{GHSER}_{Sb} - 25012.3$
$PtSb_2$	$(Pt)_{0.333}(Sb)_{0.667}$	$G_{Pt:Sb}^{PtSb_2} = +0.333\text{GHSER}_{Pt} + 0.667\text{GHSER}_{Sb} - 30727.5 + 0.1023T$
Pt_5Sb	$(Pt,Sb)_{0.833}(Pt,Sb)_{0.167}$	$G_{Pt:Pt}^{Pt_5Sb} = +\text{GHSER}_{Pt} + 5000.0$
		$G_{Pt:Sb}^{Pt_5Sb} = +0.833\text{GHSER}_{Pt} + 0.167\text{GHSER}_{Sb} - 8357.4 - 1.3948T$
		$G_{Sb:Pt}^{Pt_5Sb} = +0.833\text{GHSER}_{Sb} + 0.167\text{GHSER}_{Pt} + 18357.4 + 1.3948T$
		$G_{Sb:Sb}^{Pt_5Sb} = +\text{GHSER}_{Sb} + 5000.0$
		$^0L_{Pt,Sb:Sb}^{Pt_5Sb} = {}^0L_{Pt,Sb:Pt}^{Pt_5Sb} = -22770.3$
		$^0L_{Pt:Pt,Sb}^{Pt_5Sb} = {}^0L_{Sb:Pt,Sb}^{Pt_5Sb} = -3981.8$

[a]In J·mole^{-1} of the formula units.

Table 3. Invariant reactions of the Pt-Sb system.

Reaction	T (K)	x(Sb)			References
liq. + fcc(Pt) → Pt_5Sb	1021	0.275	0.100	0.166	[11]
	1139	0.2775	0.1240	0.1652	This work
fcc + Pt_5Sb → Pt_7Sb	898	0.104	0.155	0.125	[11]
	898	0.0330	0.1441	0.1250	This work
Pt_5Sb → Pt_7Sb + Pt_3Sb	833	0.164	0.125	0.250	[11]
	833	0.1530	0.125	0.2500	This work
liq. + Pt_5Sb → Pt_3Sb	948	0.302	0.189	0.250	[11]
	1128	0.2844	0.1703	0.2500	This work
liq. → Pt_3Sb + Pt_3Sb_2	903	0.315	0.250	0.400	[9]
	1127	0.2877	0.2500	0.4000	This work
liq. + $PtSb$ → Pt_3Sb_2	1012	0.352	0.500	0.400	[9]
	1133	0.2957	0.5000	0.4000	This work
liq. + $PtSb_2$ → $PtSb$	1319	0.475	0.667	0.500	[8]
	1197	0.3485	0.6667	0.5000	This work
liq. → $PtSb_2$	1499	0.667	0.667	–	[7]
	1497	0.667	0.667	–	This work
liq. → $PtSb_2$ + rhomb	899	0.990	0.667	1.000	[11]
	903	0.9974	0.6667	1.000	This work

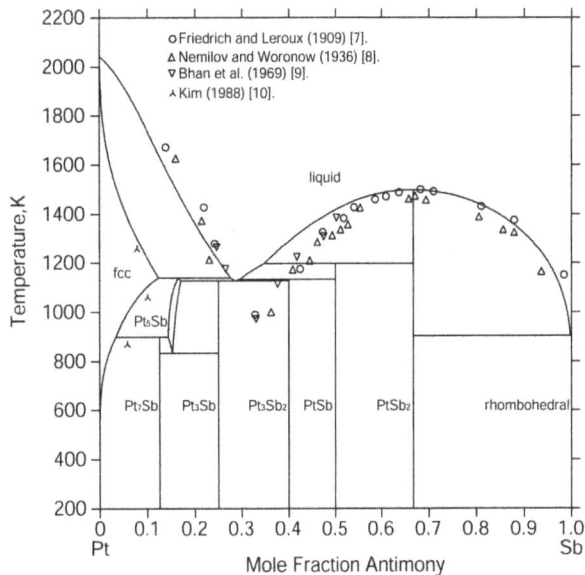

Figure 1. Calculated Pt-Sb phase diagram by the present thermodynamic escription with the experimental data measured by Friedrich and Leroux [7], Nemilov and Woronow [8], Bhan et al. [9] and Kim [10].

In view of the estimated experimental errors (about 0.01 - 0.02), 24 of the 26 experimental invariant reaction compositions in the Pt-Sb system are well reproduced.

6. Conclusion

The phase relations and the thermodynamic description of the Pt-Sb system were critically evaluated from the experimental information available in the literature. A set of consistent thermodynamic parameters were derived. With the thermodynamic description available, one can now make various calculations of practical interest or in the optimization of high-order systems.

7. Acknowledgements

This work was supported by the National Natural Science Foundation of China (Nos. 50934011, 51264012) and Jiangxi University of Science and Technology Grade Scientific Research Project (No. jxxj12032).

REFERENCES

[1]   L. W. Lin, T. Zhang, J. L. Zang and Z. S. Xu, "Dynamic Process of Carbon Deposition on Pt and Pt-Sn Catalysts for Alkane Dehydrogenation," *Applied Catalysis*, Vol. 67, No. 1, 1990, pp. 11-23.
      doi:10.1016/S0166-9834(00)84428-0

[2]   A. Hinz, B. Nilsson and A. Andersson, "Simulation of Transients in Heterogeneous Catalysis: A Comparison of the Step- and Pulse-Transient Techniques for the Study of Hydrocarbon Oxidation on Metal Oxide Catalysts," *Chemical Engineering Science*, Vol. 55, No. 20, 2000, pp. 4385-4397.

[3]   L. C. Li and Y. J. Lan, "Advance in the Catalysis for Hydrazine Decompositon," *Industrial Catalysis*, Vol. 2, No. 1, 1994, pp. 3-7. (in Chinese)

[4]   S. Gao and L. D. Schmidt, "Effect of Oxidation-Reduction Cycling on C_2H_6 Hydrogenolysis: Comparison of Ru, Rh, Ir, Ni, Pt, and Pd on SiO_2," *Journal of Catalysis*, Vol. 115, No. 2, 1989, pp. 356-364.

[5]   W. Chu, Q. G. Yan, X. Liu, Q. Li, Z. L. Yu and G. X. Xiong, "Rare Earth Promoted Nickel Catalysts for the Selective Oxidation of Natural Gas to Syngas," *Studies in Surface Science and Catalysis*, Vol. 119, 1998, pp. 855-856. doi:10.1016/S0167-2991(98)80538-7

[6]   V. P. Itkin and C. B. Alcock, "The Pt-Sb (Platinum-Antimony) System," *Journal of Phase Equilibria*, Vo. 17, No. 4, 1996, pp. 356-361.

[7]   K. Friedrich and A. Leroux, "Melting Diagram for Alloys of Platinum with Antimony," *Metallurgie*, Vol. 6, 1909, pp. 1-3. (in German)

[8]   W. A. Nemilov, N. M. Woronow and Z. Anorgm "Über Legierungen des Platins mit Antimon," *Zeitschrift für anorganische und allgemeine Chemie*, Vol. 226, No. 2, 1936, pp. 177-184. doi:10.1002/zaac.19362260211

[9]   S. Bhan, T. Godecke, K. Schubert, "Konstitution Einiger Mischungen des Platins Mit b-Elementen (B = Sn, Sb, Te)," *Journal of the Less Common Metals*, Vol. 19, No. 2, 1969, pp. 121-140. doi:10.1016/0022-5088(69)90027-7

[10] W.-S. Kim, "Phase Constitution of the Pt-Sb System," *Journal of the Korean Institute of Metals*, Vol. 26, 1988,
      pp. 378-384.

[11] P. Durussel and P. Feschotte, "Les Systèmes Binaires Pd Sb et Pt Sb," *Journal of Alloys and Compounds*, Vol. 176, No. 1, 1991, pp. 173-181.
      doi:10.1016/0925-8388(91)90023-O

[12] S. Bhan, K. Schubert, "Über die Struktur von Phasen Mit Kupfer Unterstruktur in Einigen t-b Legierungen (T= Ni, Pd, Pt; B = Ga, In, Tl, Pb, Sb, Bi," *Journal of the Less Common Metals*, Vol. 17, No. 1, 1969, pp. 73-90.
      doi:10.1016/0022-5088(69)90038-1

[13] K. Schubert, S. Bhan, T. K. Biswas, K. Frank and E. K. Pandy, "Einige Strukturdaten Metallischer Phasen," *Naturwissenschaften*, Vol. 55, No. 11, 1968, pp. 542-543.
      doi:10.1007/BF00660131

[14] P. K. Srivastava, B. C. Giessen and N. J. Grant, "A Noncrystalline Pt-Sb Phase and Its Equilibration Kinetics," *Metallurgical Transactions*, Vol. 3, No. 4, 1972, pp. 977-988. doi:10.1007/BF02647675

[15] B. T. Mattias, "Superconducting Compounds of Non-superconducting Elements," *Physical Review*, Vol. 90, No. 3, 1953, p. 487. doi:10.1103/PhysRev.90.487

[16] A. Kjekshus and K. P. Walseth, "On the Properties of the Cr(1+x)Sb, Fe(1+x)Sb, Co(1+x)Sb, Ni(1+x)Sb, Pd(1+x) Sb, and Pt(1+x)Sb Phases," *Acta Chemica Scandinavica*, Vol. 23, 1969, pp. 2621-2630.
      doi:10.3891/acta.chem.scand.23-2621

[17] A. Kjekshus, "Redetermined Lattice Constants of PtP2, PtAs2, PtSb2, and alpha-PtBi2," *Acta Chemica Scandinavica*, Vol. 14, 1960, pp. 1450-1451.
      doi:10.3891/acta.chem.scand.14-1450

[18] F. Roessler and Z. Anorg, "Synthese Einiger Erzmineralien und Analoger Metallverbindungen Durch Auflösen und Krystallisierenlassen DErselben in Geschmolzenen Metallen," *Zeitschrift für Anorganische Chemie*, Vol. 9, No. 1, 1895, pp. 31-77. doi:10.1002/zaac.18950090108

[19] E. M. Jaeger and T. J. Popema, "VIII. La Détermination Exacte des Chaleurs Spécifiques à des Températures Élevées: Sur la Règle Additive des Chaleurs Atomiques des Métaux Dans Leurs Combinaisons Binaires," *Recueil des Travaux Chimiques des Pays-Bas*, Vol. 55, No. 6, 1936, pp. 492-517. doi:10.1002/recl.19360550606

[20] A.T. Dinsdale, SGTE Pure elements (unary) database, version 4.5, 2006. (private Communication, Unpublished)

[21] O. Redlich and A. T. Kister, "Algebraic Representation of Thermodynamic Properties and the Classification of Solutions," *Industrial & Engineering Chemistry*, Vol. 40, No. 2, 1948, pp. 345-347. doi:10.1021/ie50458a036

[22] Th. Massalski, "Binary Alloy Phase Diagrams," American Society for Metals, Metal Park, 1986.

[23] L. Thomassen, "Crystallization of Binary Compounds of Metals of Platinum Group II," *Zeitschrift fur Physik B: Condensed Matter*, Vol. 4, 1929, pp. 277-287.

[24] L. Thomassen, "Crystallization of Binary Compounds of Metals of Platinum Group," *Zeitschrift fur Physik B: Condensed Matter*, Vol. 2, 1929, pp. 349-379.

[25] B. Sundman, B. Jansson and J.-O. Andersson, "The Thermo-Calc Databank System," *Calphad*, Vol. 9, No. 2, 1985, pp. 153-190. doi:10.1016/0364-5916(85)90021-5

20

Ball Milling and Annealing of Co-50 at% W Powders

A. S. Bolokang[1,2*], M. J. Phasha[2*], D. E. Motaung[3]

[1]Department of Engineering Metallurgy, University of Johannesburg, Johannesburg, South Africa
[2]Transnet Rail Engineering, Pretoria, South Africa
[3]DST/CSIR Nanotechnology Innovation Centre, National Centre of Nano-Structured Materials,
Council for Scientific and Industrial Research, Pretoria, South Africa
Email: *Amogelang.bolokang@transnet.net, *majay_phasha@yahoo.com

ABSTRACT

Broadening and height reduction of X-ray diffraction peaks were observed after cold-pressing of unmilled Co-W powder mixture. It seems the effect of cold pressing has slightly reduced the lattice parameter of W from 3.165 to 3.143 Å. Consequent annealing of unmilled compacts yielded metastable phases. Upon 10 and 20 h ball milling of Co-W powder, no alloying was obtained. Although milling did not yield significant crystal changes in W and Co ground state structures, its effect is evident during subsequent annealing. An eta phase is obtained for the first time from unmilled-annealed Co-W powder mixture in the absence of interstitial elements like carbon, while the milled counterpart yielded the rhombohedral Co_7W_6-type phase with composition deviated from stoichiometric value.

Keywords: Ball Milling; Crystal Structure; XRD, Annealing; Co-W Powder

1. Introduction

Cobalt-tungsten (Co-W) alloy is a promising material that can be used for among others coatings and corrosion resistance applications. The alloy is often manufactured as thin films produced by deposition [1-4] or magnetron sputtering spanning [5]. Due to rapid cooling involved during deposition and sputtering, formation of novel metastable and amorphous phases is attained. Electrodeposited Co-W is a potential candidate to replace Co-Cr due to better mechanical properties such as high surface sliding hardness, wear resistance and good ductility [6]. In addition, the processing of Co-W system is environmentally friendly compared to Co-Cr. Some studies reveal that a deposited film of Co-W consists of a bi-phasic structure, mainly metastable hexagonal close-packed (HCP) and amorphous phases [6,7]. In some cases, only the existence of amorphous structure is observed [1,3,8]. To the best of our knowledge, the study on the influence of ball milling (BM) on Co-W powder mixture is still lacking. BM is a versatile solid-state powder synthesizing technique used to produce alloying of powders with reduced crystalline size [9,10]. The objective of the current study is to investigate the crystal structure of unmilled and milled Co 50 at% W powder mixtures after annealing.

2. Experimental Procedure

Commercial W and Co powders of 99.5% purity were used during ball milling and cold pressing experiments. The measured particle sizes has percentage particle distribution D_{50} = 32.40 and 5.42 μm for Co and W, respectively. Milling was conducted in a high-energy ball mill at 650 rpm and 20:1 ball to powder ratio for time intervals of 10 and 20 hours (h). In order to minimize contamination, milling was performed under inert atmosphere with no process control agent (PCA) added. The changes in powder particle morphology were analysed using the LEO 1525 field-emission scanning electron microscope (FE-SEM) coupled with a Robinson Backscatter Electron Detector (RBSD). Phase evolution was traced with a Panalytical X'pert PRO PW 3040/60 x-ray diffractometer (XRD) equipped with a Cu K_α monochromated radiation source, scanning from 20° to 90° (2θ) in 0.02° step size. The crystalline sizes (D) of unpressed and cold-pressed powders were estimated by Williamson-Hall (W-H) equation as follows:

$$\beta_{hkl} \cos\theta_{hkl} = \frac{k\lambda}{D} + 4\varepsilon \sin\theta_{hkl} \qquad (1)$$

where θ is the Bragg diffraction angle, D is the average crystalline size, ε is the average internal strain, k is a constant with a value of 0.9, λ = 0.154056 nm is the wavelength for Cu $K\alpha$ radiation and β is the diffraction

*Corresponding authors.

peak width at half maximum intensity. The average internal strain (ε) is estimated from the slope of $\beta\cos\theta$ versus $4\sin\theta$ linear plot, while the average crystalline (D) is estimated from the y-intercept. Annealing of powder compacts was carried out in a Carbolite tube furnace under flowing argon gas at 800˚C and1200˚C.

3. Results and Discussions

3.1. Unmilled Co-W Powders

Shown in **Figures 1(a)-(d)** is the SEM images of unmilled Co, W and milled Co-W powder mixture for 10 and 20 h, respectively. The Co particles appear coarse compared to those of W. The mixture of fine and large, flat-round particles is evident on both 10 and 20 h milled powders. This large particles could be due to agglomeration as a result of cold-welding during milling.

The XRD patterns of the Co-W compact after cold-pressing, 800˚C and 1200˚C annealing are presented in **Figures 2(a)-(c)**. It is evident from **Figures 2(a)** and **(b)** that the diffraction peaks belonging to body centered cubic (BCC) W are more intense compared to those representing HCP and FCC (face centered cubic) Co phases. In

addition, these peaks indicate broadening.

XRD peak widening is well known to be caused by the refinement of particles and sometimes also associated with amorphous phases. However, the lattice parameter of the cold-pressed compact is 3.143 Å, smaller than that of pure W (3.165 Å). It is likely that thereduction of the lattice parameter is due to surface deformation during cold-pressing. Moreover, it has been reported that cold pressing can promote structural change on Co [11,12]. Although low melting temperature elements such as tin (Sn) and tellurium (Te) mixture were alloyed by repeated cold-pressing process [13], it is not logical to expect high melting temperature metals such as Co and W to be alloyed by cold-pressing, but their surfaces might be cold-welded. From the XRD analyses of Co-W compacts annealed at 800˚C and 1200˚C shown in **Figures 2(b)** and **(c)**, respectively, it is apparent that the peak broadening observed after cold-pressing was reversed by annealing as shown by sharp peaks. This behaviour is could be attributed to the beginning of sintering effect. In addition to the retained BCC phase with lattice parameter of 3.143 Å after annealing at 800˚C, FCC Co with lattice parameter of 3.554 Å was detected, as presented in **Table 1**.

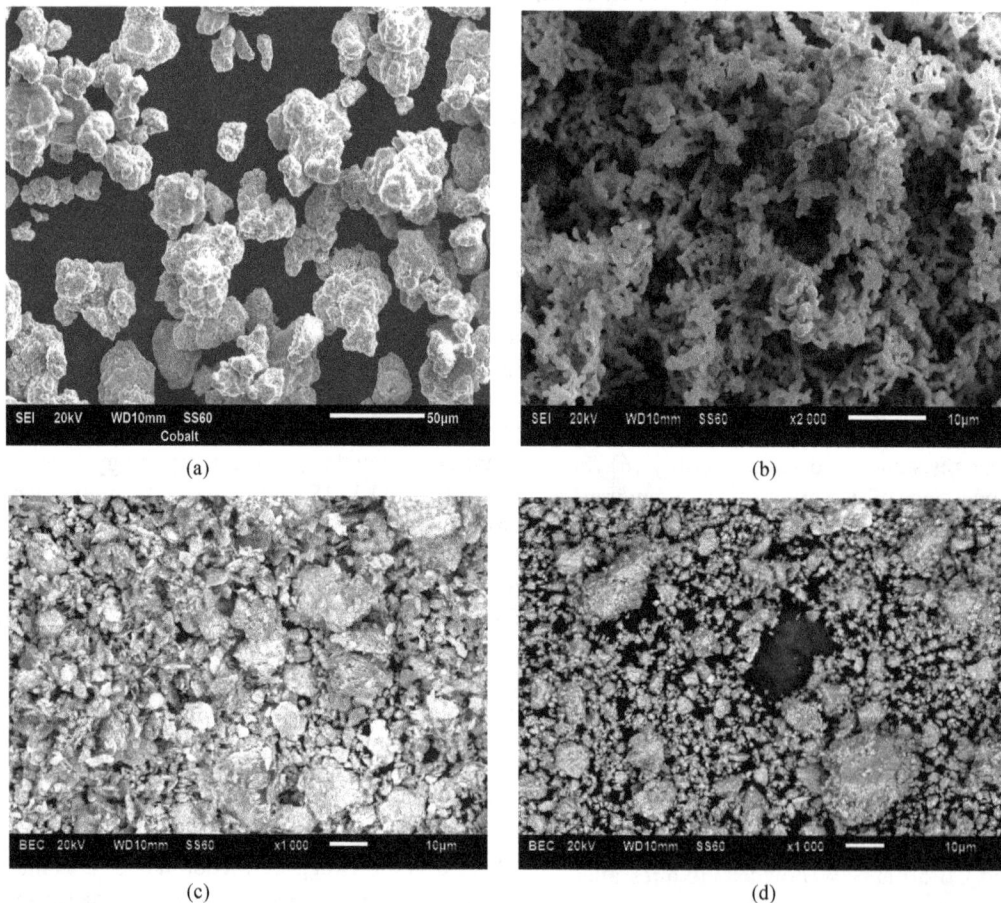

(a)

(b)

(c)

(d)

Figure 1. SEM morphology of unmilled (a) Co and (b) W unmilled powders, and of powder mixture after 10 h BM (c) and 20 h BM (d).

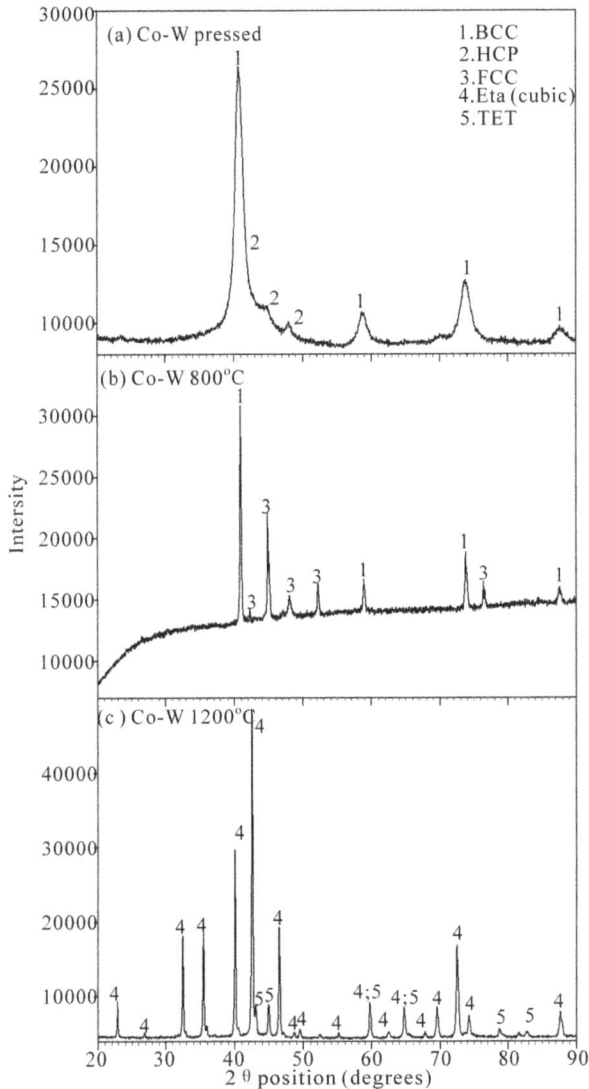

Figure 2. XRD patterns of (a) cold-pressed Co-W powder (b) sintered at 800°C and (c) 1200°C.

Table 1. Experimental XRD data of cold pressed, ball milled and annealed Co-W powders.

Sample condition	Crystal Structure	Lattice parameter (Å)	
		a	c
Co-W pressed	BCC W HCP Co	3.143 2.514	4.105
Co-W 800°C	BCC W FCC Co	3.143 3.554	
Co-W 1200°C	Eta TET	11.090 2.850	3.091
Co-W 10 h BM	BCC W HCP Co	3.147 2.514	4.105
Co-W 20 h BM	BCC W HCP Co	3.165 2.514	4.105
Co-W 10 h BM 1200°C	RHL HCP	4.751 - 4.905 2.751 - 2.764	25.670 - 23.787 4.282 - 4.127
Co-W 20 h BM 1200°C	RHL HCP	4.738 - 5.021 2.728	25.850 - 25.890 4.226

A similar FCC phase of about 3.506 Å was obtained in milled Co annealed at 800°C in previous study [14]. Upon annealing Co-W compacts at 1200°C, formation of tetragonal (TET) and FCC superstructure called eta (η) phase with space group Fd-3m # 227 were detected, as indicated in **Figure 2(c)**. Similar phases were obtained and reported as follows: 1) TET phase formed from ball milled W powder compacts annealed at 1200°C [15], 2) eta phase obtained from milled and 800°C annealed W [14], W milled and annealed at 730°C [16], milled W-Ni powder mixture annealed at 730°C and 1400°C [16] as well as from mixture of separately milled elemental Co and W powders compacted and annealed at 800°C [14]. The corresponding formation mechanisms were also provided. Although η-phase has been detected in pure W [14,16], pre-milled Co-W mixture [14], Ni-W [17] and Ni-Mo materials [18], it does not exist under equilibrium conditionsin any of the above systems including the Co-W, Ni-W and Ni-Mo systems in the absence of carbon. Furthermore, it is well known that this phase is unstable and dissociates at high temperatures [14,19]. The obtained lattice parameter (11.090 Å) of η-phase is larger than that found after reduction of cobalt-tungsten oxide (10.846 Å) [20]. The phases obtained in the current study are not found in Co-W equilibrium phase diagram shown in **Figure 1** in [14]. They are therefore regarded as metastable or rather intermediate.

3.2. Ball milled Co-W Powders

The XRD patterns of 0, 10 and 20 h ball milled Co-W powders are shown in **Figures 3(a)-(c)**. In comparison with the unmilled, the peaks of 10 h sample are slightly broader and their intensity significantly reduced. Almost similar to the compaction effect, the lattice parameter of BCC W was decreased from 3.165 to 3.147 Å with the estimated crystallite size of about 60 nm, while lattice parameters of HCP Co remained unchanged. Upon 20 h of BM, the peaks were further broadened and reduced in height. Surprisingly, the undeformed lattice of pure BCC W (3.165 Å) was recovered. This could imply that BM just reversed the deformation on the surface of W particles observed on the 10 h ball-milled powder. Similarly, crystallite size of ~60 nm was calculated. Since the crystal structure of Co did not change either, it thus follows that no alloying was induced by BM. For alloying to be achieved, high energy is required considering the melting temperatures of both Co and W. Physical properties such as thermal expansion coefficient, melting temperature, activation energy, elastic modulus and yield strength which are 13.4×10^{-6} C^{-1}, 1495°C, $Q_B = 117$ KJ/mol, 211 GPa and 345 MPa for Co, and 4.5×10^{-6} C^{-1}, 3410°C, $Q_B = 385$ KJ/mol, 411 GPa, and 550 MPa in the case of W, respectively [21], also play an important role in order for alloying to occur at approximately room temperature.

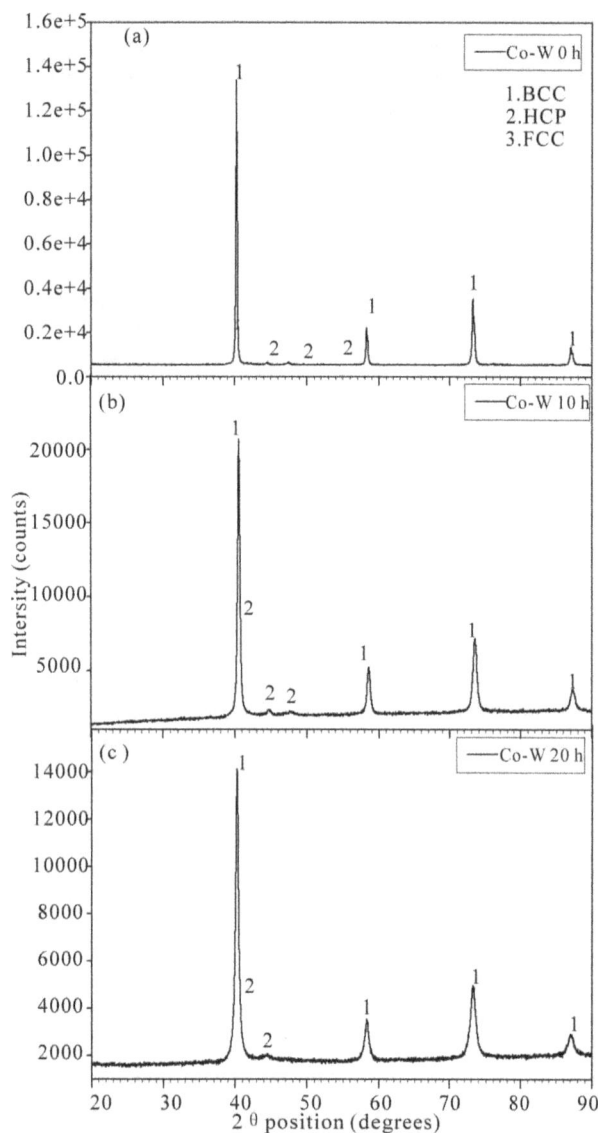

Figure 3. XRD patterns of (a) 0, (b) 10, and (c) 20 h ball milled Co-W powders.

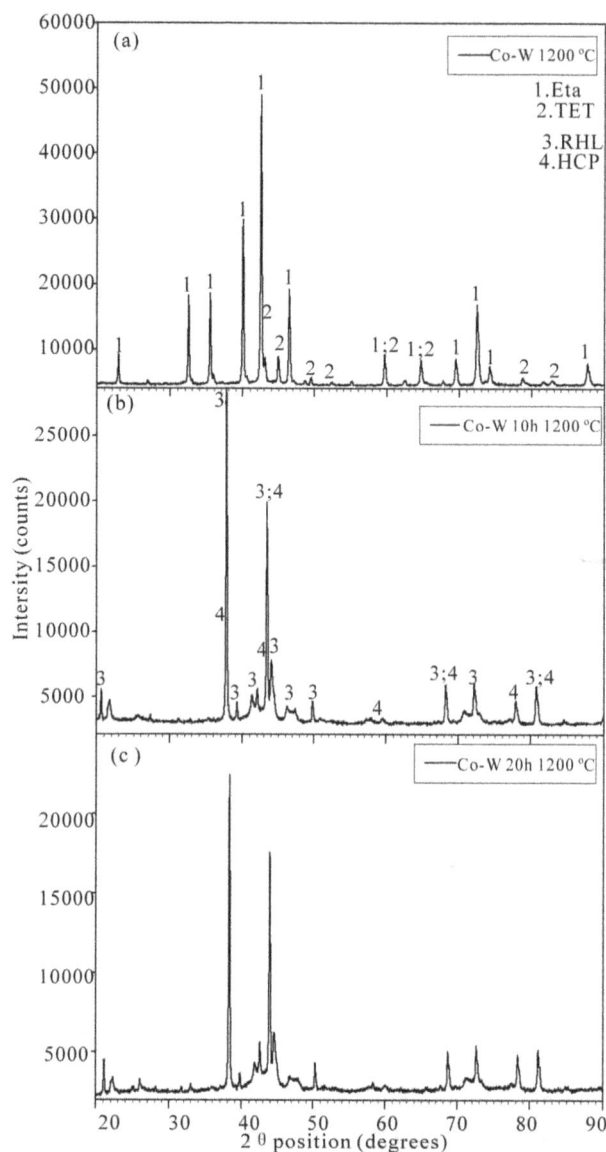

Figure 4. XRD patterns of 1200°C annealed compacted powder unmilled (a), after 10 h (b) and 20 h (c) ball milling.

Other than quicker grain refinement, the above properties imply that W is expected to have higher resistance to deformation during BM compared to Co.

Figure 4 presents the XRD patterns of 1200°C compacts, though the results for unmilled powder was only shown for comparison since it has been discussed in **Figure 2(c)**. The XRD patterns of annealed 10 and 20 h ball milled compacts show similar phases as shown in **Figures 4(b)** and **(c)**, though the peak intensities remains slightly higher for the 10 h milled compacts. Furthermore, the detected phases were alike in crystal structure, HCP and RHL (rhombohedral), though their corresponding lattice parameters varied as indicated in **Table 1**. This variation from the reported lattice parameters of equilibrium phases such as RHL Co_7W_6 (a = 4.751, c = 25.670 Å) is attributed to deviation from stoichiometric composition

as a consequence of BM and subsequent annealing. A similar HCP phase was reported for milled W powder annealed at 730°C [16], 1000°C [14] and W-Ni compact milled-annealed at 1400°C [16]. The agility of BM creates fresh surfaces of fine powder particles which promotes high diffusion rate during high temperature annealing. To date, solid state alloying of metal powders at lower temperatures (induced by BM process) is still a controversial topic, because powders produced by BM have to be shaped and sintered into products. As a result, thermal treatment dictates the final structure and properties of the product. However, mechanism of alloying elemental powders during BM is vastly different to those milled under the presence of interstitial elements such as C, N, O to form carbides, nitrides and oxides. In pure elementals, welding of fresh surfaces is highly possible while diffusion may be the process for the formation of carbides, nitrides and oxides. Therefore, actual alloying of binary alloy occurs by annealing or sintering of powders, while BM influences the process by changing the surface properties of the particles.

4. Conclusion

Despite broadening and shortening of XRD peaks, no alloying was observed upon 10 and 20 h BM of Co-W powder mixtures. Even if BM did not yield significant crystal changes in W and Co ground state structures, its effect is evident during subsequent annealing. An eta phase is obtained for the first time from unmilled-annealed Co-W powder mixture in the absence of interstitial elements like carbon, while the milled counterpart yielded the rhombohedral Co_7W_6-type phase with composition deviated from stoichiometric value. Larger deviation corresponded with longer milled products. In addition, HCP Co-W solid solutions with slightly different lattice parameters were obtained from annealed powders milled for both periods.

REFERENCES

[1] M. Donten and Z. Stojek, "Pulse Electroplating of Rich-in-Tungsten Thin Layers of Amorphous Co-W Alloys," *Journal of Applied Electrochemistry*, Vol. 26, No. 6, 1996, pp. 665-672. doi:10.1007/BF00253466

[2] D. Z. Grabco, I. A. Dikusar, V. I. Petrenko, E. E. Harea and O. A. Shikimaka, "Micromechanical Properties of Co-W Alloys Electrodeposited under Pulse Conditions," *Surface Engineering of Applied Electrochemistry*, Vol. 43, No. 1, 2007, pp. 11-17. doi:10.3103/S1068375507010024

[3] S. S. Grabchikov and A. M. Yaskovich, "Effect of the Structure of Amorphous Electrodeposited Ni-W and Ni-Co-W Alloys on Their Crystallization," *Russian Metallurgy*, Vol. 2006, No. 1, 2006, pp. 56-60. doi:10.1134/S0036029506010101

[4] Z. Guo, X. Zhu, D. Zhai and X. Yang, "Electrodeposition of Ni-W Amorphous Alloy and Ni-W-SiC Composite Deposits," *Journal of Materials Science and Technology*, Vol. 16, No. 2, 2000, pp. 323-326.

[5] C. Borgia, T. Scharowsky, A. Furrer, C. Solenthaler and R. Spolenak, "A Combinational Study on the Influence of Elemental Composition and Heat Treatment on the Phase Composition, Microstructure and Mechanical Properties of Ni-W Alloy Thin Films," *Acta Materialia*, Vol. 59, No. 1, 2011, pp. 386-399. doi:10.1016/j.actamat.2010.09.045

[6] B. Yang, G. Qin, W. Pei, Y. Ren, N. Xiao and X. Zhao, "Abnormal Saturation Magnetization Dependency on W Content for Co-W Thin Films," *Acta Metalligica Sinica*, Vol. 23, No. 1, 2010, pp. 8-12.

[7] M. Mulukutla, V. K. Kommineni and S. P. Harimkar, "Pulsed Electrodeposition of Co-W Amorphous and Crystalline Coatings," *Applied Surface Science*, Vol. 258, No. 7, 2012, pp. 2886-2893. doi:10.1016/j.apsusc.2011.11.002

[8] K. Wikiel and J. Osteryoung, "Voltammetric Study of Plating Baths for Electrodeposition of Co-W Amorphous Alloys," *Journal of Applied Electrochemistry*, Vol. 22, No. 6, 1992, pp. 506-511. doi:10.1007/BF01024089

[9] S. Bolokang, C. Bangayayi and M. Phasha, "Effect of C and Milling Parameters on the Synthesis of WC Powders by Mechanical Alloying," *International Journal of Refractory Metals and Hard Materials*, Vol. 28, No. 2, 2010, pp. 211-216. doi:10.1016/j.ijrmhm.2009.09.006

[10] K. F. Kobayashi and H. Kawaguchi, "Amorphization of Al-Cr Atomized Powder by Mechanical Alloying," *Materials Science and Engineering A*, Vol. 181-182, 1994, pp. 1253-1257. doi:10.1016/0921-5093(94)90841-9

[11] A. Karin, A. Bonefačić and D. Duževič, "Phase Transformation in Pressed Cobalt Powder," *Journal of Physics F: Metal Physics*, Vol. 14, No. 11, 1984, pp. 2781-2786. doi:10.1088/0305-4608/14/11/030

[12] A. S. Bolokang, M. J. Phasha, D. E. Motaung and S. Bhero, " Effect of Mechanical Milling and Cold Pressing on Co Powder," *Journal of Metallurgy*, Vol. 2012, 2012, Article ID: 290873. doi:10.1155/2012/290873

[13] S. D. De la Torre, K. N. Ishihara and P. H. Shingu, "Synthesis of SnTe by Repeated Cold-Pressing," *Materials Science and Engineering A*, Vol. 266, No. 1-2, 1999, pp. 37-43. doi:10.1016/S0921-5093(99)00043-X

[14] A. S. Bolokang, M. J. Phasha, D. E. Motaung and S. Bhero, "Metastable Phases in the Co-W System Traced from Elemental Co and W Powders," *International Journal of Refractory Metals and Hard Materials*, Vol. 31, 2012, pp. 274-280. doi:10.1016/j.ijrmhm.2011.12.012

[15] A. S. Bolokang, M. J. Phasha, K. Maweja and S. Bhero, "Structural Characterization of Mechanically Milled and Annealed Tungsten Powder," *Powder Technology*, Vol. 225, 2012, pp. 27-31. doi:10.1016/j.powtec.2012.03.028

[16] K. Maweja, M. J. Phasha and L. J. Choenyane, "Thermal Stability and Magnetic Saturation of Annealed Nickel-Tungsten and Tungsten Milled Powders," *International Journal of Refractory Metals and Hard Materials*, Vol. 30, No. 1, 2012, pp. 78-84. doi:10.1016/j.ijrmhm.2011.07.005

[17] C. Borgia, T. Scharowsky, A. Furrer, C. Solenthaler and R. Spolenak, "A Combinational Study on the Influence of Elemental Composition And Heat Treatment on the Phase Composition, Microstructure And Mechanical Properties of Ni-W Alloy Thin Films," *ActaMaterialia*, Vol. 59, No. 1, 2011, pp. 386-399. doi:10.1016/j.actamat.2010.09.045

[18] D. Oleszak, V. K. Portnoy and H. Matyja, "Formation of Metastable Phases in Ni-43·5 at % Mo Powder Mixtures during Mechanical Alloying and after Heat Treatment," *Philosophical Magazine*, Vol. 76, No. 4, 1997, pp. 639-649. doi:10.1080/01418639708241130

[19] N. A. Dubrovinskaia, L. S. Dubrovinsky, S. K. Saxena, M. Selleby and B. Sundman, "Thermal Expansion and Compressibility of Co_6W_6C," *Journal of Alloys and Compounds*, Vol. 285, No. 1-2, 1999, pp. 242-245.

doi:10.1016/S0925-8388(98)00932-3

[20] P. Bracconi and L. C. Dufour, "Investigation of Cobalt (II)-Tungsten (VI)-Oxide Reduction in Hydrogen," *Metallurgical Transaction A*, Vol. 7, No. 3, 1976, pp. 321-327.

[21] R, German, "Powder Metallurgy Science," Metal Powder Industry Federation, Princeton, 1994.

Microbial Recovery of Manganese Using *Staphylococcus epidermidis*

Alok Prasad Das[1*], Lala Behari Sukla[2], Nilotpala Pradhan[2]

[1]Centre of Biotechnology, Siksha 'O' Anusandhan University, Bhubaneswar, India
[2]Institute of Minerals and Materials Technology, Bhubaneswar, India
Email: *alok1503@gmail.com

ABSTRACT

Manganese minerals are widely distributed throughout the globe. The most important industrial uses of Mn are in the manufacture of steel, non-ferrous alloys, carbon-zinc batteries and some chemical reagents. Microbial recovery of manganese from low grade manganese ores using bioleaching was investigated in this paper. A bacterial strain, *Staphylococcus epidermidis* (*MTCC*-435) was collected from microbial type culture collection, IMTECH Chandigarh and used for the experiment. The experimental results for bioleaching with *S. epidermidis* showed that under pH 5.5, particle size −150 μm, pulp density 10%, temperature 35°C and agitation 200 rpm, about 80% of Mn was recovered within 20 days of incubation.

Keywords: Manganese; *Staphylococcus epidermidis*; Bioleaching

1. Introduction

The necessity for manganese ore has increased due to the escalating shortage of natural resources and increase in the manufacture of Steel, dry cell batteries and some portion of manganese oxide [1]. The 96% of global production of manganese today is from barely 7 countries viz. CIS, RSA, Brazil, Gabon, Australia, China and India in decreasing order of tonnages raised annually. The global resource base is close to 12 billion tonnes including Indian reserve of about 240 million tonnes. Indian manganese ores are preferred by many as they are generally hard, lumpy and amenable to easy reduction. The high grade ores with manganese concentrations in excess of 35% are preferred industrially, and the low grades with manganese concentration less than 35% are presently ignored because extraction methods applied to low grade manganese ore are not economically viable [2].

The development of bioleaching technology focuses on achieving effective recovery of precious metals by improving the efficiency of bioleaching microorganisms [3,4]. This relates to manganese solubilising activities of manganese-oxidizing microbes and the speciation of intermediate compounds formed during bioleaching processes [5]. The effectiveness of bioleaching is also highly dependent on the physical, chemical and biological factors in the system. The maximum yield of metal leaching may be achieved when these parameters are considered and optimized collectively [6]. In the present research work, bioleaching technology was applied for manganese recovery from low grade manganese ore.

2. Materials and Methods

2.1. Collection of Ore

The low grade manganese ore collected from Ferromanganese mines, Keonjhar district in Odisha, India. The large pieces were crushed in a jaw-crusher separately. Then, all pieces were ground to relatively smaller particles using ball mill machine. The physical and chemical properties of samples were characterized. The presence of other heavy metals was estimated by Atomic Absorption Spectroscopy methods (A. Analyst 200. Perkin Elmer). The Mn content of the ore was 20%.

2.2. Microorganism and Culture Condition

A bacterial strain, *Staphylococcus epidermidis* (*MTCC*-435) was procured from microbial type culture collection, IMTECH Chandigarh. The strain was subculture on nutrient agar medium plates containing 50 mg of MnO_2/ml to the medium. The growth of the bacterial colony was observed after 24 h of incubation at 30°C. Isolated colonies picked up with sterilized wire loop and streaked on nutrient agar medium plate containing 100 mg of MnO_2/ml. Leaching study was carried out under shaking condition

---
*Corresponding author.

inside an incubator shaker. At five day interval, pH of each flask was noted under aseptic condition & manganese estimation was done by titration method.

2.3. Effect of Process Parameters

Process optimization experiments for temperature, pH, Particle size, pulp density and carbon concentration were carried out. Leaching experiments were done in 250 ml flask containing 100 ml of sterilized minimal salt medium for a period of 20 days. For pH optimization, the five standards used were 4.5, 5.5, 6.5, 7.5 and 8.5. The respective pH was adjusted with 0.5N NaOH and 0.5N HCl. For temperature optimization the five standard temperatures considered were 25°C, 30°C, 35°C, 40°C and 45°C. Different size fractions of manganese ore ranging from +4 mm, −4 mm, +1 mm, −1 mm, +150 μm and −150 μm at 2% pulp density were considered. Pulp densities ranging from 2%, 4%, 6%, 8% and 10 % (w/v) of the sieved particle size −150 μm were taken for the experiment. Aseptic conditions were maintained throughout the experiment. Uninoculated media containing manganese ore served as control. Samples were drawn at regular intervals and were analyzed by titration method to determine the percentage of manganese leached in the liquid medium.

2.4. Manganese Bioleaching

The low grade manganese ore was used for these experiments. Effect of temperature and pH variation was studied. Leaching experiments were done in 250 ml flask containing 90 ml of sterilized Bromfield medium (BM) containing ammonium sulphate 0.027 g, dihydroxy potassium phosphate 0.025 g, yeast extract 0.1 g and sucrose 10 g.

3. Results and Discussion

3.1. Ore Characteristics

The moisture content of the collected ore was estimated to be 15%. The heavy metal content of the ore was estimate to be Fe (570 g·kg^{-1}), Si (147 g·kg^{-1}), Mn (210 g·kg^{-1}) and Zn (146 g·kg^{-1}). The phase of Mn and Fe in the collected are sample are in their oxide forms.

3.2. Process Parameters Optimization

Efficient Mn bioleaching was observed at 5.5 pH and temperature 30 - 35°C (**Figures 1(a)** and **(b)**). Interpretation of the experimental data obtained showed that there was significant rise in the leaching percentage (up to 68%) by the *Staphylococcus epidermidis* sp. Best manganese leaching solubilisation was observed at −150 μm, it was assumed that manganese extraction yield was maximum (64%) for fine particle size and due to the accessibility of the reaction areas with the particle size to the microbial strain (**Figure 2(a)**). With different pulp density of manganese ore, the highest dissolved manganese concentration was achieved (44%) with 2% (w/v). The results suggest that with increase in pulp density manganese leaching efficiency decreased which may be due to unavailability of nutritional stress to the microorganism (**Figure 2(b)**).

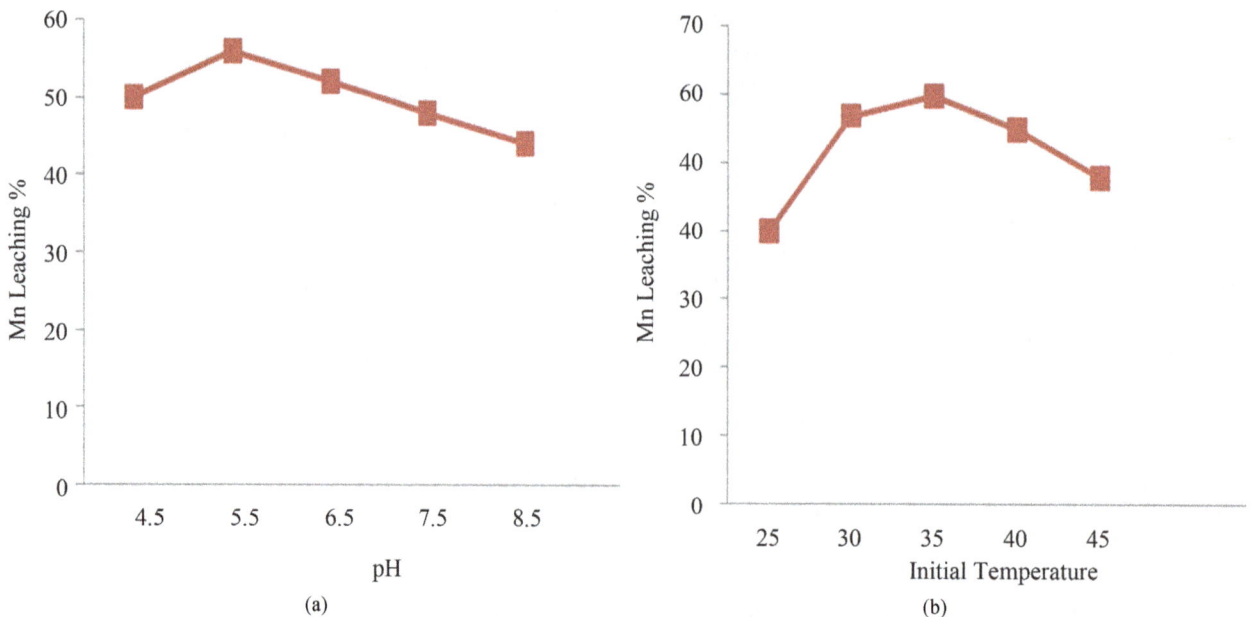

Figure 1. (a) and (b) pH and temperature optimization for Mn leaching.

3.3. Mn Bioleaching with *Staphylococcus epidermidis* (*MTCC*-435)

Manganese bioleaching efficiency of *S. epidermidis* was considerably affected by process parameters like initial pH, and temperature. At optimum pH 5.5, particle size −150 μm, pulp density 10%, temperature 35°C and agitation 200 rpm, about 80 % Mn bioleaching was achieved in 20th day with 10% sucrose as carbon source (**Figure 3**).

Best Mn bioleaching was evidenced at 5.5 pH. Mn leaching and pH relationship was not surprising because microbial production of organic acid lowers the media pH. Organic acid act as proton source helping metal leaching [7]. For optimum temperature it can be said that bacteria leaches best at its optimum growth temperature which is 30°C - 37°C. Higher temperatures (42°C) significantly reduced bacterial growth and manganese

bioleaching due to loss of viability or metabolic activity of cells [8]. Interpretation of the experimental data obtained showed that there was significant rise in the leaching percentage (up to 80%) by the *S. epidermidis* sp up to 20th day.

4. Conclusion

In this study, the bacterial species *S. epidermidis* was selected for its potential for manganese bioleaching. *S. epidermidis* is not usually pathogenic. The maximum recovery efficiency of Mn was 80% by the selected bacterial species at ph 5.5, particle size −150 μm, pulp density 10%, temperature 35°C and agitation 200 rpm with sucrose as carbon source after 20 day of incubation. The results suggest potential application of this species for environmental clean up from industrial and mining waste and act as an alternative to pyrometallurgical and

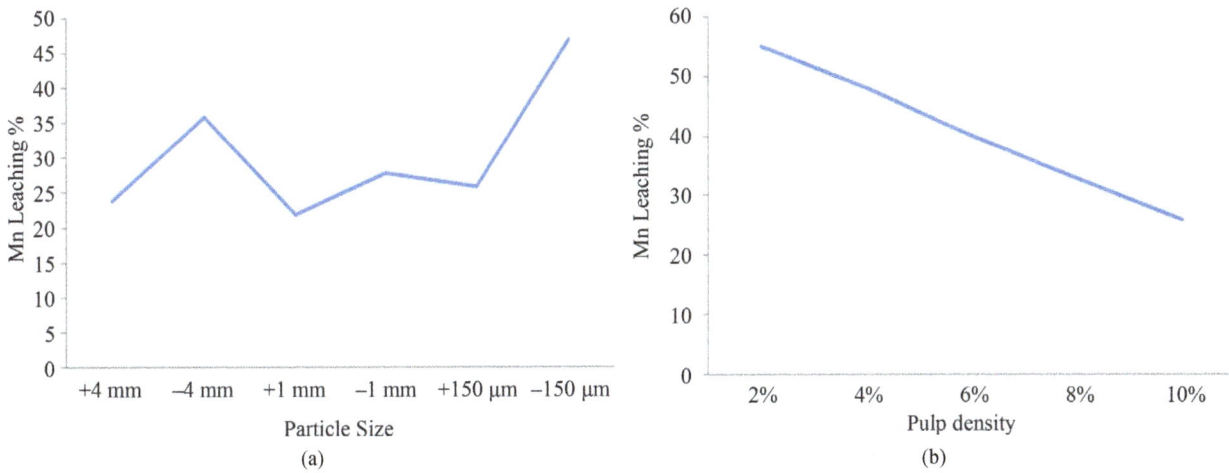

Figure 2. (a) and (b) particle size and pulp density optimization for Mn leaching.

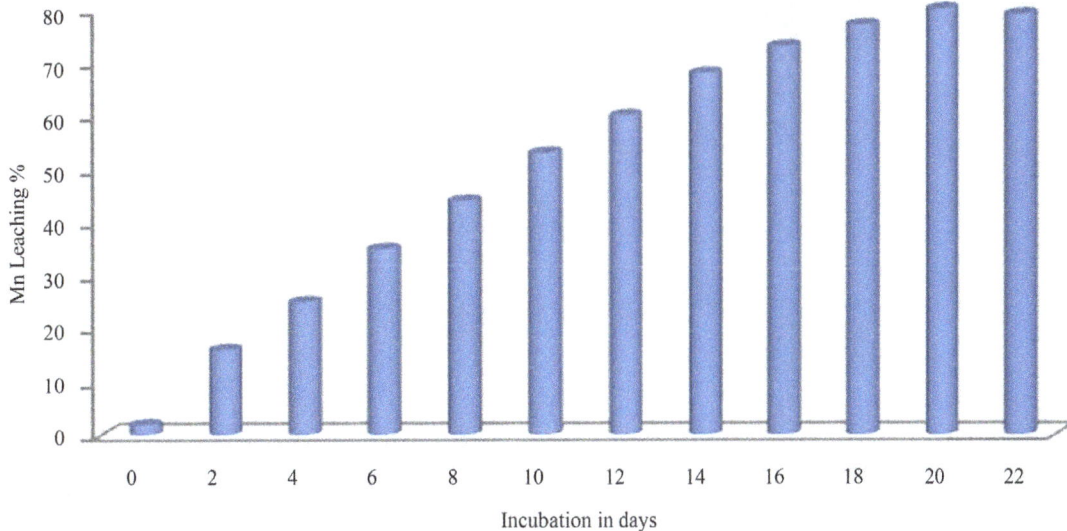

Figure 3. Mn leaching kinetics at initial pH- 5.5 particle size −150 μm, pulp density 10%, temperature 35°C and agitation 200 rpm.

hydrometallurgical techniques, to recover metals from low grade manganese ore, spent batteries and heavy metal polluted soil, economically.

REFERENCES

[1]   B. P. Xin, D. Zhang, X. Zhang, W. Feng and L. Li, "Bio-leaching Mechanism of Co and Li from Spent Lithium-Ion Battery by the Mixed Culture of Acidophilic Sulfur Oxidizing and Iron-Oxidizing Bacteria," *Bioresource Technology*, Vol. 100, No. 24, 2009, pp. 6163-6169. doi:10.1016/j.biortech.2009.06.086

[2]   C. Acharya, R. N. Kar and L. B. Sukla, "Studies on Reaction Mechanism of Bioleaching of Manganese Ore," *Minerals Engineering*, Vol. 16, No. 10, 2003, pp. 1027-1030. doi:10.1016/S0892-6875(03)00239-5

[3]   A. P. Das, L. B. Sukla, N. Pradhan and S. Nayak, "Manganese Biomining: A Review," *Bioresource Technology*, Vol. 102, No. 16, 2011, pp. 7381-7387. doi:10.1016/j.biortech.2011.05.018

[4]   J. L. Xia, Y. Yang, H. He, X. J. Zhao, C. L. Liang, L. Zheng, C. Y. Ma, Y. D. Zhao, Z. Y. Nie and G. Z. Qiu, "Sulfur Oxidation Activities of Pure and Mixed Thermo-philes and Sulfur Speciation in Bioleaching of Chalcopy-rite," *Bioresource Technology*, Vol. 102, No. 4, 2011, pp. 3877-3882. doi:10.1016/j.biortech.2010.11.090

[5]   H. B. Zhou, W. M. Zeng, Z. F. Yang, Y. J. Xie and G. Z. Qiu, "Bioleaching of Chalcopyrite Concentrate by a Moderately Thermophilic Culture in a Stirred Tank Reactor," *Bioresource Technology*, Vol. 100, No. 2, 2009, pp. 515-520. doi:10.1016/j.biortech.2008.06.033

[6]   K. Bosecker, "Bioleaching: Metal Solubilization by Microorganisms," *FEMS Microbiology Reviews*, Vol. 20, No. 3-4, 1997, pp. 591-604. doi:10.1111/j.1574-6976.1997.tb00340.x

[7]   Y. X. Huang, J. B. Cao, X. M. Li, Q. Yang, H. J. Huang, X. Liu and H. Yang, "Study on the Bioleaching Mechanism of Manganse (II) from Manganese-Electrolytic Residue by Manganese-Resistant Strain *Fusarium* sp.," *Huan Jing Ke Xue*, Vol. 32, No. 9, 2011, pp. 2703-2739.

[8]   L. S. Flavio, V. A. Oliveira, D. Guimarães, D. Adelson and A. L. Versiane, "High-Temperature Bioleaching of Nickel Sulfides: Thermodynamic and Kinetic Implications," *Hydrometallurgy*, Vol. 105, No. 1-2, 2010, pp. 103-109. doi:10.1016/j.hydromet.2010.08.006

22

A Statistical Method for Determining the Best Zinc Pregnant Solution for the Extraction by D2EHPA

Hossein Kamran Haghighi[1], Davood Moradkhani[2], Mohammad Mehdi Salarirad[1]

[1]Department of Mining and Metallurgical Engineering, Amirkabir University of Technology, Tehran, Iran
[2]Faculty of Engineering, University of Zanjan, Zanjan, Iran
Email: h.kamran.h@aut.ac.ir, dmoradkhani@gmail.com, salarim@yahoo.com

ABSTRACT

The application of D2EHPA in zinc solvent extraction has extensive background. To utilize more effectively, response surface methodology was used to optimize the concentration condition of zinc pregnant solution (ZPL) extracted by D2EHPA. In the current research, zinc, iron and manganese extraction along with separation factor of zinc-iron (Sf (Zn-Fe)) and zinc-manganese (Sf (Zn-Mn)) were considered as the response values. The optimal ZPL conditions extracted with 30% D2EHPA as the extraction solvent were as follows: Zn 21.96 g/L, Fe 382.57 ppm, Mn 1 g/L, Sf (Zn-Fe) 8.26 and Sf (Zn-Mn) 1529.82. In addition, it was found that the iron and manganese concentration were the most effective factors affecting the zinc and manganese extraction, respectively.

Keywords: D2EHPA; Response Surface Methodology; Pregnant Solution; Optimization

1. Introduction

The extraction of zinc sulfate with di-2-ethylhexyl phosphoric acid (D2EHPA) is a well-known route in zinc purification industry. According to the literatures, extraction of zinc increases from 10% to ca. 99% with increasing pH from 0.5 to 2.5 and increasing D2EHPA concentration from 5% to 40% (w/w) [1]. It is obvious that enhancement of extractant makes distribution coefficient increase; however, it is noteworthy that the high cost of the organic extractant limits the usage of D2EHPA to less than 30% (v/v or w/w) [2]; therefore, the best composition of D2EHPA in industrial zinc solvent extraction is 30% (v/v) dissolved in kerosene.

Mehdi Abad lead and Zinc mine located in Yazd, Iran with the fixed capacity of 200 M tons of sulfur and oxide ore is one of the greatest lead and zinc mines in the world. The investigations reveal that manganese and iron are the main and major impurities of Mehdi Abad ore, which consequently come to leach solution and associate with zinc ion. Therefore, evaluating the optimized concentration of impurities for solvent extraction process plays the significant role in the leaching and pre-concentration steps. These impurities have undesirable effect on the process. For instance, Mn^{2+} ions, oxidized anodically to MnO_4^- ions, depolarize the H^+ ions discharge and thus reduce the current efficiency for zinc deposition [3-7].

Furthermore the iron constitutes a severe impurity in zinc solution and must be removed before electrolysis [8]. Implementing iron (III) solvent extraction into the zinc roast-leach-electrowin flowsheet as a means of iron rejection has been under consideration for at least two decades [9].

Response surface methodology is used to reduce the number of assays necessary to optimize the process and to collect results more precise than those obtainable by traditional full factorial designs [10,11]. Accordingly, RSM has been increasingly employed to optimize solvent extraction process. However, there is little information that shows which concentration of ZPL can be optimally extracted by D2EHPA. Therefore, the optimization condition of ZPL in detail which is extracted optimally by 30% (v/v) D2EHPA is the aim of this report. In the present research, the best concentrations of iron, manganese and zinc concentration, which are significant factor in Mehdi Abad ore, were found. Furthermore, the interactions effects between ions and the most effective factors on extractions were investigated.

2. Experiment

2.1. Reagents

Analytical grade inorganic reagents used in the experiments have been illustrated in **Table 1**. The synthetic solutions were prepared with the chemicals at the target concentrations and are presented in **Table 2**. The extractant, D2EHPA, was provided from BDH in England. It was dissolved in the industrial kerosene from Tehran Refinery Company, Iran as the diluent. The metal ion concentrations in the solutions were analyzed by Perkin-Elmer AA300 model atomic absorption spectro-photometer.

2.2. Procedure of Extraction

The extraction experiments were carried out in mechani-

call agitated and thermostatic beakers. In each experiment, 50 mL of the solution containing various zinc, iron and manganese concentrations (see **Table 3**) and 50 mL of the extractant were agitated by a magnetic stirrer at a constant rate. The pH of the solution was adjusted to 2.5 by sulfuric acid and hydrogen hydroxide. After agitating the beakers for 10 min at equilibrium state, the organic phase was separated from the aqueous phase in a separator funnel. After separation, the concentrations of ions in the aqueous phase were analyzed by Perkin-Elmer AA300 model atomic absorption spectrophotometer. Concentration of metal ions calculation in the organic

Table 1. Inorganic reagents used in the experiments.

Solution/Application	Component	Supplier	Prepared Concentration
Aq. Feed	$MgSO_4 \cdot H_2O$	Fisher	See **Table 2**
Aq. Feed	$FeSO_4 \cdot 7H_2O$	Merck	See **Table 2**
Aq. Feed	$ZnSO_4 \cdot H_2O$	Merck	See **Table 2**
Aq. Feed	H_2O_2	Mojallali	3 cc per liter
pH adjusting	H_2SO_4	Mojallali	98%
pH adjusting	NaOH	Mojallali	36 %

Table 2. The coded values and corresponding actual values of the optimization parameters.

Factor	Name	Units	Type	Low Actual	High Actual
A	Zn	g/L	Numeric	15	60
B	Fe	ppm	Numeric	10	1000
C	Mn	g/L	Numeric	1	5

Table 3. The coded, experimental and predicted values for RSM design using D2EHPA as solvent.

Run	Factor 1 A: Zn g/L	Factor 2 B: Fe ppm	Factor 3 C: Mn g/L	Resp. 1 %E Zn	Resp. 2 %E Fe	Resp. 3 %E Mn	Resp. 4 Sf (Zn-Fe)	Resp. 5 Sf (Zn-Mn)
1	38.8	150.25	2.20225	100	100	31.88784	0.258231	82874.33
2	13.1625	10.83	0.9145	100	100	78.13013	1.215394	3684.115
3	15.36	754.3	1.055	99.86784	100	64.52133	0.001002	415.5133
4	15	10	5	99.8432	98.6	61.98	9.041147	390.6007
5	23.765	155.4	0.959	99.91601	99.10553	38.18561	10.73688	1925.76
6	13.11	11.17	4.354	87.12433	99.99991	99.45361	6.06E-06	0.037175
7	40.65	45	2.946	92.61993	99.99978	66.05567	2.79E-05	6.449126
8	15.81	494.3	3.276	93.52309	99.98988	99.42643	0.001461	0.083297
9	44.65	561.9	0.6347	63.91937	99.61559	9.878683	0.006836	16.1617
10	44.65	561.9	0.6347	63.91937	99.61559	9.878683	0.006836	16.1617
11	24.07	750.1	4.921	81.79477	99.94267	17.57773	0.002577	21.06741
12	28.1	7.29	1.9926	87.16014	99.98601	75.27351	0.00095	2.229862
13	32.82	764.2	2.565	80.34735	99.91364	10.72125	0.003534	34.04499
14	15.02	733.8	0.921	99.9674	99.91823	55.23344	2.509152	2485.134
15	56.2	14.58	0.9963	93.58007	99.993	50.54702	0.00102	14.261
16	28.175	394.1	2.537	83.9929	99.92641	8.671659	0.003864	55.26286
17	13.1625	10.83	0.9145	100	100	100	ignored	ignored

phase was carried out according to the concentrations of ions in the aqueous phase.

2.3. Experimental Design of RSM

To determine the optimal combination of extraction variables for the extraction ions, response surface method (RSM) was used. **Table 2** shows the coded parameters and their levels, and **Table 3** illustrates the coded, experimental and predicted values. As seen in **Table 3**, three factors (*i.e.* concentrations of three ions) as the inputted data were used to model the extraction. The values for the extraction percent of zinc, iron, manganese (%E Zn, %E Fe and %E Mn), separation factor of zinc-iron (Sf (Zn-Fe)) and zinc-manganese (Sf (Zn-Mn)) in each trial were average of duplicates. Based on the experimental data, regression analysis was done and fitted into the quadratic model as shown in Equation (1).

$$Y = A_0 + \sum_{i=1}^{k} A_i X_i + \sum_{i=1}^{k} A_{ii} X_i^2 \\ + \sum_{i=1}^{k=1} \sum_{j=i+1}^{k} A_{ij} X_i X_j + e_i \tag{1}$$

where Y represents the response, X_i and X_j are variables, k is the number of independent variables (factors), A_0 is assigned as the constant coefficient, A_{ii} and A_{ij} are interaction coefficients of linear, quadratic and the second-order terms, respectively, and ei stands for the error. Design-Expert 7.0.1.0 (Trial version, Stat-Ease Inc., Minneanopolis, MN, USA) was used for the experimental design and regression analysis of the experimental data. The Student's t-test and Fischer's F-test were used to check the statistical significance of the regression coefficient, and determine the second-order model equation, respectively. The lack of fit, the coefficient of determination (R^2) and the F-test value obtained from the analysis of variance (ANOVA) were applied to evaluate the adequacy of the model.

3. Result and Discussion

If all the aforementioned variables are assumed to be measurable, the response surface will be expressed as Equation (2):

$$Y = f(X_1, X_2, X_3, \cdots, X_i) \tag{2}$$

where Y is candidate of responses and the X_i variables are called factors. To model using RSM, a total of 18 experimental runs are required. The results inserted to Design Expert software were used to fit a model to these results. The equations of models in terms of coded factors are obtained as Equations (3) to (5) for %E Zn, %E Mn, Sf (Zn-Fe) and Sf (Zn-Mn), respectively:

For %E Zn:

$$\%E\ Zn = +81.43 - 12.09X_1 - 11.65X_2 + 6.14X_3 \\ -8.97X_1X_2 + 12.83X_1X_3 - 3.66X_2X_3 \tag{3}$$

The equation of model for iron extraction is not

significant because p-value of model is less than 0.05. This is due to high extraction percent of iron (III) in any pH ranges, which reaches above 99%.

For %E Mn:

$$\%E\ Mn = -19.31 - 27.09X_1 - 22.78X_2 - 58.09X_3 \\ -74.38X_1X_2 - 78.88X_1X_3 + 29.75X_2X_3 \\ +94.90X_1^2 + 60.53X_2^2 - 43.60X_3^2 \tag{4}$$

Selective extraction of A ion from B ion can be expressed by $Sf(A\text{-}B) = D_A/D_B$, where $D_A = [A]_{organic}/[B]_{aqueous}$ and $D_B = [B]_{organic}/[A]_{aqueous}$. The equation of model for $Sf(Zn\text{-}Fe)$ is not presented in this study because it is not significant due to p-value less than 0.05. Nevertheless, $Sf(Zn\text{-}Mn)$ has been modeled using RSM as Equationn (5).

$$Sf(Zn\text{-}Mn) = -560.69 - 1419.98X_1 \\ -387.08X_2 - 1014.53X_3 \tag{5}$$

The result of analysis of variance (ANOVA) is illustrated in **Table 4-6**.

The results of this table reveal that the prediction models of the zinc and manganese extraction percent and separation factor of zinc-manganese are significant since the p-value is less than 0.05.

The result of **Table 4** indicated that the effect of ions concentration and their interactions on the zinc extraction are not significant. As observed in this table, iron concentration has the highest effect on zinc extraction. The reason for this effect is probably because of selective extraction of iron (III) ions (*i.e.*, among other species) by D2EHPA. **Table 5** illustrates the results of Mn extraction. The effect of all factors (variables) and their interactions except zinc concentration are significant on Mn extraction. As **Table 5**, manganese concentration has the highest effect on manganese extraction. In addition, **Table 6** displays that the results of $Sf(Zn\text{-}Mn)$, the zinc and manganese concentration are only significant factors. The high value of correlation coefficient (R^2) indicates that the model has been fitted very well. If this is a response surface design which is intended to be used for modeling the design space, then the R-squared values should be rather high (perhaps above 0.60) (Design Expert 7 Help). R^2 was found to be 0.904 for %E Zn, 0.991 for %E Mn and 0.627 for $Sf(Zn\text{-}Mn)$, as shown in **Figures 1** to **3**, which are acceptable statistically.

3.1. 3D Response Surface Plots

The 3D response surface plots simulated by Design-Expert software are graphical representations in order to understand the interaction effects of variables and the

Table 4. Analysis of variance (ANOVA) of developed models for zinc extraction.

Source	Sum of Squares	df	Mean Square	F-Value	p-value Prob > F	comment
Model	1841.69	6	306.95	11.04	0.0029	significant
A-Zn	79.98	1	79.98	2.88	0.1336	
B-Fe	151.85	1	151.85	5.46	0.052	
C-Mn	40.43	1	40.43	1.45	0.2669	
AB	38.9	1	38.9	1.4	0.2754	
AC	83.32	1	83.32	3	0.127	
BC	27.87	1	27.87	1	0.35	
Residual	194.54	7	27.79			
Lack of Fit	194.54	6	32.42			
Pure Error	0	1	0			
Cor Total	2036.23	13				

Table 5. Analysis of variance (ANOVA) of developed models for manganese extraction.

Source	Sum of Squares	df	Mean Square	F-Value	p-value Prob > F	
Model	14489.83	9	1609.981	50.42903	0.0009	significant
A-Zn	127.3896	1	127.3896	3.990193	0.1164	
B-Fe	357.9424	1	357.9424	11.21174	0.0286	significant
C-Mn	1330.366	1	1330.366	41.67074	0.0030	significant
AB	2234.237	1	2234.237	69.98246	0.0011	significant
AC	1478.275	1	1478.275	46.30365	0.0024	significant
BC	1176.359	1	1176.359	36.84678	0.0037	significant
A^2	2206.262	1	2206.262	69.1062	0.0011	significant
B^2	1087.321	1	1087.321	34.05789	0.0043	significant
C^2	2547.387	1	2547.387	79.79115	0.0009	significant
Residual	127.7027	4	31.92568			
Lack of Fit	127.7027	3	42.56757			
Pure Error	0	1	0			
Cor Total	14617.53	13				

Table 6. Analysis of variance (ANOVA) of developed models for separation factor of zinc and manganese.

Source	Sum of Squares	df	Mean Square	F-Value	p-value Prob > F	
Model	1.12E + 07	3	3.75E + 06	5.6	0.0162	significant
A-Zn	6.28E + 06	1	6.28E + 06	9.39	0.012	significant
B-Fe	7.73E + 05	1	7.73E + 05	1.15	0.3078	
C-Mn	7.40E + 06	1	7.40E + 06	11.06	0.0077	significant
Residual	6.69E + 06	10	6.69E + 05			
Lack of Fit	6.69E + 06	9	7.44E + 05			
Pure Error	0	1	0			
Cor Total	1.79E + 07	13				

relationship between the variables and responses. Three dimensional (3D) plots for the aforementioned responses were molded based on the model equations for zinc and manganese extraction and separation factor of zinc-manganese as Equations (3) to (5). These plots are shown in **Figures 4** to **6**. In these figures, two variables versus responses at the center level of third variable have constructed the plots.

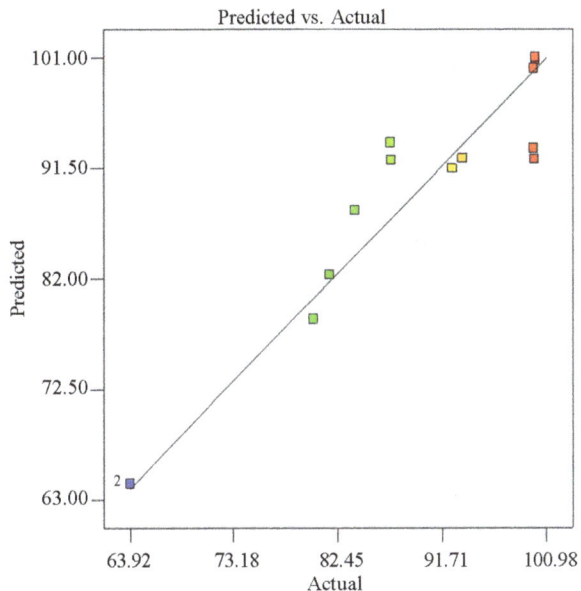

Figure 1. Relationship between predicted and actual (observed) values for zinc extraction.

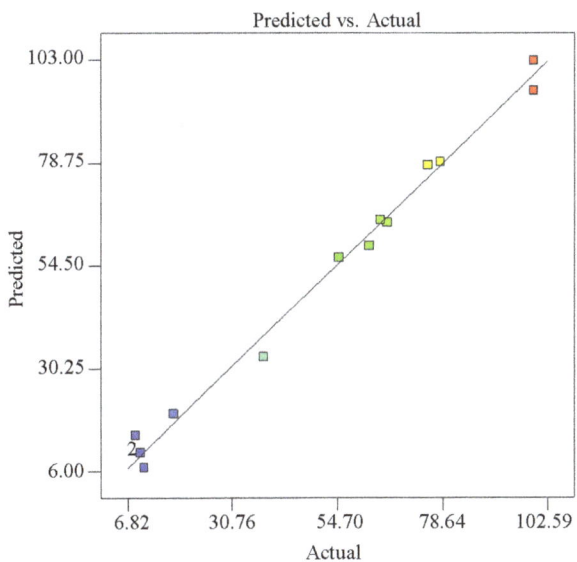

Figure 2. Relationship between predicted and actual (observed) values for manganese extraction

Figure 4(a) shows the effect of manganese and iron concentration on zinc extraction at 40°C and pH of 2.5. It expresses that increasing iron concentration in the aqueous feed decreases zinc extraction and enhancement of manganese concentration increases zinc ions extraction. Figure 4(b) demonstrates that at high levels of zinc and manganese concentrations, the zinc extraction decreases. Furthermore, in Figure 4(c), enhancement of zinc and iron ions in the aqueous phase reduces zinc extraction. It is noteworthy that high concentration of ions in aqueous phase diminishes the capability of D2EHPA, which is related to specific capacity of extractant; moreover, **Fig-**

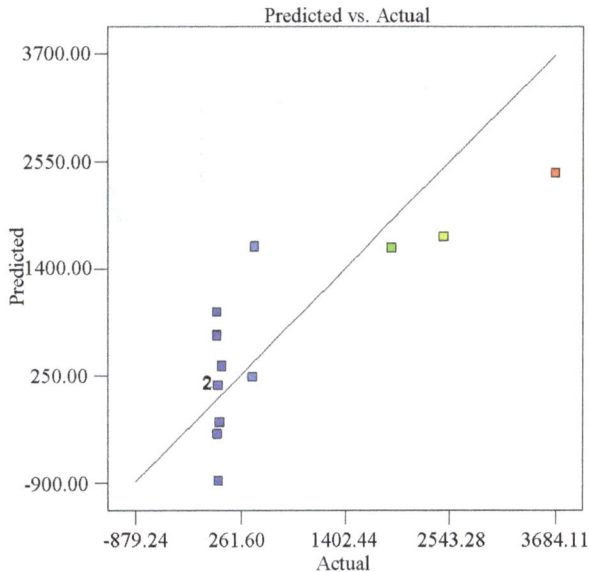

Figure 3. Relationship between predicted and actual (observed) values for separation factor of zinc and manganese.

ures 4(b) and **(c)** justify this note.

Figure 5(a) shows considerable effect of iron concentration and invariable effect of manganese concentration on the manganese extraction. **Figure 5(b)** illustrates that enhancement of manganese ions in the aqueous phase increases its extraction. In addition, this figure shows that zinc ions have approximately invariable effect on manganese extraction. In **Figure 5(c)**, the effect of zinc and iron concentrations on manganese extraction is relative. As seen in this figure, the lowest manganese extraction has occurred at the middle levels of zinc and iron concentrations. Finally, all plots of **Figure 6** shows that higher amount of ions in the aqueous phase decreases separation factor of zinc-manganese.

3.2. Optimization by RSM

The aim of optimization is to have ZPL with the lowest impurities. Therefore, the highest zinc extraction, the lowest iron and manganese extraction and the highest values of separation factors were considered for optimizing by RSM. This optimization was carried out by DX7 software and the results of the process optimization with respect to the aforementioned aim were obtained as illustrated in **Table 7**. As seen in this table, the zinc, iron and manganese extraction percent at pH of 2 and temperature of 40°C reached 93.72%, 99.20% and 11.18%, respectively. At this condition, it was found that Zn 21.22 g/L, Fe 376.08 ppm and Mn 1.00 g/L are extracted by 30% (v/v) D2EHPA dissolved in kerosene.

This result reveals that to extract more effectively by 30% D2EHPA, the best ZPL should be as the optimum condition of ions concentration. The desirability of this

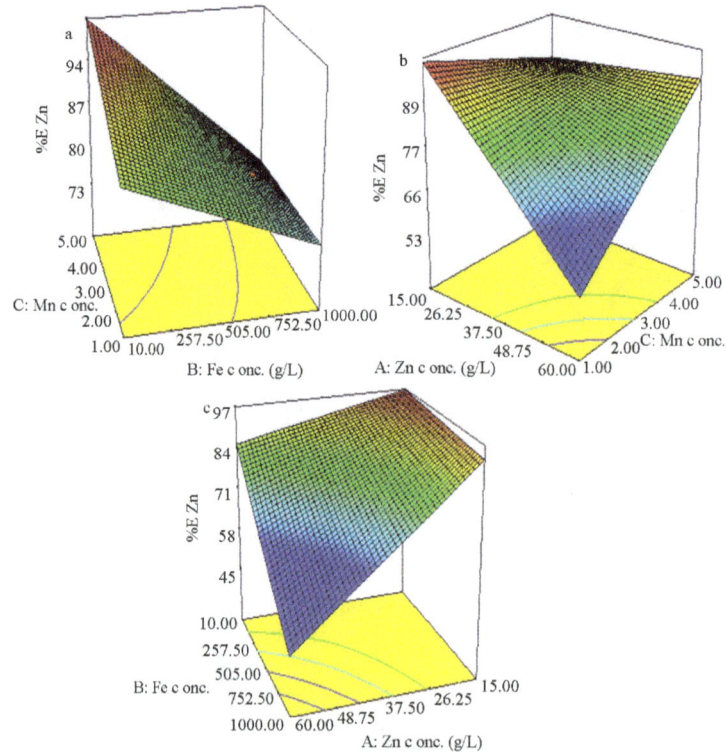

Figure 4. 3D response surface plots showing effect of two variables (factors) on zinc extraction at the center level of other variable. (a) Mn and Fe concentration (g/L and ppm, respectively). (b) Zn and Mn concentration (g/L). (c) Zn and Fe concentration (g/L and ppm, respectively).

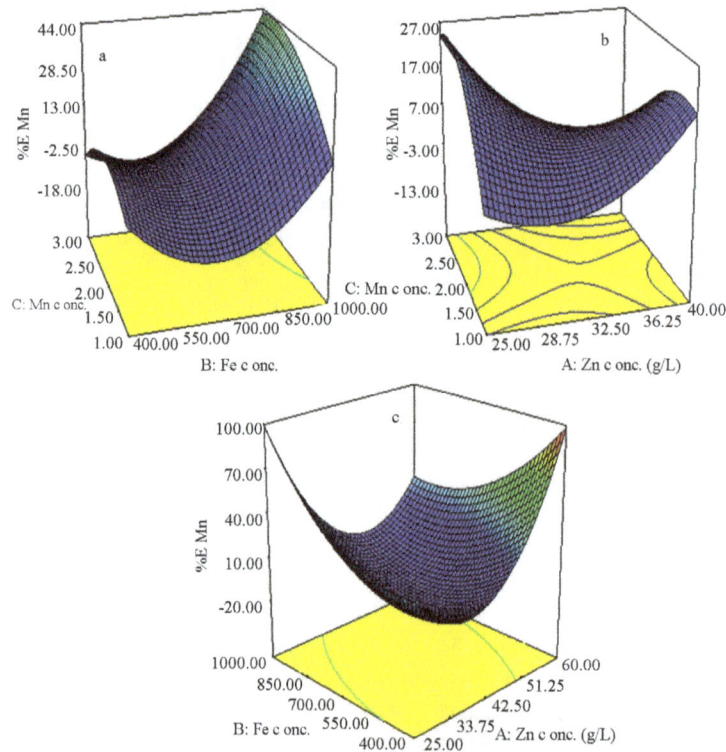

Figure 5. 3D response surface plots showing effect of two variables (factors) on manganese extraction at the center level of other variable. (a) Mn and Fe concentration (g/L and ppm, respectively). (b) Zn and Mn concentration (g/L). (c) Zn and Fe concentration (g/L and ppm, respectively).

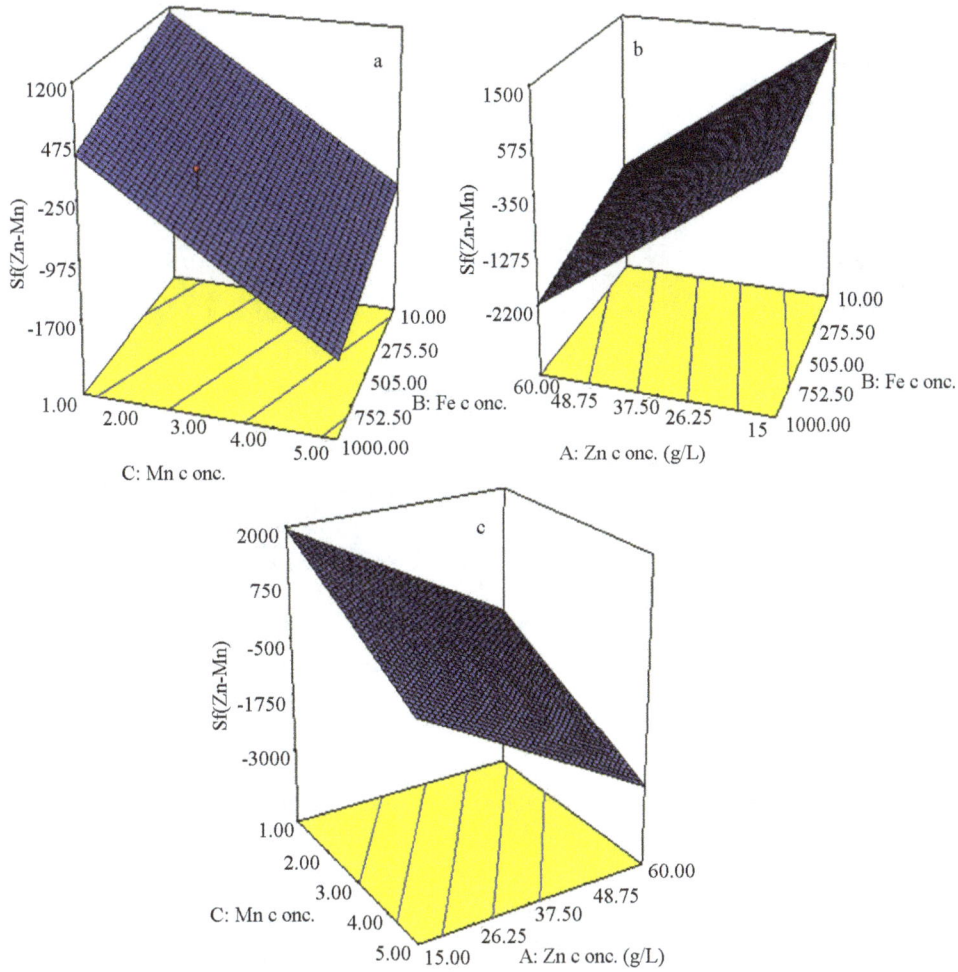

Figure 6. 3D response surface plots showing effect of two variables (factors) on separation factor of zinc-manganese at the center level of other variable. (a) Mn and Fe concentration (g/L and ppm, respectively). (b) Zn and Fe concentration (g/L and ppm, respectively) (c) Zn and Mn concentration (g/L).

Table 7. Results of process optimization and optimum levels of variable.

Name	Goal	Zn (g/L)	Fe (ppm)	Mn (g/L)	%E Zn	%E Fe	%E Mn	Sf (Zn-Fe)	Sf (Zn-Mn)
Zn	is in range								
Fe	is in range								
Mn	is in range								
E Zn	maximize	21.22	376.08	1	93.72	99.20	11.18	8.10	1582.39
E Fe	minimize								
E Mn	minimize								
Sf (Zn-Fe)	maximize								
Sf (Zn-Mn)	maximize								

optimum condition achieved 0.67, which is statistically acceptable. **Figure 7** shows the desirability of the optimum condition. As seen in this figure, at the optimum condition of iron concentration factor and the lowest levels of zinc and manganese concentration, the desirability of the model is high.

Design-Expert® Software

Desirability

1

0

X_1 = A: Zn
X_2 = C: Mn

Actual Factor
B: Fe = 376.08

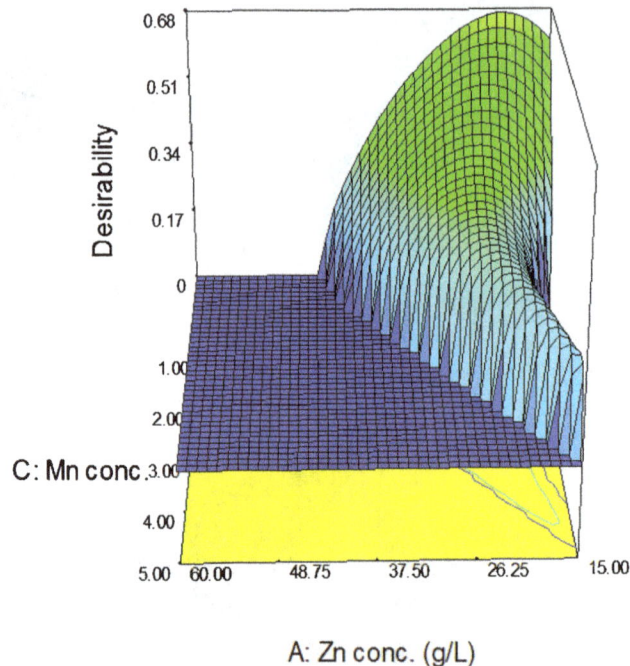

Figure 7. Response surface plot showing effect of ions concentration on desirability of optimum condition.

4. Conclusions

1) At the optimum condition, the zinc, iron and manganese extraction percent at pH of 2 and temperature of 40°C reached 93.72%, 99.20% and 11.18%, respectively.

As a result, the best aqueous feed for extraction by 30%

2) As ANOVA tables indicate, iron extraction and Sf(Zn-Fe) are not significant responses to model.

3) Iron and manganese concentration had the highest effect on the zinc and manganese extraction, respectively.

REFERENCES

[1]   E. Vahidi, F. Rashchi and D. Moradkhani, "Recovery of Zinc from an Industrial Zinc Leach Residue by Solvent Extraction Using D2EHPA," *Minerals Engineering*, Vol. 22, No. 2, 2009, pp. 204-206. http://dx.doi.org/10.1016/j.mineng.2008.05.002

[2]   M. Bolourfroush, M. Oliyazadeh and K. Gharibi, "Investigation of Zinv Solvent Extraction by D2EHPA," *Iranian Journal of Mining Engineering (IRJME)*, Vol. 2, No. 4, 2008, pp. 21-28.

[3]   A. G. Pecherskaya and V. V. Stender, *JPC*, Vol. 9, No. 920, 1950.

[4]   V. V. Stender and A. G. Pecherskaya, *Non-Fer. Met*, Vol. 45, No. 4, 1950.

[5]   U. F. Turomoshina and V. V. Stender, *JPC*, Vol. 166, No. 2, 1955.

[6]   P. Zaidler and V. V. Stender, *JPC*, Vol. 17, No. 282, 1944.

[7]   I. Ivanov and Y. Stefanov, "Electroextraction of Zinc from Sulphate Electrolytes Containing Antimony Ions and Hydroxyethylated-butyne-2-diol-1,4: Part 3. The Influence of Manganese Ions and a Divided Cell," *Hydrometallurgy*, Vol. 64, No. 3, 2002, pp. 181-186. http://dx.doi.org/10.1016/S0304-386X(02)00039-7

[8]   M. R. C. Ismael and J. M. R. Carvalho, "Iron Recovery from Sulphate Leach Liquors in Zinc Hydrometallurgy," *Minerals Engineering*, Vol. 16, No. 1, 2003, pp. 31-39. http://dx.doi.org/10.1016/S0304-386X(02)00039-7

[9]   F. Principe and G. P. Demopoulos, "Comparative Study of Iron(III) Separation from Zinc Sulphate-Sulphuric Acid Solutions Using Organophosphorus Extractants, OPAP and D2EHPA: Part II. Stripping," *Hydrometallurgy*, Vol. 79, No. 3-4, 2005, pp. 97-109. http://dx.doi.org/10.1016/j.hydromet.2005.06.006

[10] Y. Sun, *et al.*, "Optimizing the Extraction of Phenolic Antioxidants from Kudingcha made from Ilex Kudingcha C. J. Tseng by Using Response Surface Methodology," *Separation and Purification Technology*, Vol. 78, No. 3, 2011, pp. 311-320. http://dx.doi.org/10.1016/j.seppur.2011.01.038

[11] C.-H. Tan, *et al.*, "Extraction and Physicochemical Properties of Low Free Fatty Acid Crude Palm Oil," Food Chemistry, Vol. 113, No. 2, 2009, pp. 645-650. http://dx.doi.org/10.1016/j.foodchem.2008.07.052

23

Thermal Studies and Analytical Applications of a Newly Synthesized Composite Material "Polyaniline Stannic Molybdate"

Sajad Ahmad Ganai[1*], Javid Ahmad Banday[1], Abid Hussain Shalla[2], Tabassum Ara[1]

[1]Department of Chemistry, National Institute of Technology, Srinagar, Kashmir
[2]Islamic University of Science and Technology, Awantipora, India
Email: *sajadali16@gmail.com

ABSTRACT

Polyaniline stannic molybdate—an organic-inorganic composite material, was prepared via sol-gel mixing of organic polymer polyaniline into matrices of inorganic precipitate of stannicmolybdate. The composite material synthesized at pH 1.2 showed an ion exchange capacity 1.8 meq/g for Na^+ ions. Ion exchange capacity, pH titration and distribution studies were carried out to determine the preliminary ion exchange properties of the material. The distribution studies showed the selectivity of Hg(II) ions by this material. The effect of temperature on the ion exchange capacity of the material at different temperatures had been studied. The sorption behavior of metal ions was also explored in different surfactant media.

Keywords: Organic-Inorganic Hybrid Material; Polyaniline Stannic Molybdate; Synthesis; Thermal Studies

1. Introduction

Ion exchangers are always of large scale interest because of their higher selectivity and resistance at higher temperature. These materials have found large scale applications in water treatment and pollution control. An ion exchanger may be organic, inorganic or hybrid depending upon the nature of matrix of which it is made up of. Out of these, hybrid ion exchangers have received more attention because both organic as well as inorganic ion exchangers have some limitations. Organic ion exchangers are very sensitive to exposure to high radiation doses, which cause significant changes in their capacity and selectivity, presumably through hydrolysis of their functional groups, chain scission, and changes in their crosslinking. The inorganic ion exchangers have several superior qualities required for the treatment of industrial waste effluents compared to organic resins. In order to obtain a combination of these advantages associated with polymeric and inorganic materials as ion-exchangers, attempts have been made to develop polymeric-inorganic composite ion-exchangersby incorporation of inorganic monomers in the organic matrix. A composite ion-ex-

changer must be an interesting material, as it should possess the mechanical stability due to the presence of organic polymeric species and the basic characteristics of an inorganic ion-exchanger regarding its selectivity for some particular metal ions [1]. It was therefore considered to synthesize such hybrid ion-exchangers with a good ion-exchange capacity, high stability, reproducibility and selectivity for metal ions [2,3], indicating its useful environmental applications.

The work deals with the synthesis, ion exchange properties and thermal stability of polyaniline stannic molybdate, prepared by the incorporation of conducting polymer, i.e. polyanniline into the matrices of inorganic ion exchanger stannic molybdate.

2. Experimental

2.1. Reagents and Instruments

The main reagents used for the synthesis were aniline, stannic chloride pentahydrate and potassium persulphate (LobaChemePvt Ltd (India). All other chemicals and reagents used were of analytical grade. The main instruments used during the study were UV-vis spectrophotometer (Elico EI 301E, India), a PW 1148/89 based

X-ray diffractometer (Phillips, Holland), a water bath incubator shaker (MSW-275, India) and muffle furnace (G1-111, India).

2.2. Preparation of Reagent

The solutions of 0.1 M Stannic chloride pentahydrate were prepared in demineralized water (DMW) and solutions of 10% (v/v) of double distilled aniline and 0.1 M potassium persulphate were prepared in 1.0 M HCl.

2.3. Preparation of Polyaniline Stannic Silicate

Polyaniline gels were prepared by mixing different volume ratio of 10% aniline and 0.1 M potassium persulphate with continuous stirring by a magnetic stirrer (**Table 1**). Green colored polyaniline gels were obtained by keeping the solution below 10°C for half an hour [4]. A precipitate of stannic (IV) molybdate was prepared at room temperature (25°C ± 2°C) by adding mixture of 0.1 M sodium molybdate solution gradually to an aqueous solution of 0.1 M stannic chloride pentahydrate in different volume ratios. The white precipitates were obtained. The desired pH of the mixture was adjusted by adding HCl with continuous stirring. The gels of polyaniline were then added to the white inorganic precipitate of stannic(IV) molybdate and mixed thoroughly with constant stirring [5-7]. The resultant green color gels so obtained were kept for 24 h at room temperature (25°C ± 2°C) for digestion. The supernatant liquid was filtered off by suction. The gel was washed with demineralized water (DMW) till the filtrate was neutral. The product was dried in an oven at 50°C. The dried material was broken into small granules on immersion in DMW. The granules were converted into H^+ form by placing in 1 M HNO_3 solution for 24 h. The excess acid from the material was removed after several washings with DMW and finally dried at 50°C ± 2°C. In this way a number of samples of polyaniline stannic molybdate were prepared in different experimental conditions described in **Table 1**. It was decided to study sample S-9 in detail on the basis of highest ion exchange capacity for Na^+ ions.

2.4. Column Ion-Exchange Capacity

To determine the ion-exchange capacity, column process was used. 1.0 g dry cation-exchanger in H^+ form was packed in a column (1.0 cm id) fitted with glass wool at the bottom. Metal nitrates as eluents were used to elute the H^+ ions completely from the cation-exchanger column. The effluent was titrated against a standard solution of 0.1 M NaOH.

2.5. Thermal Stability

In order to determine the effect of heating temperature on ion-exchange capacity, 1.0 g sample of the material (S-9) in H^+ form was heated at different temperatures in a muffle furnace for 1h and Na^+ ion-exchange capacity was determined by standard column process as mentioned above.

2.6. Distribution Studies

The distribution coefficients (Kd) for various metal ions were determined in different concentrations of surfactant

Table 1. Synthesis of polyaniline stannic molybdate in different conditions.

Sample	Mixing volume ratio				pH of the mother liqor with inorganic precipitate	Sodium ion exchange capacity (meq/g)	Appearance of beads after drying at 50°C
	A	B	C	D			
S1	1	1	1.0	1.0	0.5	0.37	Green
S2	2.0	2.0	1.0	1.0	0.5	0.6	Green
S3	1.0	2.0	1.0	1.0	0.5	0.85	Green
S4	1.0	2.0	1.0	1.0	0.5	0.7	Dirty White
S5	1.0	2.0	1.0	1.0	0.7	0.74	Dark Green
S6	1.0	2.0	1.0	1.0	0.8	0.90	Dark Green
S7	1.0	2.0	1.0	1.0	0.9	1.44	Dark Green
S8	1.0	2.0	1.0	1.0	1.0	1.62	Dark Green
S9	1.0	2.0	1.0	1.0	1.2	1.8	Dark Green
S10	1.0	2.0	1.0	1.0	1.4	1.5	Dark Green
S11	1.0	2.0	1.0	1.0	1.6	1.4	Dark Green

A: 0.1M Stannic chloride pentahydrate; B: 0.1M sodium molybdate; C: 10% Aniline in 1.0M HCl; D: 0.1M K2S2O8 in 1.0M HCl.

(SDS).The effect of temperature on the distribution coefficient was also studied. Polyaniline stannic molybdate (0.3 g) in H^+ form was put into 100 ml conical flasks each containing 30 mL solution of 0.1 M concentration of metal ions. The mixture was continuously shaken for a definite period of time at 30°C, 45°C, 60°C and 75°C (**Table 2**). The amount of metal ions present in the solution was determined by titrating it against disodium salt of EDTA (0.01 M) using standard procedures.

3. Results and Discussion

In this study of preparation and characterization of organic-inorganic composite cation exchange material, a number of samples of organic-inorganic electrically conducting composite stannicmolybdate providing a new class of hybrid ion-exchangers were prepared by sol-gel mixing of organic conducting polymer , polyaniline with the inorganic precipitate of stannic molybdate (**Table 1**). Among them, sample S-9 possessed enhanced Na^+ ion exchange capacity (1.8 meq/g) and better thermal stability. Polyaniline gel was prepared by oxidation coupling using $K_2S_2O_8$ in acidic aqueous medium.

The effect of temperature on the reaction seems to be very pronounced. Aniline underwent oxidation coupling only below 10°C very effectively, leading to formation of polyaniline with fairly good yield.

However sample S9 of polyaniline stannic molybdate exhibited better granulometric and mechanical properties and reproducible behavior as is evident from **Table 1**. The samples prepared in various batches did not show any appreciable deviation in their ion exchange capacities. The ion exchange capacity of the composite cation exchanger for alkali metal ions and alkaline earth metal ions increases according to the increase in the hydrated ionic radius increases (**Table 3**). It was observed that on heating the material at different temperatures, ion exchange capacity of the dried sample material (S9) was changed as the temperature increased as shown in **Table 4**. From the comparative study of Na^+ ion exchange capacity of polyaniline stannic molybdate with those of other ion exchangers, as shown in **Table 5**, it is apparent that this composite cation exchanger has more ion exchange capacity than others.

Table 2. Effect of temperature on theKd($mL \cdot g^{-1}$) of metal ions in 0.1 M SDS.

Metal ions	Temperature (°C)			
	30	45	60	70
Mg^{2+}	18	15	11	4
Cd^{2+}	11	19	13	8
Zn^{2+}	35	49	31	19
Mn	27	35	24	22

Table 3. Ion exchange capacity of polyaniline stannic molybdate (sample S-9) for different metal ions.

Exchanging ions	Ionic radii (A°)	Hydrated ionic radii (A°)	Ion exchange capacity (meq/g)
Na^+	0.97	2.76	1.8
K^+	1.33	2.32	1.3
Mg^{2+}	0.78	7.00	1.2
Ba^{2+}	1.43	5.90	0.9

Table 4. Effect of temperature on the ion-exchange capacity of polyaniline stannic molybdate.

Temperature °C	color	Retention of ion exchange capacity (%)
50	Green	100
100	Green	94
200	Green	78
300	Light green	63
400	Brown	52
500	White	41
600	white	35

Table 5. Comparison of ion exchange capacity of polyaniline stannic molybdate with those of other cation exchangers.

Ion exchange material	Na^+ exchange capacity
Polyanilinesn(II) arsenophosphate [6]	1.58
PolyanilineSn(II) tungstate [7]	0.75
PolyanilineSn(II)arsenate [8]	0.87
PolyanilineSn(II) tungstoarsenate [9]	1.67
Sn(IV) tungstoarsenate [7]	1.06
PolyanilineZr(IV) tungstophosphate [10]	1.46
PolyanilineSn(II) molybdate	1.8

In order to explore the potentiality of polyaniline stannic molybdate in separation of metal ions, distribution studies of some metal ions were achieved in various concentrations of SDS. It is apparent from the data in **Tables 2** and **6** that the Kd values varied with temperature and conc. of contacting solvents. The Kd values of metal ions increases up to 45°C and then decreases, Thus 45°C is the optimum temperature for the Kd value determination. It was also observed from kd values that the Hg(II) was highly adsorbed in all solvents while remaining metal ions were poorly adsorbed. The high uptake of Hg(II) in all solvents demonstrate not only the ion exchange properties but all the adsorption and ion selective characterization of the material. The utility of this compound has been demonstrated by achieving some binary separations (Mg^{2+}-Hg^{2+}, Cd^{2+}-Hg^{2+} and Mn^{2+}-Hg^{2+}) of metal ions on its column. The separations were found to be quantitative and the results obtainedwere quite precise and accurate as is evident from **Table 7**. Hg^{2+} has also been selectively separated from a synthetic mixture of metal ions. The salient features of selective separation are summarized in **Table 8**. Thus we can say that this composite material is highly selective for Hg(II) and can be used for the separation of mercury from waste effluxents.

4. Conclusion

PolyanilineSn(IV) molybdate, an organic-inorganic com-

Table 6. The distribution coefficients of metal ions in SDS ($KdmLg^{-1}$) at 30˚C ± 2˚C.

Metal ions	SDS (Surfactant)				
	0.1 M	0.01 M	0.001 M	0.0001 M	0.00001 M
Mg^{2+}	8	15	17	31	29
Cd^{2+}	11	25	29	34	28
Hg^{2+}	137	145	152	159	143
Zn^{2+}	35	37	45	52	33
Mn^{2+}	27	32	53	66	62

Table 7. Quantitative Separations of metal ions in binary mixtures on a column of polyaniline stannic molybdate.

Separation achieved	Amount loaded, mL	Amount found, mL	Recovery %	Eluent used	Volume of eluent, mL
Mg^{2+}	2.42	2.36	97.40	0.1M SDS	70.00
Hg^{2+}	20.05	20.05	100.00	0.00001M SDS	100.00
Cd^{2+}	11.24	11.21	99.73	0.1M SDS	90.00
Hg^{2+}	20.05	19.10	95.26	0.00001M SDS	110.00
Mn^{2+}	5.49	5.37	97.81	0.1M SDS	80.00
Hg^{2+}	20.05	19.98	99.65	0.00001M SDS	90.00

Table 8. Selective separation of Hg^{2+} from a synthetic mixture oh Hg^{2+}, Cd^{2+}, Mn^{2+} and Mg^{2+} on a column of polyaniline stannic molybdate.

Amount loaded, mg	Amount found, mg	% Recovery	Volume (mL) of 0.00001M SDS used for complete elution of Hg^{2+} mL
2.50	2.39	95.6	110
5.0	4.9	98.0	130

posite cation exchange material represents a new class of ion exchange materials that combines good characteristics of both organic and inorganic components within a single composite. It is quite clear from the results that the composite cation exchanger has enhanced ion exchange capacity and thermal stabilities. The selective behavior of polyanilineSn(II) molybdate is important from environmental pollution chemistry point of view, where an effective separation method is needed for the removal of Hg(II) ions.

5. Acknowledgements

The authors are highly thankful to the H.O.D. Chemistry (Dr. S. A. Shah) and Director, National Institute of Technology (NIT) Hazratbal, Srinagar for providing the research facilities.

REFERENCES

[1] K. G. Varshney, N. Tyal and U. Gupta, "Acrylonitrile Based Cerium (IV) Phosphste as a New Mercury Selective Fibrous Ion-Exchanger: Synthesis, Characterization and Analytical Applications," *Colloidal and Surfaces A*, Vol. 145, No. 2-3, 1998, pp. 71-81.

[2] H. Zhang, J. H. Pang, D. Wang and Z. Jiang, "Sulphonated Poly(acryleneether nitrile ketone) and Its Composite with Phosphotungstic Acid as Materials for Proton Exchange," *Journal of Membrane Science*, Vol. 264, No. 1-2, 2005, pp. 56-64.

[3] W. A. Siddiqui and S. A. Khan, "Synthesis, Characterization and Ion Exchange Properties of a New and Novel Organic-Inorganichybridcation Exchanger: Poly(methyl methacrylate) Zr(IV) Phosphate," *Colloids and Surfaces A: Physicochemical and Engineering Aspects*, Vol. 295, No. 1-3, 2007, pp. 193-219. doi:10.1016/j.colsurfa.2006.08.053

[4] I. M. El-Naggar, E. S. Zakaria, I. M. Ali and M. Khalil, Chemical Studies on Synthetic Polyaniline Titanotungstate and Its Uses to Reduce Cesium from Solutions and Polluted Milk," *Journal of Environmental Radioactivity*, No. 112, 2012, pp. 108-117.

[5] A. A. Khan and Inamuddin, "Applications of Hg(II) Sensitive Polyaniline Sn(IV) Phosphate Compositecation Exchange Material in Determination of Hg^{2+} from Aqueous Solutions and in Making Ion Selective Membraneelectrode," *Sensors and Activators B*, Vol. 120, No. 1, 2006, pp. 10-18. doi:10.1016/j.snb.2006.01.033

[6] R. Niwas, A. A. Khan and K. G. Varshney, "Synthesis and Ion Exchange Behaviour of Polyaniline Sn(IV) Arsenophosphate: A Polymeric Ion Exchanger," *Colloids and Surfaces*, Vol. 150, 1999, pp. 7-14

[7] A. A. Khan and M. M. Alam, "Synthesis, Characterization and Analytical Applications of a New and Novel Organic-Inorganic Composite Material as Acation Exchanger and Cd(II) Ion-Selective Membrane Electrode: Polyaniline Sn(IV) Tungstoarsente," *Journal of Reactive and Functional Polymers*, Vol. 55, No. 3, 2003, pp. 277-290. doi:10.1016/S1381-5148(03)00018-X

24

Microstructural Evolution in Cu-Mg Alloy Processed by Conform

Lianpeng Song[1], Yuan Yuan[1,2], Zhimin Yin[1]

[1]School of Materials Science and Engineering, Central South University, Changsha, China
[2]China Railway Construction Electrification Bureau Group Kang Yuan New Materials Co., LTD., Jiangyin, China
Email: songlp@csu.edu.cn

ABSTRACT

The objective of this study is to investigate the possibility of continuous extrusion forming (Conform process) and microstructural evolution the of Cu-Mg alloy. The results indicate that Conform process can break as-cast grains and refine the structure, meanwhile. This process can improve the degree of the structure homogeneity. The TEM and EBSD techniques were used to investigate the morphology, grain size and misorientation of the samples at cavity entrance and cavity export. Refined structures after shear deformation include broken grains and subgrains formed by dislocation reconstruction. Due to the relatively high deformation temperature, dynamic recrystallization occurred during deformation. The subgrain rotation nucleation took place, and grain boundary migration resulted in grain growth. However, the coarse grains were refined by anneal twins.

Keywords: Cu-Mg Alloy; Microstructural Evolution; Conform; Dynamic Recrystallization

1. Introduction

Conform continuous extrusion forming process has been most widely used in the efficient continuous production of soft aluminum alloys rod, tube, sections and cored products, as well as copper wires, rectangular sections and a wide range of copper profiles [1-3], since the introduction of this methodology by the Atomic Energy Authority in UK (UKAEA) early in 1970s, due to its superiority of pre-heating-free, energy-saving, high extrusion ratio and production efficiency, production with large length and high homogeneity and so on [4,5].

Similar to ECAP (equal channel angular pressing), severe plastic deformation technique was performed using a tool steel die with two channels intersecting at an inner angle of 90° [6-11]. The schematic of the Conform process is shown in **Figure 1**. In principle, the feedstock is fed into the profiled groove of the extrusion wheel by means of a coining roll and the groove closed by a close fitting shoe. The material is prevented from continuing its passage around the wheel by means of an abutment [12,13]. As a result, high temperatures and pressures are developed in the material, which becomes plastic and subsequently emerges from the machine through an extrusion die [14].

The microstructural evolution and grain refinement during ECAP have been well documented [15]. Previous studies indicate that grain refinement is predominately operated by accumulation, interaction, tangling and spatial rearrangement of dislocations [16-18]. However, due to the large deformation heat produced by metal plastic deformation process of Conform, the microstructural evolution has not been reported. The main purpose of the present study is to research the microstructural evolution during Conform process.

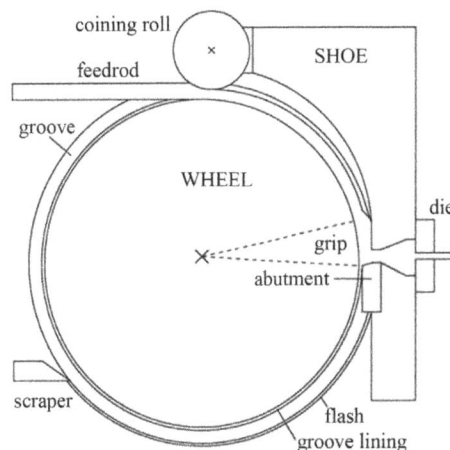

Figure 1. A schematic view of Conform process.

2. Materials and Experimental Procedures

The Cu-0.45 Mg alloys were prepared by melting high purity cathode copper, high purity magnesium in a 3000 kg industrial inductive furnace, then the ingots were cast by the upward casting system. The rods were extruded by TLJ500 Conform machine which was manufactured by Dalian Railway Institute. The parameters of extrusion process were extrusion temperature of 650°C, wheel angular velocity of 4.0 r/min. **Figure 2** shows the photograph of copper block in cavity obtained in the interrupted extrusion production. The copper samples at cavity entrance and export were researched.

To evaluate the microstructural evolution during the process, transmission electron microscopy (TEM) observations were performed on a transmission electron microscope FEI TECNAIG^2 at 200 kV. Thin foil samples were electropolished in a twin-jet instrument using a mixture of 30% hydrogen nitrate with alcohol as the balance. The SEM examination was conducted with a KYKY-2800 type scanning electron microscope under control voltage 20 kV, and equipped electron backscatter diffraction (EBSD) was utilized to investigate the fracture behavior and quantitative analysis of grain and sub-grains.

3. Results

3.1. Microstructure Developed during the Conform Process

The microstructure of Cu-Mg alloy rod before Conform process is shown in **Figure 3(a)**. It can be found that coarse equiaxed grains and dendritic structure exist in the as-cast Cu-Mg alloy, the average grain size is about 2000 μm. **Figure 3(b)** shows the microstructure of Cu-Mg alloy after Conform process. It can be found that the structure was consisted of fine grains. The evolution of microstructure shows that the as-cast structure is transformed into equiaxial and fine grains. So Conform process is an effective method to refine the as-cast grain

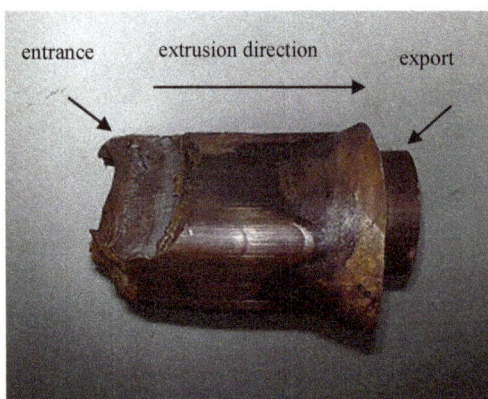

Figure 2. A photograph of copper block in cavity.

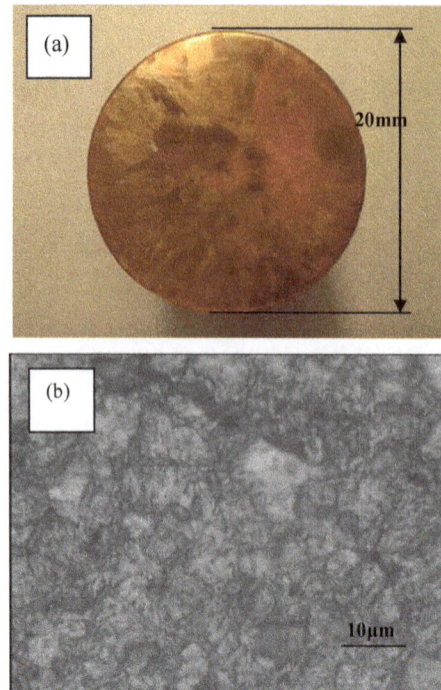

Figure 3. Optical micrographs of Cu-Mg alloy (a) before Conform process (b) after Conform process.

structure in copper, and can be successfully applied to the continuous deformation of Cu-Mg alloy.

3.2. EBSD Characterization of Cu-Mg Alloys

In order to characterize the homogeneity of microstructure during Conform process on a smaller scale, the advanced EBSD technique was used to investigate the morphology, grain size and misorientation distributions of samples ar cavity entrance and cavity export. From the corresponding EBSD images in **Figures 4(a)** and **(b)**, it can be seen that the microstructure of sample at the cavity entrance is consisted of high-angle grains, and the grain sizes are about 10 μm. However, the grain boundaries are not clearly observed (**Figure 4(a)**), which are caused by substructures consisting primarily of subgrains of small misorientation after shear deformation [19]. The grain boundaries can be be clearly observed for the sample at the cavity export, and a lot of small-size grains and twins are shown in **Figure 4(b)**. These grains showed well-defined grain boundaries, suggesting the presence of a large fraction of high angle grain boundaries. It's worth mentioning that recrystalllization has occurred in die cavity, possibly due to the heat generated during the deformation.

Grain size distributions of corresponding samples are shown in **Figure 5(a)**. When subgrain sizes are considered, the grain sizes of the sample at entrance range from 0.3 to 10.3 μm after shear deformation, with average grain size of 4.94 μm (**Figure 5(a)**). After annealing at

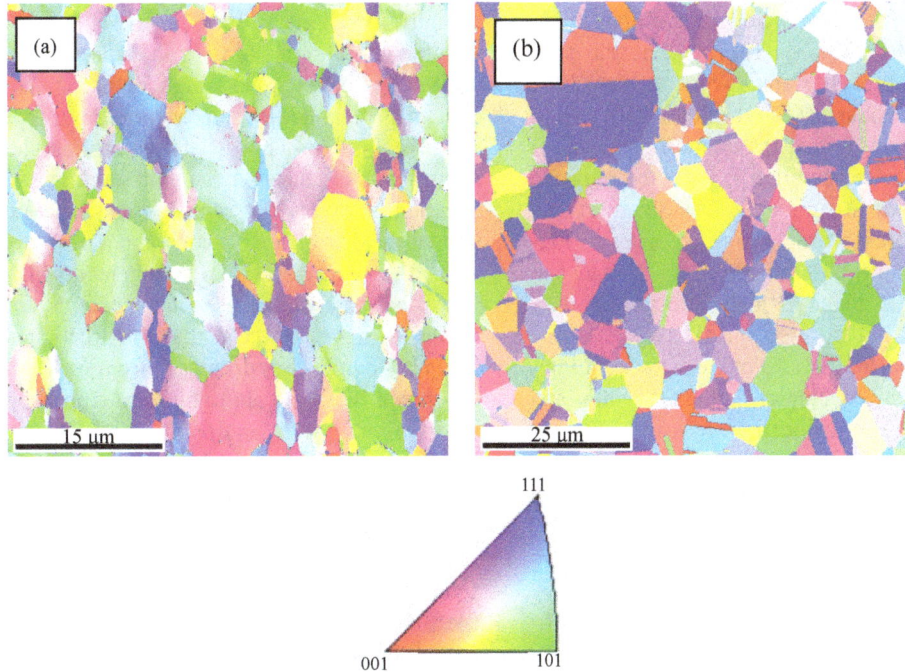

Figure 4. EBSD maps concerning microstructures of Cu-Mg alloy after Conform (a) sample at the cavity entrance (b) sample at the cavity export.

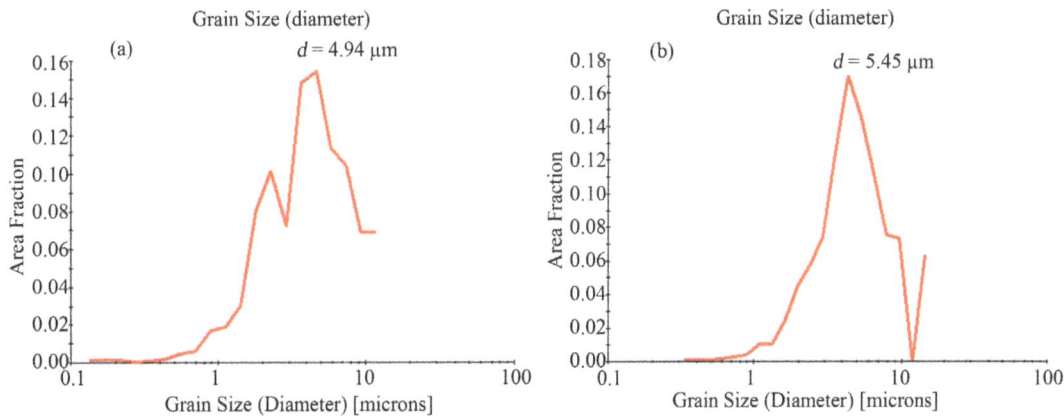

Figure 5. Grain size distributions of Cu-Mg alloy alloys obtained by EBSD (a) cavity entrance's sample (b) cavity export's sample

the cavity, the average grain size changes to 5.45 μm. It's worth noting that some grains have grown up at the cavity.

The distributions of the misorientation angles of corresponding samples are presented in **Figures 6(a)** and **(b)**, respectively. The volume fraction of low-angle grain boundaries that powerfully substantiates the formation of the relatively homogeneous fine microstructures after shearing deformation is about 60%. However, from the data, the corresponding fraction of the low-angle grain boundaries of the extruded product is about 27%, which is essentially lower than that of sample at the cavity entrance. From this difference, we can conclude that some

low-angle grain boundaries had transformed to high-angle grain boundaries during deformation. Meanwhile, the volume fraction of misorientation angles of 60° has increased up to 33.8% for the sample at the cavity export.

4. Discussion

According to the EBSD data, the grains have been refined to 4.94 μm after shear deformation at the cavity entrance. There are two reasons for the refinement. One is severe plastic deformation that increases the misorientation of geometrically necessary boundaries and incidental boundaries leading to the formation of high-angle grain boundaries, as well as smaller grains inside the original coarse grains, this is the well-known dislocation subdivision mechanism [20-23]. The other is the subgrains forming in the broken grains, which are separated by low-angle grain boundaries (**Figure 7(a)**). The second reason can be explained by that the dislocation distribution produced by shear deformation is apt to be replaced by energetically favorable subgrains. The subgrain formation evidences in broken grains were observed in **Figure 7(b)**. Lots of dislocation entangle in position A and form dislocation-tangle zone. At high deformation temperatures, annealing of deformed metals results in the decrease of the number of dislocations in the cell's interior while the dislocations tangle in the cell walls and change into more regular dislocation networks. The reconstruction was observed in position B. Then the dislocation walls gradually transform into subboundaries such as dense-dislocation wall (position D) and uncondensed dislocation wall (position C), and cells transform into subgrains free of dislocations. Thermal energy assists in creating "clean" boundaries. Meanwhile, it is to be noted that the deformation time for subgrain formation is ex-

tremely low ($-2.5 \times 10 - 5$ s) [24], that's why a large number of subgrains form in short time.

Due to the relatively high deformation temperature, dynamic recrystallization occurred during deformation (**Figure 8**). The TEM examination has shown that the grain boundaries become smooth, and the density of dislocation is low. Lots of anneal twins have nucleaned and grown up during heat treatment. These evidences confirmed the occurrence of dynamic recrystallization due to the effect of heat and pressure, which is consistent with that shown in **Figure 6** that the fraction of low-angle grain boundaries has decreased from 61% to 27%.

Anneal twins are apt to be formed in main slip system (twin plan: (111), angle: 60°) [25-28]. According to EBSD data calculation, the fraction of twins formed in main slip system is 20.3%, that's why misorientation angles of 60° increased greatly. Meanwhile, there is no change at the angles of 38.9°, which means there is no twin formed in second slip system (twin plan (110), angle 38.9°). In **Figure 9**, a new grain (T) formed at the triple point between grains E, F and G. The location of

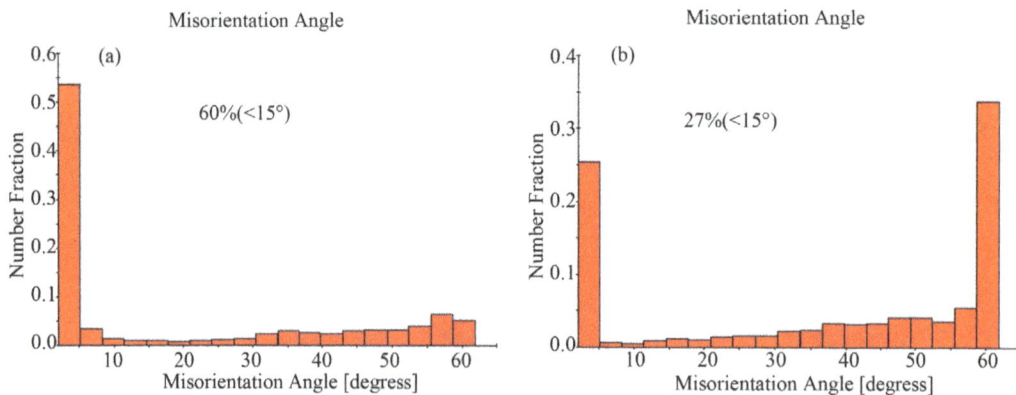

Figure 6. Misorientation angle distributions of Cu-Mg alloy alloys obtained by EBSD (a) sample at the cavity entrance (b) sample at the cavity export.

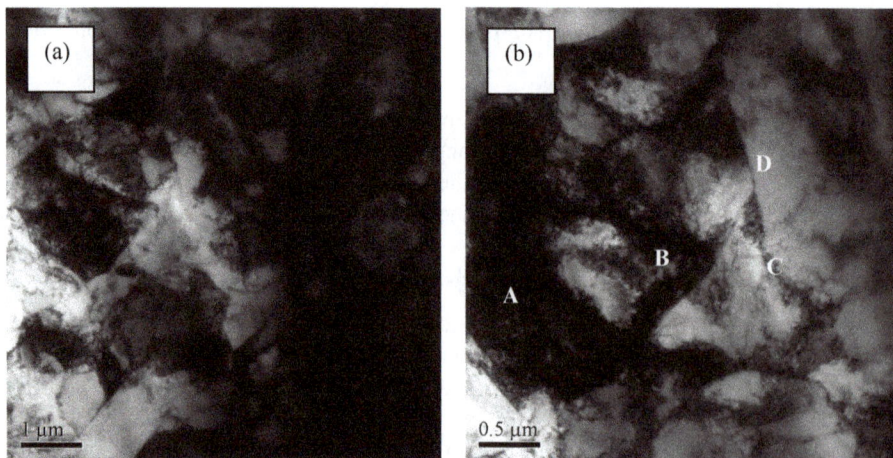

Figure 7. Bright-field TEM images of cavity entrance's sample (a) TEM micrographs of microstructures (b) TEM images of subgrain formation.

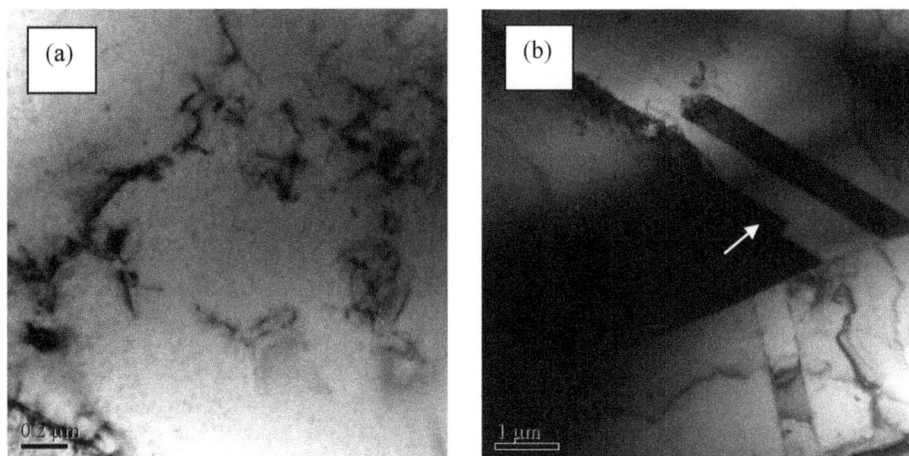

Figure 8. Bright-field TEM images of sample at the cavity export (a) dislocation in grain (b) twins and ledge of boundary migration.

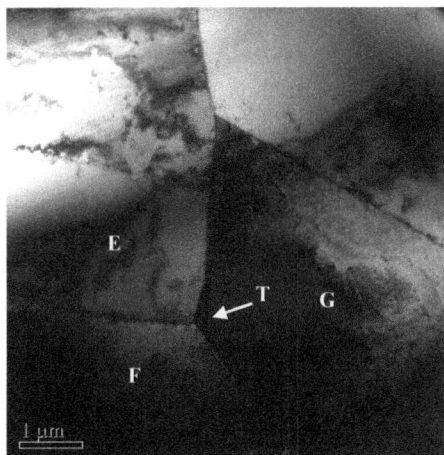

Figure 9. A new grain T formed at the triple point between grains E, F and G.

the new grains in the structure appears to be more favorable for their growth. According to Ref. [29], the grain (T) is twinned with grain G, and the energy of the boundary GT is lower than that of GF and GE, because the energy of the coherent twin boundary GT is very low, there may be a reduction in total boundary energy despite the extra boundary area created. So the anneal twins will refine the coarse-grain structure.

5. Conclusions

Continuous extrusion forming (Conform process) was an effective method to refine the cast grain structure in copper and can be successfully applied to Cu-Mg alloy. From the TEM and EBSD observation, the grains have been refined to 4.94 μm after shear deformation. However, due to the relatively high deformation temperature, dynamic recrystallization occurs during deformation. The average grain size changes to 5.45 μm.

By comparing to the microstructural evolution in

Conform process, broken grains which composed of dislocations formed during shear deformation, meanwhile, subgrains formed by dislocation reconstruction.

Due to the relatively high deformation temperature, dynamic recrystallization occurred during deformation. The fraction of the low-angle grain boundaries decreased from 60% to 27%. According to EBSD data calculation, the fraction of twins formed in main slip system (twin plan: (111), angle: 60°) is 20.3%. That's why misorientation angles of 60° increased greatly.

6. Acknowledgements

The authors would like to thank School of Materials Science and Engineering of Central South University for the financial support and provision of research facilities used in this work.

REFERENCES

[1] C. Etherington, "CONFORM—A New Concept for the Continuous Extrusion Forming of Metals," *Transactions of the American Society of Mechanical Engineers, Journal of Engineering for Industry*, Vol. 96B, No. 3, 1974, pp. 893-900.

[2] S. Harper, "Special Extrusion Processes for Non-Ferrous Metals," *The Metallurgist and Materials Technologist*, Vol. 12, No. 5, 1980, pp. 257-260.

[3] T. Reinikainen, A. S. Korhonen, K. Anderson and S. Kivivuori, "Computer-Aided Modelling of a New Copper Extrusion Process," *Annals of CIRP*, Vol. 42, No. 1, 1993, pp. 265-268. doi:10.1016/S0007-8506(07)62440-8

[4] D. Green, "Continuous Extrusion-Forming of Wire Section," Journal of the Institute of Metals (London), Vol. 100, 1972, pp. 295-300.

[5] C. Etherington, "Continuous Extrusion Forming of Metals," *Transactions of the American Society of Mechanical Engineers, Series B*, Vol. 96, No. 3, 1968, pp. 893-900.

[6] Z. Horita, T. Fujinami, M. Nemoto and T. G. Langdon,

"Improvement of Mechanical Properties for Al Alloys Using Equal-Channel Angular Pressing," *Journal of Materials Processing Technology*, Vol. 117, No. 3, 2001, pp. 288-292. doi:10.1016/S0924-0136(01)00783-X

[7] W. Z. Han, S. D. Wua, S. X. Li and Y. D. Wang, "Intermediate Annealing of Pure Copper during Cyclic Equal Channel Angular Pressing," *Materials Science and Engineering A*, Vol. 483-484, 2008, pp. 430-432. doi:10.1016/j.msea.2006.10.179

[8] A. Azushima, R. Kopp, A. Korhonen, D. Y. Yang and F. Micari, "Severe Plastic Deformation (SPD) Processes for Metals," *CIRP Annals: Manufacturing Technology*, Vol. 57, 2008, pp. 716-735. doi:10.1016/j.cirp.2008.09.005

[9] A. Mishra, V. Richard, F. Grégori, R. J. Asaro and M. A. Meyers, "Microstructural Evolution in Copper Processed by Severe Plastic Deformation," *Materials Science and Engineering A*, Vol. 410-411, 2005, pp. 290-298. doi:10.1016/j.msea.2005.08.201

[10] M. Kawasakia, Z. Horitab and T. G. Langdona, "Microstructural Evolution in High Purity Aluminum Processed by ECAP," *Materials Science and Engineering A*, Vol. 524, No. 1-2, 2009, pp. 143-150. doi:10.1016/j.msea.2009.06.032

[11] A. A. Gazder, F. Dalla Torre, C. F. Gu, C. H. J. Davies and E. V. Pereloma, "Microstructure and Texture Evolution of bcc and fcc Metals Subjected to Equal Channel Angular Extrusion," *Materials Science and Engineering A*, Vol. 415, No. 1-2, 2006, pp. 126-139. doi:10.1016/j.msea.2005.09.065

[12] I. V. Alexandrov and R. Z. Valiev, "Nanostructures from Severe Plastic Deformation and Mechanisms of Large-Strain Work Hardening," *Nanostructured Materials*, Vol. 12, No. 5-8, 1999, pp. 709-712. doi:10.1016/S0965-9773(99)00223-8

[13] T. Manninen, T. Katajarinne and P. Ramsay, "Analysis of Flash Formation in Continuous Rotary Extrusion of Copper," *Journal of Materials Processing Technology*, Vol. 177, No. 1-3, 2006, pp. 600-603. doi:10.1016/j.jmatprotec.2006.04.051

[14] L. P. Lu, X. B. Yun, J. Y. Yang and B. Y. Song, "Study on Deforming Behavior of Copper-Magnesium Alloy Wire in Extending Continuous-extrusion Process," *Hot Working Technology*, Vol. 39, 2010, pp. 92-95.

[15] S. Qu, X. H. An, H. J. Yang, C. X. Huang, G. Yang, Q. S. Zang, Z. G. Wang, S. D. Wu and Z. F. Zhang, "Microstructural Evolution and Mechanical Properties of Cu-Al Alloys Subjected to Equal Channel Angular Pressing," *Acta Materialia*, Vol. 57, No. 5, 2009, pp. 1586-1601. doi:10.1016/j.actamat.2008.12.002

[16] P. B. Prangnell, A. Gholinia, V. M. Markushev, "The Effect of Strain Path on the Rate of Formation of High Angle Grain Boundaries during ECAE," In: T. C. Lowe and R. Z. Valiev, Eds., *Investigations and Applications of Severe Plastic Deformation*, Kluwer Academic Publisher, Dordrecht, 2000, pp. 65-71. doi:10.1007/978-94-011-4062-1_9

[17] Y. Iwahashi, Z. Horita, M. Nemoto and T. G. Langdon, "The Process of Grain Refinement in Equal-Channel Angular Pressing," *Acta Materialia*, Vol. 46, No. 9, 1998, pp. 3317-3331. doi:10.1016/S1359-6454(97)00494-1

[18] Y. Iwahashi, Z. Horita, M. Nemoto and T. G. Langdon, "An Investigation of Microstructural Evolution during Equal-Channel Angular Pressing," *Acta Materialia*, Vol. 45, No. 11, 1997, pp. 4733-4741. doi:10.1016/S1359-6454(97)00100-6

[19] F. J. Humphreys, "Review Grain and Subgrain Characterisation by Electron Backscatter Diffraction," *Journal of Materials Science*, Vol. 36, 2001, pp. 3833-3854. doi:10.1023/A:1017973432592

[20] R. Z. Valiev, R. K. Islamgaliev and I. V. Alexandrov, "Bulk Nanostructured Materials from Severe Plastic Deformation," *Progress in Materials Science*, Vol. 45, No. 2, 2000, pp. 103-189. doi:10.1016/S0079-6425(99)00007-9

[21] R. Z. Valiev and T. G. Langdon, "Principles of Equal-Channel Angular Pressing as a Processing Tool for Grain Refinement," *Progress in Materials Science*, Vol. 51, No. 7, 2006, pp. 881-981. doi:10.1016/j.pmatsci.2006.02.003

[22] Y. Iwahashi, Z. Horita, M. Nemoto and T. G. Langdon, "An Investigation of Microstructural Evolution during Equal-Channel Angular Pressing," *Acta Materialia*, Vol. 45, No. 11, 1997, pp. 4733-4741. doi:10.1016/S1359-6454(97)00100-6

[23] Y. Iwahashi, Z. Horita, M. Nemoto and T. G. Langdon, "The Process of Grain Refinement in Equal-Channel Angular Pressing," *Acta Materialia*, Vol. 46, No. 9, 1998, pp. 3317-3331. doi:10.1016/S1359-6454(97)00494-1

[24] A. Mishra, V. Richard, F. Gregori, R. J. Asaro and M. A. Meyers, "Microstructural Evolution in Copper Processed by Severe Plastic Deformation," *Materials Science and Engineering A*, Vol. 410-411, 2005, pp. 290-298. doi:10.1016/j.msea.2005.08.201

[25] R. E. Reed-Hill, J. P. Hirth and H. C. Rogers, "Deformation Twinning," Gordon and Breach, New York, 1964, p. 7.

[26] T. H. Blewitt, R. R. Coltman and J. K. Redman, "Low-Temperature Deformation of Copper Single Crystals," *Journal of Applied Physics*, Vol. 28, No. 6, 1957, pp. 651-660. doi:10.1063/1.1722824

[27] J. W. Christian and S. Mahajan, "Deformation Twinning," *Progress in Materials Science*, Vol. 39, No. 1-2, 1995, pp. 1-157. doi:10.1016/0079-6425(94)00007-7

[28] A. Rohatgi, S. K. Vecchio and T. G. Gray III, "The Influence of Stacking Fault Energy on the Mechanical Behavior of Cu and Cu-Al Alloys: Deformation Twinning, Work Hardening, and Dynamic Recovery," *Metallurgical and Materials Transactions A*, Vol. 32, No. 1, 2001, pp. 135-145. doi:10.1007/s11661-001-0109-7

[29] F. J. Humphreys and M. Hatherly, "Recrystallization and Related Annealing Phenomena," Elsevier Ltd., Oxford, 2004, pp. 261-266.

25

Extraction of Palladium from Acidic Chloride Media into Emulsion Liquid Membranes Using LIX 984N-C[®]

Satit Praipruke[1], Korbratna Kriausakul[2*], Supawan Tantayanon[2,3]

[1]Program of Petrochemistry, Faculty of Science, Chulalongkorn University, Bangkok, Thailand
[2]Green Chemistry Research Laboratory, Department of Chemistry, Faculty of Science, Chulalongkorn University, Bangkok, Thailand
[3]National Center of Excellence for Petroleum, Petrochemicals and Advanced Materials (NCE-PPAM), Bangkok, Thailand
Email: *korbratna.k@chula.ac.th

ABSTRACT

The extraction of palladium from hydrochloric acid solutions into emulsion liquid membranes (ELMs) using LIX 984N-C as the extractant was investigated. The influential factors and the total capacities of palladium extraction were determined by a batch method. The behavior of palladium extraction by ELMs under the operational conditions—pH of the external feed phase, surfactant and extractant concentration, internal stripping phase concentration, treat ratio and agitation speed were reported. Using LIX 984N-C, palladium was effectively extracted from the external acidic chloride feed phase into the internal receiving phase of W/O emulsions. More than 92% of palladium could be extracted at a feed pH of 2 with 3% Span 80, 9% LIX 984N-C and 7M HCl at a stirring speed of 300 rpm.

Keywords: Palladium Extraction; LIX 984N-C; Emulsion Liquid Membrane

1. Introduction

Recently, much attention has been paid on the development of an efficient treating process for industrial wastes because they have a probability of destroying the global environment [1,2]. Palladium, one of the precious group metals (PGMs) is a major pollutant from the chemical industry which utilizes palladium as catalysts as well as electrical and corrosion resistant materials. Solvent extraction is a common hydrometallurgical process for the recovery of palladium from most industrial plants. The studies of palladium extraction using pyridinecarboxamides [3], thiosulfate, ACORGA[®] CLX50 [4], hexadecylpyridinium [5], phosphonium [6], NH_4^+-dibenzyldiaza-18-crown-6 [7], Cyanex 471 [8], hydroxyoxime [9], Cyanex 301-immobilized material [10], α-benzoin-oximes and Cyphos[®] IL/toluene [11], imidazolium nitrate immobilized resin [12] as extractants have been reported. However, the metal extraction rates by solvent extraction are generally low and the inventories of the organic solvent and metal extractant are substantial for the separation process. Emulsion liquid membranes (ELMs) [13], on the other hand a promising technology for the separation of heavy metal ions from aqueous effluent streams, has re-

moved the limitations of solvent extraction by combining extraction and stripping in a single operation. The reduction of metal concentration in the feed stream can be achieved to very low level. The treatment of aqueous streams contaminated with palladium by ELMs using a bi-functional surfactant has been reported [14,15]. In recent years, Cognis Inc. [16] has introduced oxime derivatized extractants for copper, such as LIX 63, LIX 65, LIX 84I-C, LIX 860N-IC and LIX 984N-C, showing faster kinetics, easier phase disengagement and stronger copper extraction. A combination of salicylaldoximes with extractive strength and fast kinetics and ketoximes with proven excellent physical performance and stability led to the development of LIX 984N-C, which was found to effectively extract copper from aqueous acidic medium [17,18]. With similar coordination chemistry of palladium with phenolic oxime ligands to that of copper [19], this gives an insight for the use of LIX 984N-C for palladium extraction.

In the present work, the extraction of palladium from aqueous hydrochloric acid medium containing a low concentration of palladium (10 ppm) by ELMs using LIX 984N-C as an extractant is carried out. The behavior palladium extraction and optimum operating condition by ELMs are also investigated.

---

*Corresponding author.

2. Experimental

2.1. Reagents

The chemical structures of the carrier (extractant) and the surfactant are used in this study. The carrier, LIX 984N-C obtained from Cognis Ireland is an equi-volume mixture of LIX 860N-IC (5-nonyl salicylaldoxime) and LIX 84I-C (2-hydroxy-5-nonyl-acetophenone-oxime) (**Figure 1(a)**) in a high flash point hydrocarbon diluent and used as such. Emulsifier, Span 80 (sorbitan monooleate) (**Figure 1(b)**) was supplied by Kao Industrial, Thailand. The commercial extractant and surfactant were used as received. The organic solvent, Exxol™ D80 (dearomatized aliphatic hydrocarbon), supplied by ExxonMobil Chemical (Thailand) Co., Ltd., was used as the membrane material.

2.2. [Pd(II)-LIX 984N-C] Complex Formation

Equivalent volumes (20 mL) of aqueous acidic feed solution (pH 2) containing 10 ppm of Pd(II) and the organic phase of LIX 984N-C in Exxol D80 were allowed to contact each other for 60 min at room temperature (30 ± 1°C). After gravitational settling the aqueous and organic phases were separated for analyses. A Perkin-Elmer model AAnalyst 100 AAS was used to determine the Pd(II) concentration in the aqueous portion. UV spectra of Pd(II) in aqueous feed solution were recorded before and after the extraction process. [Pd(II)-LIX 984N-C] complex in the organic phase was also identified by a UV spectrometer equipped with a deuterium-halogen light source DH 2000 and USB 2.0 fiber optic model USB4000 UV-VIS detector.

2.3. Pd(II) Extraction into ELMs

Batch extractions were conducted in a glass-mixer-settler (7" dia.) fitted with baffles, a stop-cock for easy sampling and a variable speed 30 mm turbine impeller. The primary W/O emulsion was prepared by gradually dripping HCl solution into the oil phase in a beaker using a high speed homogenizer (IKA, Ultra Turrax model T10 basic) at 12,000 rpm for about 30 min. The resultant milky white emulsion was then dispersed into the aqueous feed phase containing 10 ppm Pd(II) using a turbine mixer at 250 to 300 rpm. Samples were drawn at fixed intervals (5 min.) and the emulsion and aqueous phases were allowed to settle in measuring test tubes. The concentration of Pd(II) in the aqueous phase was determined by standard procedures using Perkin-Elmer model AAnalyst 100 AAS. The parametric variation of the components in the external aqueous feed phase, the membrane phase and internal aqueous phase are described in **Table 1**.

Palladium extraction into ELMs is represented by C_e/C_{e0} as a function of time, where C_e is the instantaneous concentration and C_{e0} is the initial concentration of palladium in the external aqueous feed solution.

3. Results and Discussion

3.1. [Pd(II)-LIX 984N-C] Complex Formation

The UV spectra of Pd(II) in aqueous solution that a prominent peak at 244 nm with a maximum absorbance of 0.84 indicates the initial Pd(II) in the aqueous phase (**Figure 2(a)**) and after extraction the absorbance was reduced to 0.17 with a broadening of the peak (**Figure 2(b)**). The

Carrier

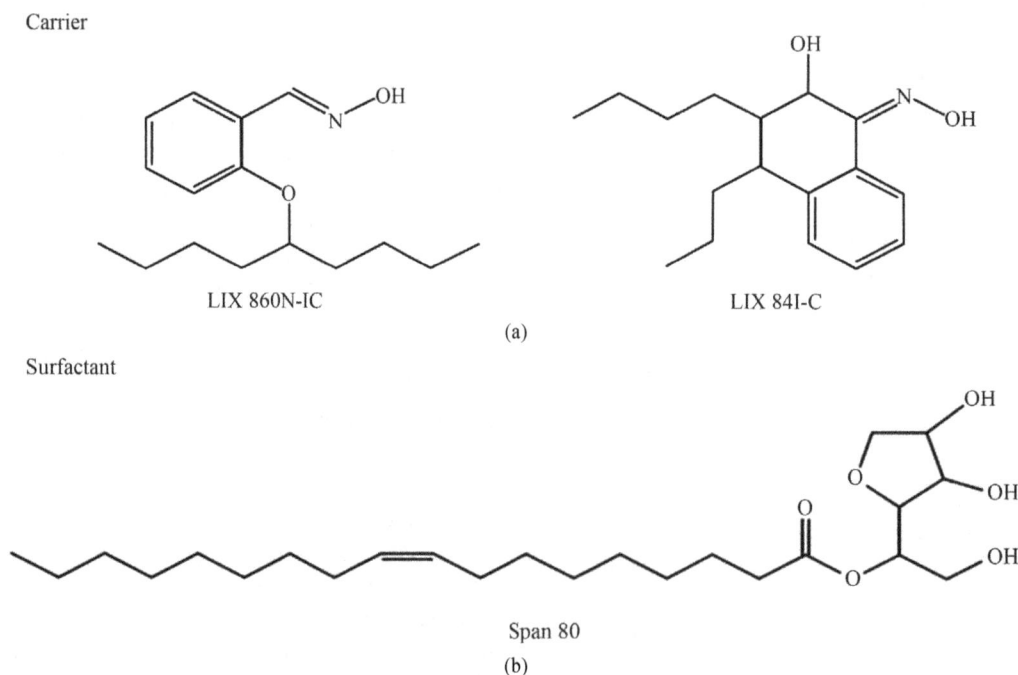

LIX 860N-IC

LIX 84I-C

(a)

Surfactant

Span 80

(b)

Figure 1. Chemical structure of (a) Carrier and (b) Surfactant.

Table 1. Parametric variation of the components in the external aqueous feed phase, the membrane phase and internal aqueous phase for Pd(II) extraction by ELMs.

Parameters	Range used	Remarks
External aqueous feed phase (W_{III})	$V_e = 300\ mL$	
(a) palladium concentration	$C_e = 10$ ppm	
(b) with buffer pH	pH = 1 - 3	Adjusted pH by hydrochloric solution
Oil Membrane phase (O_{II})	$V_m = 25$ mL	
(a) solvent	Exxol D80	De-aromatized aliphatic hydrocarbon
(b) carrier	LIX 984N-C®	
	$C_c = 5 - 12$ wt%	
(c) surfactant	Span 80	
	$C_s = 2.5 - 3.5$ wt%	
Internal aqueous phase	$V_i = 25$ mL	
(a) stripping acid	HCl	
	$C_i = 3 - 7$ M	
Treat ratios	$T_r = 0.125 - 0.250$	
Agitation speed	250 - 350 rpm	
The ratio of the phases	1:1	The emulsion is water-in-oil (W/O) type

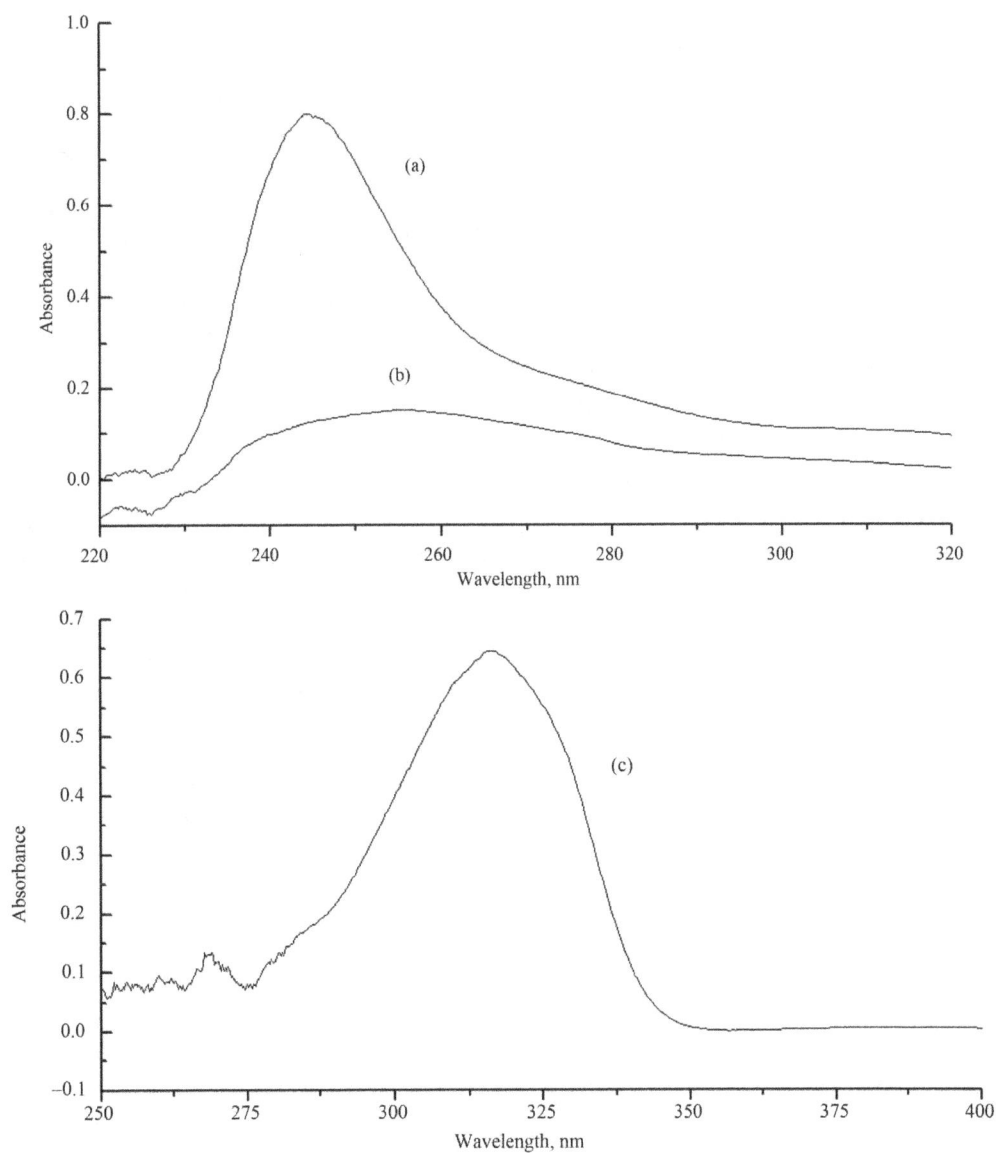

Figure 2. UV absorption spectra (a) Palladium in feed solution before extraction; (b) Palladium after extraction and (c); [Pd(II)-LIX 984N-C] complex in organic phase.

presence of the [Pd(II)-LIX 984N-C] complex in the organic phase [20] was evident by the prominent peak at 319 nm with an absorbance of 0.65 (**Figure 2(c)**).

3.2. Permeation of Pd(II) through ELMs

When the W/O emulsion is dispersed in the feed phase, several emulsion globules are formed generating a large surface area for the interfacial reactions and mass transfer. Palladium exists mainly as Pd^{2+} ions in an external feed solution containing hydrochloric acid with a low H^+ ion concentration. At the interface of an emulsion globule and the feed phase the following reaction takes place, leading to a formation of a PdR_2 complex in the organic phase:

$$2HR_{(org.)} + Pd^{2+}_{(aq.)} \ \square \ PdR_{2(org.)} + 2H^+_{(aq.)}$$

where HR represents the carrier in the organic phase. The PdR_2 complex diffuses through the organic membrane phase to the inner stripping phase, where the Pd^{2+} ion gets stripped from the carrier into the internal phase of emulsion droplets by the following chemical reaction:

$$2H^+_{(aq.)} + PdR_{2(org.)} \ \square \ Pd^{2+}_{(aq.)} + 2HR_{(org.)}$$

This type of carrier facilitated transport mechanism has been recognized already as a counter transport [21,22] as schematically shown in **Figure 3**.

The counter transport mechanism for Pd(II) extraction by ELMs is confirmed by the result of the Pd^{2+} and H^+ concentration in the external feed phase, shown in **Table 2**. A depletion of palladium concentration in the external aqueous feed solution from 10.00 to 1.742 ppm and a simultaneous fall of pH from 2.00 to 1.75 were found to take place during the 60 minutes of extraction.

3.3. Effect of Parametric Variations on Palladium Extraction

In all experimental runs, the internal phase volume fraction of the extracting emulsion was 1:1 and initial palladium

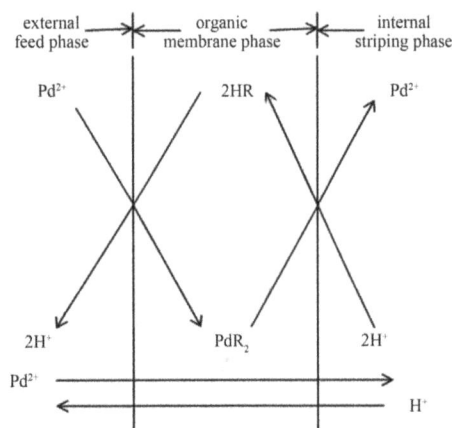

Figure 3. Schematic diagram of permeation mechanism of Pd(II) with ELMs.

Table 2. Pd(II) and H^+ concentration in the feed solution during extraction process by ELMs.

Run	Time (min)	pH	Concentration	
			Pd(II), ppm	H^+
1	0	2.00	10.00	0.010
2	5	1.93	5.397	0.012
3	10	1.89	4.629	0.013
4	15	1.86	4.007	0.014
5	20	1.85	3.291	0.014
6	25	1.83	3.199	0.015
7	30	1.78	2.821	0.017
8	35	1.77	2.373	0.017
9	40	1.77	2.243	0.017
10	45	1.77	1.995	0.017
11	50	1.75	1.908	0.018
12	55	1.75	1.745	0.018
13	60	1.75	1.742	0.018

concentration (C_{e0}) in the external feed phase 10 ppm. Other operating parameters were varied to study their effect: surfactant concentration, C_s (2.5 - 3.5 wt%); external feed pH, (1 - 3); carrier LIX 984N-C concentration, C_c (3 - 12 wt%); internal acid concentration, C_i (3 - 7 M); agitation speed (250 - 300 rpm) and treat ratio, T_r (1:4 - 1:8). The amount extracted at any time was expressed as a ratio of the instantaneous concentration to initial concentration of palladium (C_e/C_{e0}).

3.3.1. Effect of Surfactant Concentration

The influence of surfactant concentration in the membrane phase of the emulsion on the extraction rates was investigated at a concentration range of 2.5% to 3.5% Span 80 by weight. **Figure 4** indicates significant increase in the initial extraction rate within 5 minutes, due to a low surface tension of the membrane phase, resulting in small-sized emulsion droplets, allowing faster mass transfer for palladium extraction. A low concentration of Span 80 (2.5%) renders the membrane weak and the emulsion became unstable. Increasing the concentration of Span 80 (3.5%) enhances mass transfer resistance due to the increase in viscosity of the membrane phase [17]. Both results led to lower total extraction efficiency. With 3.0% Span 80 by weight maximum palladium extraction could be achieved. Therefore, the surfactant concentration of 3.0% was chosen for further studies.

3.3.2. Effect of the Carrier Concentration

The effect of LIX 984N-C concentration was studied in the range of 5% to 12% by weight of the membrane phase at an external feed pH of 2, while maintaining all other operating conditions constant. **Figure 5** shows an increase in the initial rate and total amount extracted with increasing carrier concentration from 5.0% to 9.0% by

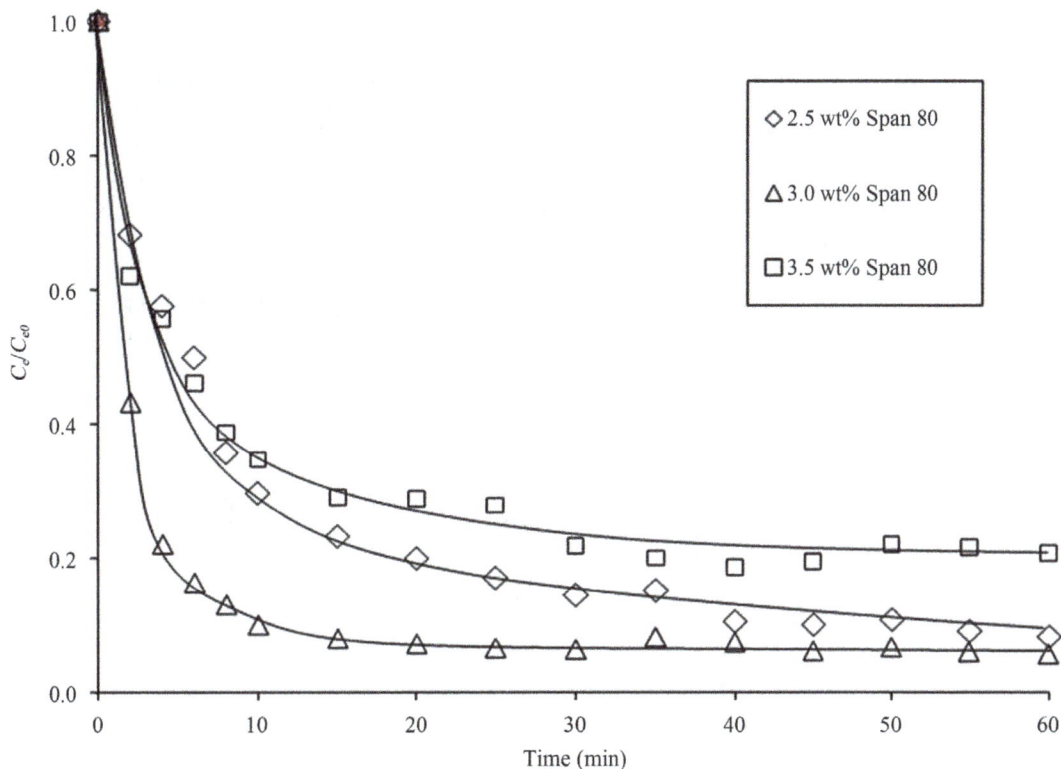

Figure 4. Effect of surfactant concentration on palladium extraction.

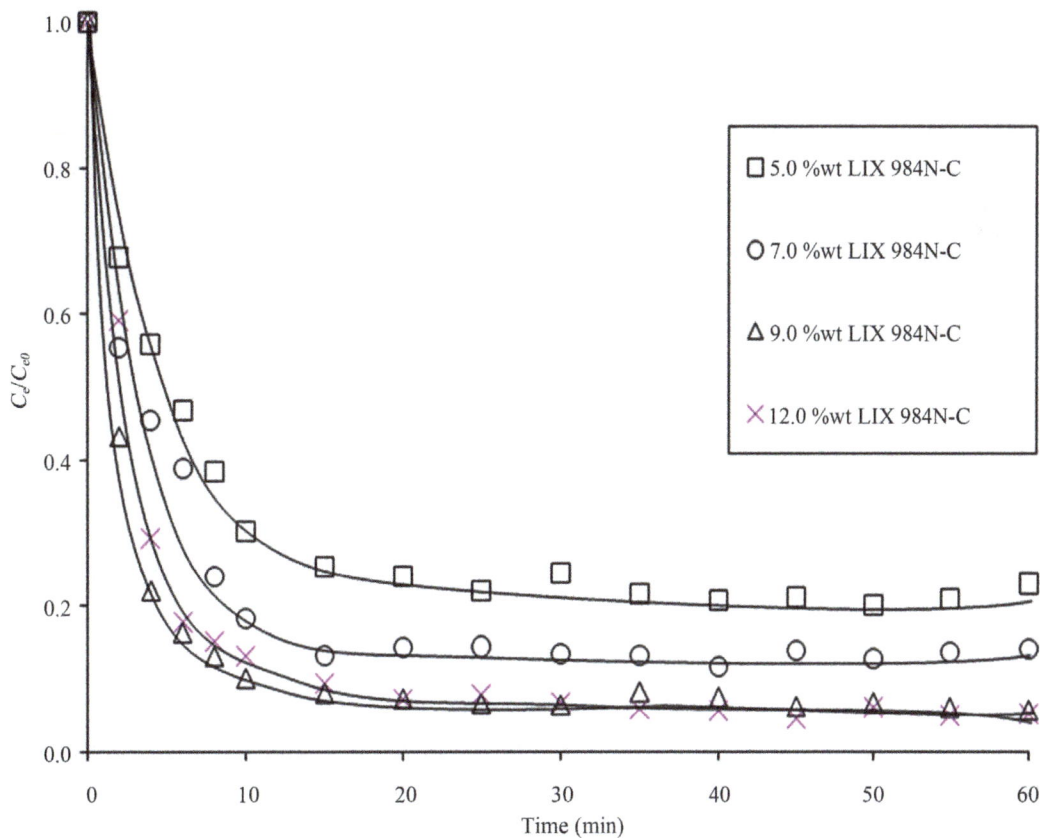

Figure 5. Effect of LIX 984N-C concentration on palladium extraction.

weight. Increasing the carrier concentration in the membrane phase phenomenologically increases the interfacial solute concentration and hence the driving force for the extraction leading to a high extraction rate [18]. Further increasing the LIX 984N-C concentration to 12% by weight did not show any change and the final amount extracted remains the same. This demonstrates that changing the carrier concentration does not change the final equilibrium conditions of the system, but affects how fast the equilibrium is reached. Moreover, the carrier acts as a thinner at high concentration due to the higher viscosity in the membrane phase leading to enhanced stability of the emulsion [21,22]. Therefore, the carrier concentration of 9.0% was used to form the W/O emulsion.

3.3.3. Effect of Acid Concentration in the Internal Stripping Phase

The experiments were conducted with HCl concentrations of 3 M, 5 M and 7 M at a phase pH of 2. From **Figure 6**, it is evident that both the initial extraction rate and total Pd(II) extraction depend on the stripping acid concentration, increasing with higher HCl molarity as a result of the difference in H^+ concentration between the external and internal phases acting as a driving force in the ELM process [23] and the ability of the internal phase for stripping the [Pd(II)-LIX 984N-C] complex at the membrane/internal stripping phase. At 7 M HCl the

chemical reaction was found to be very fast, with the extraction equilibrium being achieved in about 15 minutes. [22], using hydroxyoxime extractant and hydrochloric acid medium, found the stripping of palladium incomplete with hydrochloric acid concentrations below 6 M. The optimum concentration for the internal stripping phase regent of 7.0 M HCl is thus chosen.

3.3.4. Effect of Feed pH

The palladium extraction was carried out with the acidity of the external feed solution being adjusted from pH 1 to 3 by hydrochloric acid. **Figure 7** indicates that the palladium extraction is strongly pH dependent. With a feed pH of 1 and 3, only 10% of palladium was extracted. However, almost 100% of palladium could be extracted within 20 minutes at a feed pH of 2. [18] also found similar behavior for copper extraction at pH 1 and 2 using LIX 984N-C, but with a higher amount extracted and no further change at higher pH.

3.3.5. Effect of Treat Ratio

The ratio of the emulsion phase to the external aqueous feed phase, treat ratio (R_{ew}), is a measure of the emulsion holdup in the system. **Figure 8** shows the extraction behavior for five diffent R_{ew} values. The rate and amount of palladium extraction were found to increase with the R_{ew} value during the 25 minutes of contact time. When the

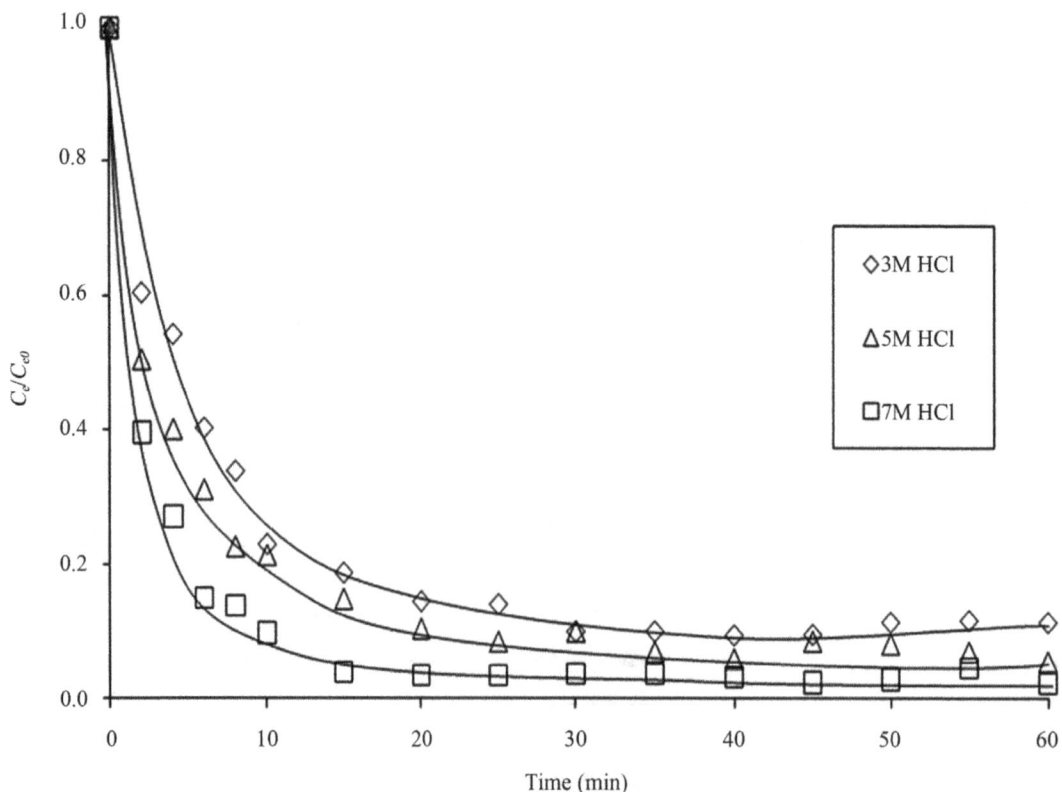

Figure 6. Effect of stripping acid concentration on palladium extraction.

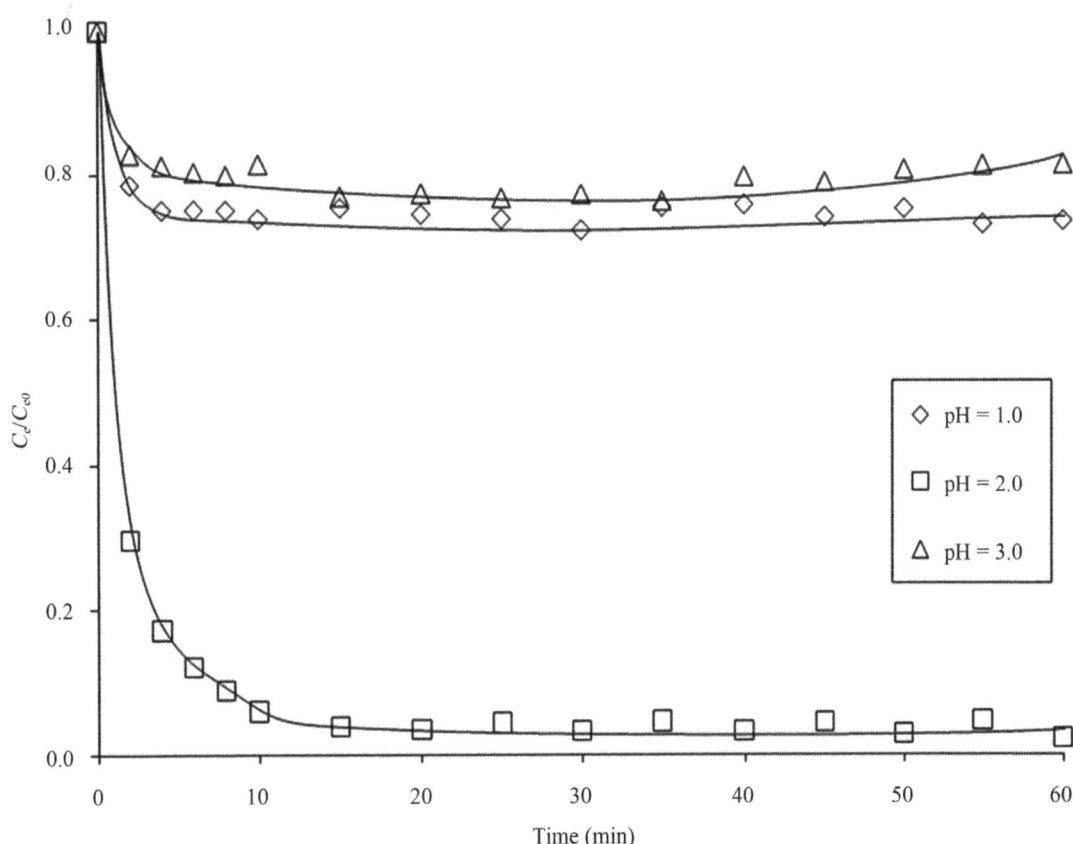

Figure 7. Effect of external feed pH on palladium extraction.

emulsion holdup is low (R_{ew} = 0.125, 0.140), the total amount extracted after 60 minutes was less than 85%. Increasing the emulsion holdup (R_{ew} = 0.200, 0.250), led to larger extent of extraction (92%), back extraction was later observed due to globule breakage, depicted as a rise in C_e/C_0 versus t curve. With R_{ew} value of 0.167 the extension behavior was found to be optimal and is chosen for the treat ratio.

3.3.6. Effect of Agitation Speed

A proper agitation speed was determined by conducting experiments with a variation of speed from 250 rpm to 350 rpm as shown in **Figure 9**. The initial rate of palladium extraction was found to increase with increasing agitation speeds from 250 to 350 rpm, due to smaller emulsion globules, which provide a larger mass transfer area, forming as a result of the shear force of agitation [24]. Total palladium extraction capacities with agitation speeds of 250 and 275 rpm were less (80%) than with 300 rpm (92%). A further increase in the agitation speed to 350 rpm, resulted in a decrease of the amount of palladium extracted after 15 minutes with a total extraction of less than 70%. At high agitation speeds the emulsion globules coalesce resulting in enlargement of the globules, indicating two competing processes in the solute transport: the diffusion of the solute through the emul-

sion membranes into the internal phase and the leakage of the internal solution due to globule breakage.

4. Conclusion

A new efficient carrier, LIX 984N-C, has been introduced for the extraction of palladium from acidic chloride media by an ELM method. A stable W/O emulsion was formulated with 7 M HCl solution, 3% Span 80 and 9% by weight LIX 984N-C in Exxol D80. The behavior of palladium extraction by ELMs under the operational conditions; pH of the external feed phase, treat ratios and agitation speed were investigated. The feed pH was found to be the most prominent factor, resulting in the highest initial rate of palladium extraction at a pH of 2. More than 92% of palladium could be extracted under the optimum conditions with treat ratio of 1:6 and agitation speed of 300 rpm. The permeation mechanism of palladium extraction by ELMs was proved to be counter transport involving a [Pd(II)-LIX 984N-C] complex. It is ascertained that LIX 984N-C allows efficient extraction of dilute palladium solutions by ELMs relative to the conventional solvent extraction method.

5. Acknowledgements

The authors would like to thank ExxonMobil Chemical

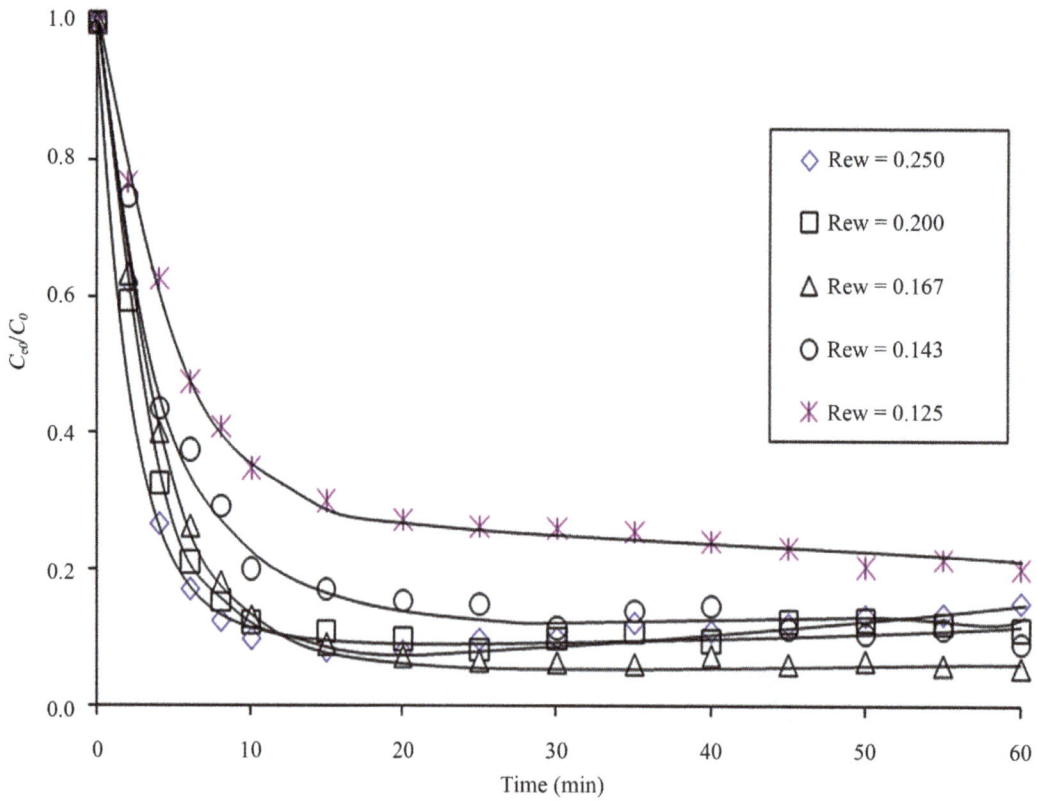

Figure 8. Effect of treat ratio on palladium extraction.

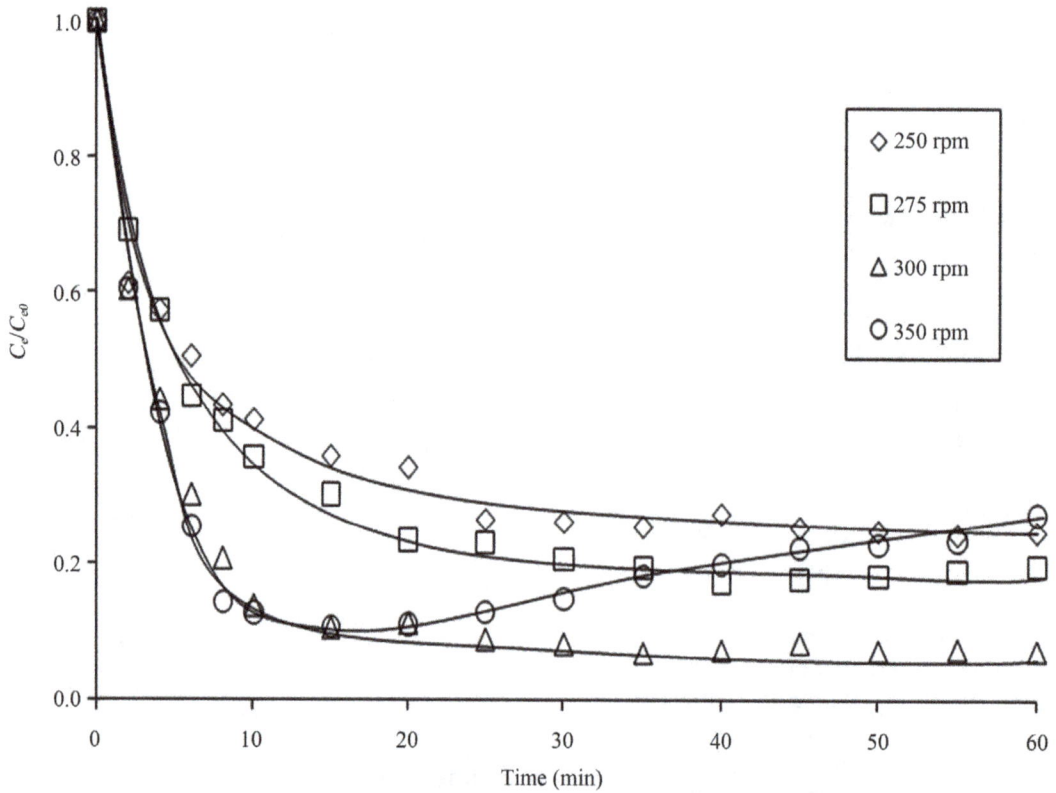

Figure 9. Effect of agitation speed on palladium extraction.

(Thailand) Limited, Bangkok, Thailand, for the provision kerosene and Cognis Inc., Ireland for the provision of LIX 984N-C® reagents. This work was supported by Program of Petrochemistry and Polymer Science, and National Center of Excellence for Petroleum, Petrochemicals and Advanced Materials (NCE-PPAM), Chulalongkorn University.

REFERENCES

[1] S. Shen, T. Pan, X. Liu, L. Yuan, Y. Zhang, J. Wang and Z. Guo, "Adsorption of Pd(II) Complexes from Chloride Solutions Obtained by Leaching Chlorinated Automotive Catalysts on Ion Exchange Resin Diaion WA21J," *Journal of Colloid and Interface Science*, Vol. 345, No. 1, 2010, pp. 12-18. doi:10.1016/j.jcis.2010.01.049

[2] A. K. Chakravarti, S. B. Chowdhury and D. C. Mukherjee, "Liquid Membrane Multiple Emulsion Process of Separation Copper(II) from Waste Water," *Colloids and Surfaces A: Physicochemical and Engineering Aspects*, Vol. 166, No. 1-3, 2000, pp. 7-25. doi:10.1016/S0927-7757(99)00452-5

[3] A. Dakshinamoorthy, A. G. C. Nair, S. K. Das, R. K. Singh and S. Prakash, "A Rapid and Selective Separation of Palladium," *Journal of Radioanalytical and Nuclear Chemistry*, Vol. 162, No. 1, 1992, pp. 155-162. doi:10.1007/BF02039936

[4] C. Nowottny, W. Halwachs and K. Schugerl, "Recovery of Platinum, Palladium and Rhodium from Industrial Leaching Solutions by Reactive Extractaction," *Separation Science and Technology*, Vol. 12, No. 2, 1997, pp. 135-144.

[5] G. Khayatian and M. Shamsipur, "Highly Selective and Efficient menbrane Transport of Palladium as Pdcl42—Ino Using NH4+-Dibenzyldiaza-18-Crown-6," *Separation and Purification Technology*, Vol. 16, No. 2-3, 1999, pp. 235-241.

[6] N. Alizadeh, S. Salimi and A. Jabbari, "Transport Study of Palladium through a Bulk Liquid Membrane Using Hexadecylpyridinium as Carrier," *Separation and Purification Technology*, Vol. 28, No. 3, 2002, pp. 173-180. doi:10.1016/S1383-5866(02)00003-5

[7] I. Szczepanska, A. Borowiak-Resterna and M. Wisniewski, "New Pyridinecarboxamides for Rapid Extraction of Palladium from Acidic Chloride Media," *Hydrometallurgy*, Vol. 68, No. 1-3, 2003, pp. 159-170. doi:10.1016/S0304-386X(02)00201-3

[8] C. Fantas, V. Salvado and M. Hidalgo, "Selective Enrichment of Palladium from Spent Automotive Catalysts by Using a Liquid Membrane System," *Journal of Membrane Science*, Vol. 223, No. 1-2, 2003, pp. 39-48. doi:10.1016/S0376-7388(03)00288-6

[9] E. Guaibal and T. Vincent, "Palladium Recovery from Dilute Effluents Using Biopolymer-Immobilized Extractant," *Separation Science and Technology*, Vol. 41, No. 11, 2006, pp. 2533-2553.

[10] Z. Hubicki, M. Leszczynska, B. Lodyga and A. Lodyga, "Palladium(II) Removal from Chloride and Chloride-Nitrate Solutions by Chelating Ion-Exchangers Containing N-Donor Atoms," *Mineral Engineering*, Vol. 19, No. 13, 2006, pp. 1341-1347. doi:10.1016/j.mineng.2006.01.004

[11] J. H. Brits and D. A. Deglon, "Palladium(II) Stripping rates in PGM Refining," *Hydrometallurgy*, Vol. 89, No. 3-4, 2007, pp. 253-259. doi:10.1016/j.hydromet.2007.07.012

[12] M. Regel-Rosocka, M. Wisniewski, A. Borowiak-Resterna, A. Cieszynska and A. M. Sastre, "Selective Extraction of Palladium(II) from Hydrochloric acid Solutions with Pyridinecarboxamides and ACORGA®CLX50," *Separation and Purification Technology*, Vol. 53, No. 3, 2007, pp. 337-341.

[13] N. N. Li, "Separating Hydrocarbons with Liquid Membranes," US Patent No. 3410794, 1968.

[14] T. Kakoi, N. Horinouchi, M. Goto and F. Nakashio, "Selective Recovery of Palladium from a Simulated Industrial Waste Water by Liquid Surfactant Membrane Process," *Journal of Membrane Science*, Vol. 118, No. 1, 1996, pp. 63-71. doi:10.1016/0376-7388(96)00102-0

[15] T. Kakoi, M. Goto and F. Nakashio, "Separation of Platinum and Palladium(II) by Liquid Surfactant Membranes Utilizing a Novel Bi-Functional Surfactant," *Journal of Membrane Science*, Vol. 120, No. 1, 1996, pp. 77-88. doi:10.1016/0376-7388(96)00137-8

[16] Cognis, "Mining Chemicals Technology". http://www.cognis-us.com/

[17] B. Sengupta, R. Sengupta and N. Subrahmanyam, "Copper Extraction into Emulsion Liquid Membranes Using LIX 984N-C," *Hydrometallurgy*, Vol. 81, No. 1, 2006, pp. 67-73.

[18] B. Sengupta, R. Sengupta and N. Subrahmanyam, "Process Intensification of Copper Extraction Using Emulsion Liquid Membranes: Experimental Search for Optimal Conditions," *Hydrometallurgy*, Vol. 84, No. 1-2, 2006, pp. 43-53. doi:10.1016/j.hydromet.2006.04.002

[19] A. G. Smith, P. A. Tasker and D. J. White, "The Structures of Phenolic Oximes and Their Complexes," *Coordination Chemistry Reviews*, Vol. 241, No. 1-2, 2003, pp. 61-85. doi:10.1016/S0010-8545(02)00310-7

[20] A. Dakshinamoorthy and V. Venugopall, "Solvent Extraction Studies on the Complexation of Palladium with Alpha Benzoin Oxime," *Journal of Radioanalytical Nuclear Chemistry*, Vol. 266, No. 3, 2005, pp. 425-429. doi:10.1007/s10967-005-0927-y

[21] E. L. Cussler, D. F. Evan and M. A. Matesich, "Theoretical and Experimental Basis for a Specific Counter Transport System in Membranes," *Science*, Vol. 172, No. 3981, 1971, pp. 377-379. doi:10.1126/science.172.3981.377

[22] M. T. A. Reis and J. M. R. Carvalho, "Modelling of Zinc Extraction from Sulphate Solutions with Bis(2-Ethylhexyl) Thiophosphoric Acid by Emulsion Liquid Membranes," *Journal of Membrane Science*, Vol. 237, No. 1-2, 2004, pp. 97-107. doi:10.1016/j.memsci.2004.02.025

[23] A. Kargari, T. Kaghazchi and M. Soleimani, "Role of Emulsifier in the Extraction of Gold (III) Ions from Aqueous Solutions Using the Emulsion Liquid Membrane Technique," *Desalination*, Vol. 162, 2004, pp. 237-247. doi:10.1016/S0011-9164(04)00047-5

[24] M. V. Rane and V. Venugopal, "Study on the Extraction of Palladium(II) and Platinum(IV) Using LIX 84I," *Hydrometallurgy*, Vol. 84, No. 1-2, 2006, pp. 54-59. doi:10.1016/j.hydromet.2006.04.005

26

Efficacy of Bacterial Adaptation on Copper Biodissolution from a Low Grade Chalcopyrite Ore by *A. ferrooxidans*

Abhilash[*], Kapil Deo Mehta, Bansi Dhar Pandey

CSIR-National Metallurgical Laboratory, Jamshedpur, India

Email: [*]biometnml@gmail.com

ABSTRACT

A low-grade ore containing ~0.3% Cu, remains unutilized for want of a viable process at Malanjkhand Copper Project (MCP), India in which copper is present as chalcopyrite associated with pyrite in quartz veins and granitic rocks. In order to extract copper from this material, bioleaching has been attempted on bench scale using *Acidithiobacillus ferrooxidans* (*A. ferrooxidans*) isolated from the native mine water. The enriched culture containing *A. ferrooxidans* when adapted to the ore and employed for the bioleaching at 5% (w/v) pulp density, pH 2.0 and 25°C with three particle sizes viz.150 -76 μm, 76 - 50 μm and <50 μm, resulted in recovery of 38.31%, 29.68% and 47.5% Cu respectively with a maximum rise in redox potential (E_{SCE}) from 530 to 654 mV in 35 days. Under similar conditions, the unadapted strains gave a recovery of 44.0% for <50 μm size particles with a rise in E_{SCE} from 525 to 650 mV. On using unadapted bacterial culture directly in shake flask at pH 2.0 and 35°C temperature and 5% (w/v) pulp density (PD) for <50 μm size particles, 72% Cu bio-dissolution was achieved in 35 days. Copper biorecovery increased to 75.3% under similar conditions with a rise in bacterial count from 1×10^7 cells/mL to 1.13×10^9 cells/mL in 35 days. The higher bio-recovery of copper with the adapted bacterial culture may be attributed to the improved iron oxidation (Fe^{2+} to Fe^{3+}) exhibiting higher E_{SCE} as compared to that of unadapted strains.

Keywords: Bioleaching; Low-Grade Ore; Chalcopyrite; *Acidithiobacillus ferrooxidans*

1. Introduction

With the depletion of high grade resources of ores, there is a need to process low grade/discarded ores and tailings to meet the current demand of metals. The existing conventional processes are not suitable to recover the metals from such resources due to high energy consumption and environmental pollution. Biohydrometallurgy is one such approach which is environmentally benign and can be used to recover metals economically with the application of microbes [1]. The understanding of bio-leaching process began with the isolation of a mesophilic and acidophilic bacteria from mine waters by Colmer and Hinke (1947) and its usage in bioleaching of copper ores [2]. The bacteria such as *Acidithiobacillus ferrooxidans* (*A. ferrooxidans*) have been often used for dissolution of copper from different low grade ores and tailings [3-5]. Various researchers have been actively involved in elucidating the role of bacteria in bioleaching while catalyzing the oxidation of metal sulfides [6-11]. Two mechanisms of bacterial action for sulfide oxidation with *A. ferrooxidans* have been suggested: 1) direct mechanism involving

bacterial oxidation through enzymatic reaction, and 2) indirect mechanism, where the bacteria generate Fe(III) by the oxidation of Fe(II), which in turn oxidizes the metal sulfides [10].

The bioleaching of copper from a low grade chalcopyrite ore is reported to involve both direct and indirect leaching mechanism [11]. The direct mechanism proceeds through the attachment of *A. ferrooxidans* (*Ac. Tf*) on the surface of minerals viz., $CuFeS_2$ and FeS_2 (pyrite) to oxidize the metals.

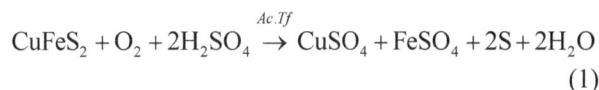

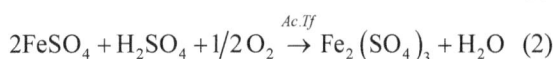

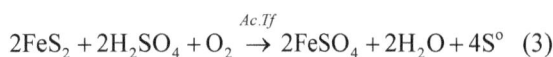

$$CuFeS_2 + O_2 + 2H_2SO_4 \xrightarrow{Ac.Tf} CuSO_4 + FeSO_4 + 2S + 2H_2O \quad (1)$$

$$2FeSO_4 + H_2SO_4 + 1/2O_2 \xrightarrow{Ac.Tf} Fe_2(SO_4)_3 + H_2O \quad (2)$$

$$2FeS_2 + 2H_2SO_4 + O_2 \xrightarrow{Ac.Tf} 2FeSO_4 + 2H_2O + 4S° \quad (3)$$

The dissolution of chalcopyrite also involves oxidative ferric reaction with the mineral, which essentially represents indirect leaching mechanism:

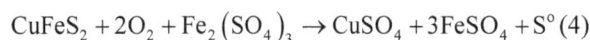

$$CuFeS_2 + 2O_2 + Fe_2(SO_4)_3 \rightarrow CuSO_4 + 3FeSO_4 + S° \quad (4)$$

This, apart from dissolution of the metal ions, produces ferrous iron and elemental sulfur ($S°$). It is this ferrous iron and the elemental sulfur that form the substrate for microbial growth according to reaction:

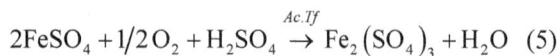

$$2FeSO_4 + 1/2 O_2 + H_2SO_4 \xrightarrow{Ac.Tf} Fe_2(SO_4)_3 + H_2O \quad (5)$$

and:

$$S° + 2O_2 + 2H^+ \xrightarrow{Ac.Tf} H_2SO_4 \quad (6)$$

The ferric iron thus formed is hydrolyzed in aqueous solution if pH is higher.

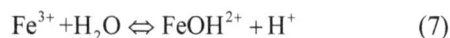

$$Fe^{3+} + H_2O \Leftrightarrow FeOH^{2+} + H^+ \quad (7)$$

$$Fe^{3+} + 2H_2O \Leftrightarrow Fe(OH)_2^+ + 2H^+ \quad (8)$$

Reaction (5) increases the pH, but the reaction (7) and (8) reduce and stabilizes it. So the extent of ferric iron hydrolysis is dependent on pH.

In India, the demand for copper is also increasing with industrialisation and to meet the requirement of the metal, the low grade ores/tailings located in Malanjkhand (M.P.), Khetri (Rajasthan) and Singhbhum (Jharkhand) regions can be an important source. The Malanjkhand Copper Project (MCP) is the largest open-cast mine having copper deposits of 22 million T of low grade ores (~0.3%) remaining untapped because no conventional processing technology has been found suitable. Biohydrometallurgical processing can be explored to assess the technical feasibility and it has so been projected to recover copper from this low grade reserve [12-14]. The R & D studies carried out on bench scale for bio-leaching of copper from the lean MCP ore at NML, Jamshedpur are presented in this paper with the aim of optimizing the process parameters using enriched culture of *A. ferrooxidans*, initially derived from mine water of MCP and also with the adapted isolates to compare their efficiency.

2. Materials and Methods

2.1. Copper Ore

Lean grade copper ore (~0.3% Cu) obtained as lumps from the mines of Malanjkhand Copper Project, M.P was crushed, ground and sieved to obtain different size fractions. Representative samples were then prepared by coning and quartering method for ore and each fraction to get sieve analysis and respective chemical analysis by using atomic absorption spectrophotometer [15]. Chemical analysis of the ore is given in **Table 1** whereas chemical analysis of sieve fractions is given in **Table 2**. It may be seen that the composition of the different fractions are almost same.

2.2. Phases in the Copper Ore

The low grade Malanjkhand copper ore is mostly a granitic

Table 1. Chemical analysis of copper ore from MCP (%).

Cu	Ni	Fe	Co	Zn	S	TiO$_2$	SiO$_2$
0.32	0.23	3.91	0.05	0.05	2.8	0.60	68.2

Table 2. Chemical analysis of different sieve fractions of copper ore.

Particle size (μm)	Fraction retained (%)	Cumulative Fraction retained (%)	Composition (%)		
			Cu	Ni	Fe
150 - 76	3.33	3.33	0.321	0.228	3.90
76 - 50	34.92	38.25	0.316	0.225	3.92
<50	61.75	100	0.320	0.230	3.91

rock and the preliminary mineralogical studies indicated the presence of chalcopyrite in the cracks and fissures of quartz veins. The petrological analysis for distribution of phases (**Figure 1**) showed the presence of chalcopyrite in the form of irregular grains, and pyrite (brighter than chalcopyrite) was observed within feldspar grains. The bulk ore was slightly pink in appearance due to high percentage of granite in the ore body.

XRD phase identification analysis of the ore (**Table 3**) showed quartz (SiO$_2$), chalcopyrite (CuFeS$_2$), pyrite (FeS$_2$) as the major phases whereas bornite was the minor phase.

2.3. Bacteria Used and Bioleaching Experiments

The micro-organism used in this study was a culture of *Acidithiobacillus ferrooxidans* (*A. ferrooxidans*), derived by successive enrichment of mine water of MCP using ferrous sulfate as the substrate in 9 K nutrient media at pH 2.0. The isolated microbial culture from the mine water was grown and specific colonies were successively re-plated and sub-cultured to get the enriched culture containing *A. ferrooxidans*. The bacterium was identified for its shape and size by staining as straight rods with variable motility and examining the cell count microscopically. The enriched culture thus derived was used in subsequent bioleaching experiments [14,16]. In general, bioleaching tests were carried out in 500 mL conical flasks with 200 mL of total solution. Experiments were carried out using unadapted culture and culture adapted over the ore. Leaching solution was inoculated with 10% (v/v) of active and enriched culture and a known quantity of sized ore was added. In case of sterile/control experimental sets, mercuric chloride (0.02 g/L) was added as a bactericide. For the experiments with adapted isolates, the enriched culture was initially adapted over ore of <150 μm size at 5% (w/v) pulp density, pH 2.0 and 35°C, and the culture was further grown in large volumes for the use in further experiments.

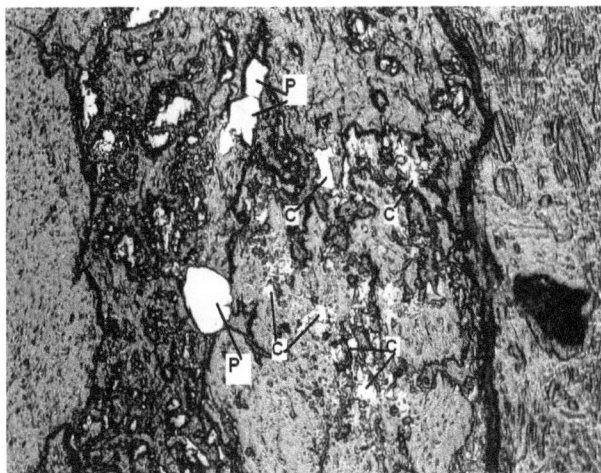

Figure 1. Petrological micrograph of copper ore (at 100×) [C-chalcopyrite, P-pyrite].

Table 3. XRD Phase analysis of copper ore.

Sample	Phases	
	Major Phases	*Minor Phases*
Ore	Chalcopyrite, Pyrite, Silica	Bornite

Flasks containing ore slurry were incubated at a temperature of 35°C ± 0.2°C and pH 2.0 in incubator shaker with orbital motion at 120 rpm unless otherwise stated. During experiments, samples were mostly taken at 5 days intervals for chemical analysis and pH of the leach solution was maintained on alternate days (initial pH maintained with 10N H_2SO_4). Redox potential (E_{SCE}) was also measured against saturated calomel electrode and reported as such in the text. A 0.5 mL supernatant solution was withdrawn for analysis of metals by AAS (Model: GBC-980BT). The iron(II) concentration was determined by titrating against 0.05 N potassium dichromate solution. All the inoculated sets had their corresponding sterile sets prepared under the same conditions. Upon termination of the leaching experiments, the solid residues were dried and samples were taken for chemical analysis and XRD phase identification. Cell count was done using Petroff Hauser's Counting Chamber and enumerated using biological microscope. The bioleaching experiments were mostly repeated for 3 times and the copper recoveries were found to vary within ±0.5 to 1.0%.

3. Results and Discussion

In the present work, extensive bench scale studies on bioleaching of copper from lean grade copper ore of Malanjkhand Copper Project (MCP) are reported for optimisation of the critical parameters. The details are presented and discussed below.

3.1. Bioleaching Experiments Using Unadapted Culture

Acidithiobacillus ferrooxidans isolated from the mine water and sub-cultured in presence of ferrous sulfate was used in the form of enriched culture in bio-leaching experiments.

3.1.1 Effect of pH

Bioleaching of copper ore was carried out using enriched culture at pH ranging from 1.5 - 2.5 at a pulp density of 5% (w/v) and 25°C. **Figure 2** shows the maximum (40%) bio-recovery of copper at pH 2.0 in 35 days which was mainly governed by increase in bacterial oxidation under this condition. Further increase in time had no effect on recovery of copper. During this period, low copper recovery of 17% was, however, observed at pH 2.0 in control/sterile leaching experiment. This may be correlated with high bio-dissolution of iron (45%) at pH 2.0 in comparison to 21% recovery in control leaching (**Figure 3**). **Figure 3** also describes the trend of Fe(III) concentration

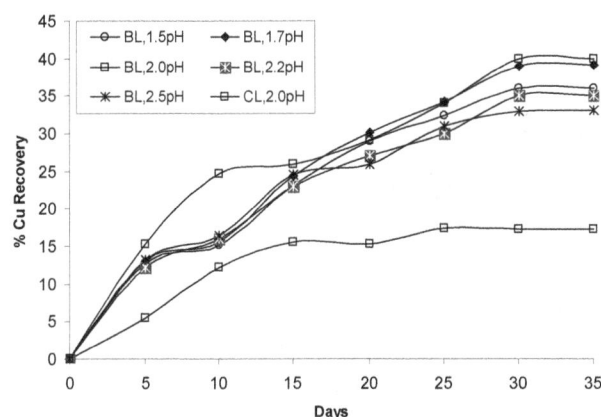

Figure 2. Recovery of copper at varying pH using unadapted culture [T: 25°C, pulp density:5% (w/v), particle size: <50 μm].

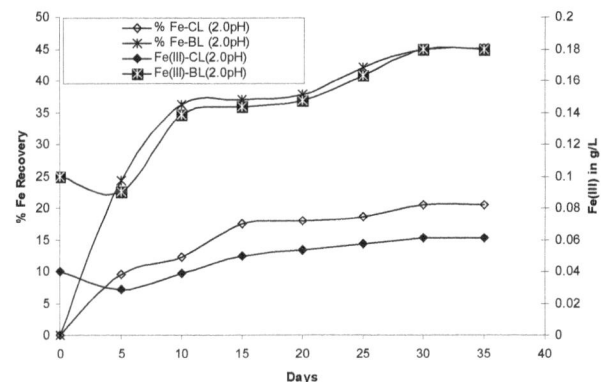

Figure 3. Iron recovery (%) with variation in Fe(III) ion concentration (g/L) using unadapted culture at 35°C, <50 μm particles and pH 2.0. [BL-Bio-leaching, CL-Chemical Leaching].

encountered during the bio-leaching/control experiments. It may be mentioned that iron dissolved in the form of ferrous ion from the pyrite/chalcopyrite present in the ore would be converted to ferric ions during bio-leaching with the help of *A. ferrooxidans*. The iron(III) present in the solution has thus improved the copper bio-dissolution process from 5th day onwards (**Figure 2**) through indirect mechanism. With a rise in pH to 2.5, the biorecovery of copper decreased to 32% due to hydronium jarosite precipitation on the surface of the ore particles which was identified by XRD phase analysis (**Table 3**) of leach residue [17]. Acid requirement was also an important aspect to be taken in to consideration which was at a minimum level (1.25 mL of 10 N H_2SO_4 for 200 mL solution) at pH 2.0 due to higher bacterial activity; whereas it was 75% higher acid that was consumed in control experiments at the same pH.

3.1.2. Effect of Pulp Density (PD)

Bio-dissolution of copper was investigated by varying pulp density in the range 5% - 20% (w/v) at pH 2.0 and 35°C with <50 μm size particles inoculating 10% (v/v) enriched culture/isolate from Cu-mine water without adaptation in 200 mL leaching medium while shaking at 100 rpm. From the results shown in **Figure 4**, it was clear that bio-recovery of copper was much high (72%) at lower pulp density of 5% (w/v) in 35 days as compared to 32% Cu dissolved in chemical leaching. The bio-dissolution of copper decreased with increase in pulp density. This may be attributed to the deficiency of oxygen availability and increased concentration of metal ions causing toxicity to bacterial growth at higher pulp densities. At 5% (w/v) PD, redox potential increased from 522 mV in 5 days to 661 mV in 35 days for bioleaching whereas the corresponding E_{SCE} were 312 mV and 401 mV in control leaching at this pulp density. These data signified that high metal dissolution was governed by high redox potential owing to the high ferric iron concentration in the solution. These studies further showed increase in copper recovery with an increase in temperature from 25°C (**Figure 2**) to 35°C under similar conditions. This may be attributed to improved biochemical action of bacteria—*A. ferrooxidans* at 35°C.

3.1.3. Effect of Particle Size

Studies on effect of particle sizes on bio-leaching of copper are shown in **Figure 5**. It may be seen that increasing fineness from 150 - 76 μm to <50 μm increased the metal recovery from 39% to 72% in 35 days at 5% (w/v) PD, pH 2.0 and 35°C. This could be mainly due to the availability of increased surface area of the ore leading to better permeation of leachant to oxidize the copper sulfide present in the ore. Finer particles were increaseingly exposed to lixiviant that dissolved copper from the chalcopyrite phase.

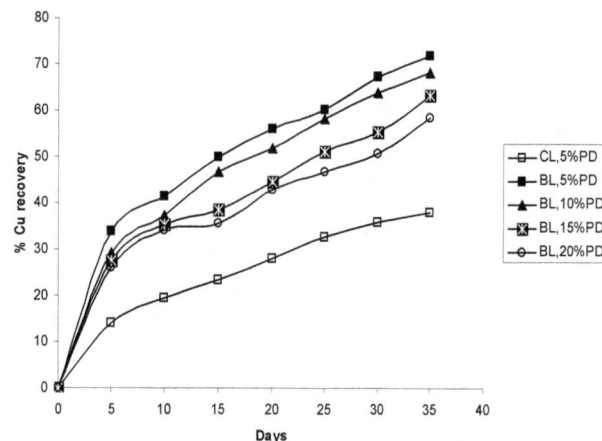

Figure 4. Copper recovery at different pulp densities using unadapted culture at 35°C, <50 μm particles, pH 2.0. [BL-Bio-leaching, CL-Chemical Leaching].

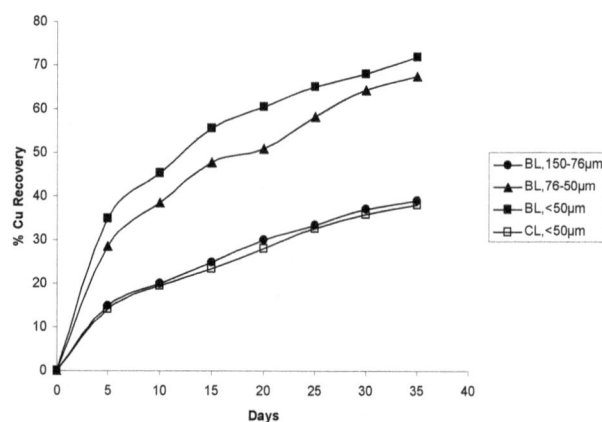

Figure 5. Effect of particle size on bioleaching of copper ore with unadapted bacteria [T:35°C, pH 2.0, 5% (w/v) PD; BL: Bio-leaching].

3.1.4. Effect of Temperature

In general, chemical reaction rate increases with rise in temperature, but the bacterial culture containing *A. ferrooxidans* being mesophilic in nature, the experiments were conducted only up to 35°C to ensure sustenance of the bacteria. Effect of temperature on bioleaching at 5% PD using 10% (v/v) enriched culture at pH 2.0 with <50 μm size particles is depicted in **Figure 6**. Maximum copper recovery was found to be 72% at 35°C in 35 days as against 44% at 25°C, whereas recovery under control leaching was found to be only 32% at 35°C (**Figure 2**). Higher bio-dissolution of the metal may be correlated with higher E_{SCE} that was attained from 316 to 661 mV in 35 days in bio-leaching as compared to that of control leaching which varied from 312 to 401 mV during the corresponding period. The bacterial population in bio-leaching at 35°C increased from 6×10^7 to 9.8×10^8 cells/mL in 35 days showing strong oxidizing conditions for improved metal recovery.

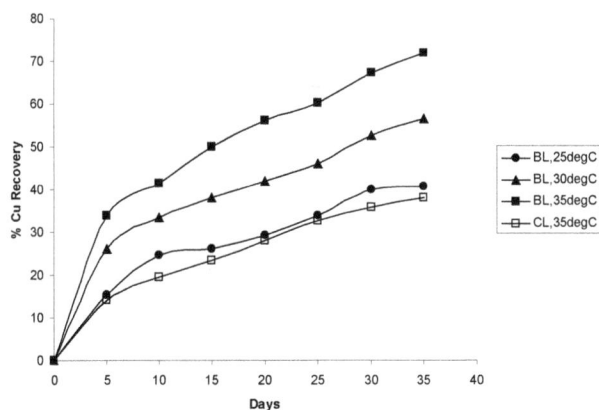

Figure 6. Effect of temperature on bioleaching of copper ore with unadapted culture [PD 5% (w/v), pH 2.0, <50 μm particles; BL: Bio-leaching, CL: Chemical Leaching].

3.2. Bioleaching Experiments Using Adapted Culture

As described earlier, the bacteria isolated from mine water was adapted on 5% (w/v) ore at pH 2.0 and 35°C and the adapted enriched culture after further sub-culturing was used in bio-leaching.

3.2.1 Effect of pH

Bioleaching of copper from the ore was carried out using adapted culture of isolate at varying pH from 1.5 - 2.5 at a pulp density of 5% (w/v) and 25°C and results are depicted in **Figure 7**. The bio-recovery was found to be maximum (41%) in 35 days at pH 1.7 and 2.0, although initial rate of copper biodissolution was faster at pH 1.7. Acid consumption was slightly more at pH 1.7 (1.8 mL 10N H_2SO_4) as compared to that of pH 2.0 (0.5 mL 10N H_2SO_4). This indicated that bio-chemical oxidation at these pH values was pronounced and the lower pH of 1.5 was not that effective for bio-oxidation with 36% Cu recovery. High recovery at pH 2.0 with the adapted culture was mainly governed by increase in bacterial oxidation which was demonstrated by high Fe(III) concentration of 0.26 g/L in 35 days as compared to 0.18 g/L iron(III) in case of un-adapted culture (**Figure 3**). With a rise in pH above 2.0, the recovery of copper decreased due to the jarosite formation on the ore surface. A recovery of 17% copper was attained in control experiments at pH 2.0 during the same period.

3.2.2 Effect of Particle Size

Effect of particle size on bio-leaching of copper was investigated for the three size range at 5% (w/v) PD using 10% (v/v) enriched culture of adapted bacteria in 200 mL leaching medium at pH 2.0 and 25°C. As shown in **Figure 8**, maximum copper recovery (47.5%) was obtained with <50 μm size material using adapted culture which was higher than that of the recovery of 40% with non-adapted

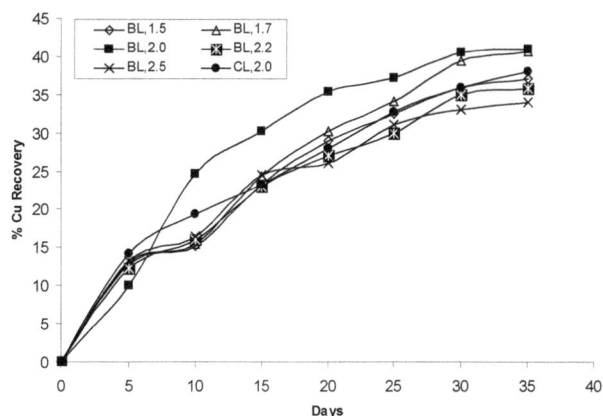

Figure 7. Recovery of copper during bioleaching with adapted culture at different pH [PD: 5% (w/v), T: 25°C, particle size: <50 μm].

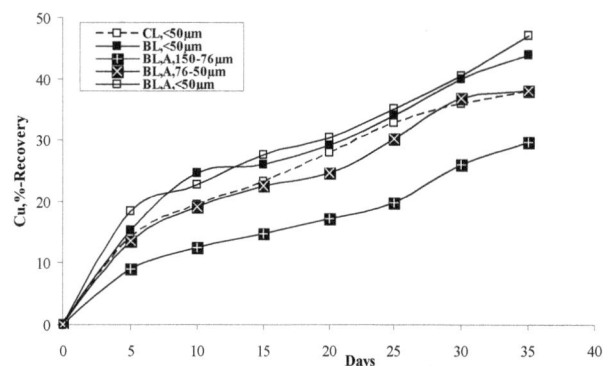

Figure 8. Cu recovery during bioleaching with adapted culture at different particle size of ore [pH: 2.0, PD: 5% (w/v), Temp: 25°C, BL-A: Bioleaching with adapted culture, BL: Bioleaching without adapted culture].

culture (**Figure 2**). This may be attributed to the fact that metal ion tolerance of adapted strains contributed more to the bioleaching of metal in comparison to the unadapted culture. Copper bio-recovery of 29.68% and 38.31% were obtained with 150 - 76 μm and 76 - 50 μm size ore particles respectively in 35 days. In control experiment for <50 μm size ore, recovery of copper was 32% in 35 days. Maximum rise in E_{SCE} of 654 mV was noticed for the bioleaching with adapted bacteria in 35 days with <50 μm size particles which resulted in high metal dissolution, because of favoured biochemical oxidation of the finer size particles.

3.2.3 Effect of Pulp Density

Recovery of copper at different pulp density at 25°C temperature, pH 2.0 and <50 μm size particles along with the change in E_{SCE} (mV) values in 35 days has been presented in **Figures 9-10**. The control experiments were also run under similar conditions at 5% (w/v) PD. The maximum copper recovery was found to be 47.5% and 44% with adapted and un-adapted culture as mentioned

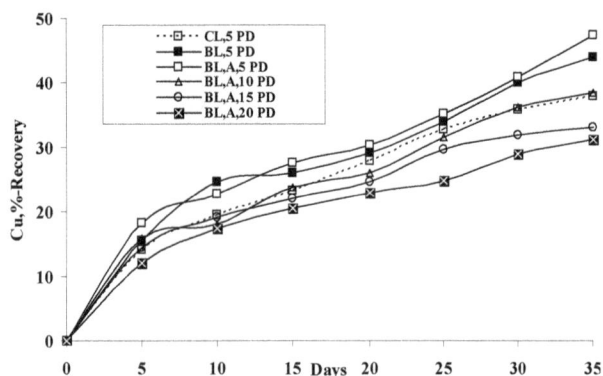

Figure 9. Effect of pulp density on bioleaching of copper with adapted culture [T: 25°C, particle size: <50 μm, pH: 2.0].

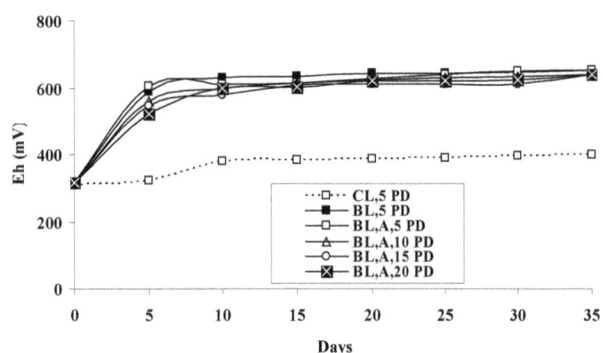

Figure 10. Change in redox potential during bioleaching at different pulp densities [T: 25°C, particle size: <50 μm, pH: 2.0].

earlier, whereas 32% copper was leached out in sterile/control experiments at 5% (w/v) PD in 35 days. Copper bio-recovery of 38.5%, 33.04% and 31.19% was obtained with the adapted culture at 10, 15 and 20% PD respectively (**Figure 9**); this indicated decreased metal recovery at higher pulp densities. At 5% (w/v) pulp density, maximum redox potential of the solution was found to be 390 mV for the chemical leaching, whereas it was 652 and 654 mV in 35 days for leaching with non-adapted and adapted cultures respectively (**Figure 10**). The higher redox potential may be correlated with the high copper recovery with adapted culture at this pulp density.

3.2.4. Effect of Temperature

Effect of temperature on bio-dissolution of copper using the enriched culture was investigated in the range 25°C - 35°C at 5% (w/v) PD and pH 2.0 and results are reported in **Figure 11**. Copper bio-recovery was maximum (75.3%) with the adapted culture in leaching as compared to the recovery (72%) with non-adapted bacteria, which may also be attributed to the metal ion tolerance of the adapted strains. It may be mentioned that strains can tolerate only up

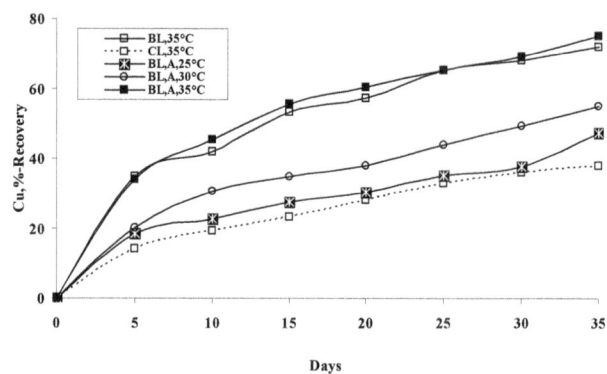

Figure 11. Effect of temperature on copper bio-recovery with adapted culture [Particle size: <50 μm, pH 2.0, PD: 5% (w/v)].

to a certain critical concentration of metal ions under normal conditions, generally ≈1 g/L for Cu, which on adaptation rises to 55 g/L [5,18]. It is the tolerance limit of *A. ferrooxidans* which is enhanced through adaptation leading to the increased metal dissolution. Bio-recovery of copper increased from 47.5% - 75.3% with increase in temperature from 25°C to 35°C. At 35°C temperature, the redox potential varied between 316 to 661 mV and 318 to 668 mV in bio-leaching experiments with unadapted and adapted cultures whereas it varied between 312 to 401 mV in control/chemical leaching in 35 days. During bioleaching, increase in bacterial growth was observed from 6×10^7 to 9.8×10^8 and 9.6×10^7 to 11.3×10^8 cells/mL with unadapted and adapted bacterial culture respectively. It was interesting to see that the bacterial population increased with time resulting in improved metal bio-recovery. Apparently the higher cell population in case of leaching with adapted culture resulted in better metal dissolution. A slightly higher redox potential (668 mV) and cell counts (11.3×10^8) favoured improved metal dissolution (75.3%) in case of adapted isolates as a consequence of metal tolerance of bacteria on ore [18].

3.3. XRD Phase Analysis of Leach Residue

The XRD phase analysis of the residue (**Table 3**) obtained during bio-leaching at 35°C and pH 2.5 with the adapted culture showed that hydronium jarosite $[H_3OFe_3(SO_4)_2(OH)_6]$ and silica were present as major phases and chalcopyrite and pyrite as the minor phases. Chalcopyrite and pyrite were found as major phases in the ore.

During the bioleaching, jarosite was formed as,

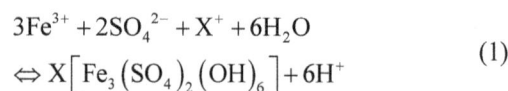

$$3Fe^{3+} + 2SO_4^{2-} + X^+ + 6H_2O$$
$$\Leftrightarrow X\left[Fe_3(SO_4)_2(OH)_6\right] + 6H^+ \tag{1}$$

Precipitation of jarosite was lower at pH 2; but at pH 2.5, its precipitation increased resulting in lower biorecovery.

4. Conclusions

Bioleaching of copper from the low grade chalcopyrite ore of MCP, India by enriched culture containing *Acidithiobacillus ferrooxidans* (*A. ferrooxidans*) is effective for extraction of the metal.

1) The maximum bioleaching of copper was 75.3% and 72% with the enriched culture adapted on the ore and with the native culture without adaptation respectively.

2) The optimum condition is worked out to be: pH 2.0, 35°C temperature and 5% (w/v) pulp density for the bioleaching process to leach maximum copper metal (75.33%) from the ore of <50 µm size particles in 35 days.

3) Additionally, the enriched culture containing *A. ferrooxidans* possesses the capacity to generate sulfuric acid through sulfur oxidation to maintain the pH in leaching solution and allow further copper dissolution through the action of Fe(III) ions.

4) The higher copper recovery with the enriched adapted culture may be correlated with the presence of higher iron level in solution and high redox potential (668 mV) acquired under the optimum conditions as compared to un-adapted culture.

5. Acknowledgements

The authors are thankful to the Director, NML Jamshedpur for giving permission to publish the paper. Thanks are also due to the Malanjkhand Copper Project (Hindustan Copper Limited) for providing the ore sample and to Dr. A. K. Upadhyay of Analytical Chemistry Centre, CSIR-NML, Jamshedpur for chemical analysis.

REFERENCES

[1]   A. E. Troma, "New Trends in Biohydrometallurgy," In: R. W. Smith and M. Misra, Eds., *Mineral Bioprocesing*, The Minerals, Metals & Materials Society, Pennsylvania, 1991, p. 43.

[2]   A. R. Colmer and M. E. Hinkle, "The Role of Microorganisms in Acid Mine Drainage," *Science*, Vol. 106, No. 2751, 1947, pp. 253-256. doi:10.1126/science.106.2751.253

[3]   G. Rossi, "Biohydrometallurgy," McGraw-Hill Book Company, Hamburg, 1990.

[4]   K. A. Natarajan, "Electrochemical Aspects of Bioleaching of Base-Metal Sulphides," In: H. L. Ehrlich and C. L. Brierley, Eds., *Microbial Mineral Recovery*, McGraw-Hill Book Company, New York, 1990, p. 79.

[5]   A. Das, J. M. Modak and K. A. Natarajan, "Studies on Multi-Metal Ion Tolerance of Thiobacillus Ferrooxidans," *Minerals Engineering*, Vol. 10, No. 7, 1997, pp. 743-749. doi:10.1016/S0892-6875(97)00052-6

[6]   M. P. Silverman, "Mechanism of Bacterial Pyrite Oxidation," *Journal of Bacteriology*, Vol. 94, 1967, pp. 1046-1051.

[7]   L. A. Brickett, R. W. Hammack and H. M. Edenborn, "Comparison of Methods Used to Inhibit Bacterial Activity in Sulfide Ore Bioleaching Studies," *Hydrometallurgy*, Vol. 39, No. 1-3, 1995, pp. 293-305. doi:10.1016/0304-386X(95)00043-G

[8]   A. Schippers and W. Sand, "Bacterial Leaching of Metal Sulfides Proceeds by Two Indirect Mechanisms via Thiosulfate or via Polysulfides and Sulfur," *Applied and Environmental Microbiology*, Vol. 65, No. 1, 1999, pp. 319-321.

[9]   L. Patnaik, R. N. Kar and L. B. Sukla, "Infuluence of pH on Bioleaching of Copper and Zinc from Complex Sulphide Concentrate Using Thiobacillus Ferrooxidans," *Transactions of the Indian Institute of Metals*, Vol. 54, No. 4, 2001, pp. 139-144.

[10]   G. S. Hansford and T. Vargas, "Chemical and Electrochemical Basis of Bioleaching Processes," *Hydrometallurgy*, Vol. 59, No. 2-3, 2001, pp. 135-145. doi:10.1016/S0304-386X(00)00166-3

[11]   W. Sand, T. J. Gehrke and A. Schippers, "(Bio)Chemistry of Bacterial Leaching—Direct vs. Indirect Bioleaching," *Hydrometallurgy*, Vol. 59, No. 2-3, 2001, pp. 159-175. doi:10.1016/S0304-386X(00)00180-8

[12]   A. D. Agate, K. M. Paknikar and N. J. Khinvasara, "Scale-Up Leaching of Malanjkhand Copper Ores—A Case Study," In: J. Salley, R. G. L. Mcready and P. L. Wichlaez, Eds., *Biohydrometallurgy*, CANMET, 1989, p. 577.

[13]   D. Pradhan, S. Pal, L. B. Sukla, G. Roy Choudhary and T. Das, "Bioleaching of Low Grade Copper Ore Using Indigenous Microorganism," *Indian Journal of Chemical Technology*, Vol. 15, 2008, pp. 558-592.

[14]   S. C. Pal, K. D. Mehta, B. D. Pandey and T. R. Mankhand, "Biomineral Processing for Extraction of Copper Metal from Lean Ore of Malanjkhand Copper Project," *International Conference on Emerging Trends in Mineral Processing and Extractive Metallurgy* (*ICME*-2005), Bhubaneswar, India, pp. 246-255.

[15]   F. Habashi, "Chalcopyrite: Its Chemistry and Metallurgy," McGraw-Hill, New York, 1978.

[16]   D. Bevilaqua, A. L. L. C. Leite, O. Garcia Jr. and O. H. Tuovinen, "Oxidation of Chalcopyrite by *Acidithiobacillus ferrooxidans* and *Acidithiobacillus thiooxidans* in Shake Flasks," *Process Biochemistry*, Vol. 38, No. 4, 2002, pp. 587-592. doi:10.1016/S0032-9592(02)00169-3

[17]   M. B. Stott, H. R. Watling, P. D. Franzmann and D. Sutton, "The Role of Iron-Hydroxy Precipitates in the Passivation of Chalcopyrite during Bioleaching," *Minerals Engineering*, Vol. 13, No. 10, 2000, pp. 1117-1127. doi:10.1016/S0892-6875(00)00095-9

[18]   Y. Rodriguez, A. Ballester, M. L. Blazquez, F. Gonzalez and J. A. Muñoz, "New Information on the Chalcopyrite Bioleaching Mechanism at Low and High Temperature," *Hydrometallurgy*, Vol. 71, No. 1-2, 2003, pp. 47-56. doi:10.1016/S0304-386X(03)00173-7

27

Extraction Kinetics of Ni(II) in the Ni^{2+}-SO_4^{2-}-Ac^- (Na^+, H^+)-Cyanex 272 (H_2A_2)-Kerosene-3% (v/v) Octan-1-ol System Using Single Drop Technique

Ranjit Kumar Biswas[*], Aneek Krishna Karmakar, Muhammad Saidur Rahman

Department of Applied Chemistry and Chemical Engineering, Rajshahi University, Rajshahi, Bangladesh

Email: [*]rkbiswas694@gmail.com

ABSTRACT

The kinetics of extraction of Ni(II) in the Ni^{2+}-SO_4^{2-}-Ac^- (Na^+, H^+)-Cyanex 272 (H_2A_2)-kerosene-3% (v/v) octan-1-ol system using the single falling drop technique have been reported. The flux of Ni^{2+} transfer (F) at 303 K in presence of 3% (v/v) octan-1-ol (de-emulsifier) can be represented as:

$$F_f \left(\text{kmol/m}^2\text{s} \right) = 10^{-3.7} \left[Ni^{2+} \right] \left[H_2A_2 \right]_{(o)}^{0.5} \left(1 + 10^{6.35} \left[H^+ \right] \right)^{-1} \left(1 + 6.3 \left[SO_4^{2-} \right] \right)^{-1} \left(1 + 0.55 \left[Ac^- \right] \right)^{-1}$$

. Depending on reaction parameters, the activation energy (E_a) and enthalpy change in activation ($\Delta H^{\neq}$) varies within 17 - 58 kJ/mol and 17 - 67 kJ/mol, respectively. Entropy change in activation ($\Delta S^{\neq}$) is always negative. Based on the empirical flux equation, E_a and $\Delta S^{\neq}$ values, mechanisms of extractions in different parametric conditions are proposed. At low $\left[SO_4^{2-} \right]$ and [Ac^-], and pH, the chemical controlled step is: $Ni^{2+} + A^- \rightarrow NiA^+$; and this reaction occurs via an S_N2 mechanism. But in most parametric conditions, the process is under intermediate control; and at high $\left[SO_4^{2-} \right]$ and [Ac^-], and pH, the extraction process is under diffusion control.

Keywords: Kinetics; Cyanex 272; Sulphate; Kerosene; Ni^{2+}; Single Drop Technique

1. Introduction

Cobalt has no natural deposit as its mine; and all nickel deposits contain invariably small proportion of cobalt. In order to obtain purified nickel and to isolate cobalt, it is necessary to separate Co(II) from Ni(II). The Co^{2+}/Ni^{2+} separation is a challenge to hydrometallurgists, who extract nickel following 1) leaching of ores, 2) purification of leach solution and 3) either reduction by hydrogen or electrolysis of purified solution. The purification of leach solution by solvent extraction is complicated by the difficult separation of Co^{2+} from Ni^{2+}.

Previously, organo-phosphorous extractants like D2EHPA [1-10], Cyanex 272 [1-6,11-16], EHEHPA or PC 88A [1-5,17,18], M2EHPA [9], TBP [2,8,9], Cyanex 301 [4,7,11,16,19-21] and Cyanex 302 [4,7,11,14], TOPS 99 [12,22], TIBPS [22], etc. have been used for Ni^{2+}/Co^{2+} separation. A few works [2-4,7,17,19] are available on extraction equilibrium of Ni^{2+}. Recently, the ex-

traction equilibrium of Ni^{2+} in the Ni^{2+}-SO_4^{2-}-Ac^- (Na^+, H^+)-Cyanex 272-kerosene-3% (v/v) n-octan-1-ol system (where, 3% (v/v) n-octan-1-ol in a de-emulsifier) has been reported from Authors' Laboratory [23]. The chemical structure of the active component of Cyanex 272 is [11]:

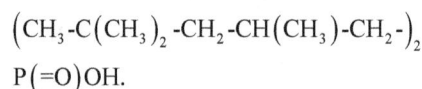

$$\left(CH_3\text{-}C\left(CH_3 \right)_2\text{-}CH_2\text{-}CH\left(CH_3 \right)\text{-}CH_2\text{-} \right)_2$$
$$P\left(=O \right)OH.$$

It is reported that equilibration time is only 2 min; and

$$\log{^C}D = 10^{-11.16} + 2pH_{(eq)} + \log \left[H_2A_2 \right]_{(o,eq)}$$
$$- \log \left(1 + 6.92 \left[SO_4^{2-} \right] - \log \left[Ac^- \right] \right)$$

when, $[H_2A_2]_{(o,eq)} \leq 0.05$ mol/L and

$$\log{^C}D = 10^{-11.56} + 2pH_{(eq)} + 3\log \left[H_2A_2 \right]_{(o,eq)}$$
$$- \log \left(1 + 6.92 \left[SO_4^{2-} \right] - \log \left[Ac^- \right] \right)$$

---

[*]Corresponding author.

when, $[H_2A_2]_{(o,eq)} \geq 0.10$ mol/L. These equations have suggested, respectively, the extraction equilibrium reactions as: $Ni^{2+} + H_2A_{2(o)} \square \ NiA_{2(o)} + 2H^+$ and

$Ni^{2+} + 3H_2A_{2(o)} \square \ NiA_{2(o)} \cdot 2H_2A_2 + 2H^+$.

Although the kinetics of Ni^{2+} extraction by non-phosphorous based extractants [24-28], have been reported, there is no report on the extraction kinetics of Ni^{2+} by organophosphorous extractants except the works of Dresinger and Cooper [29,30] who have used either D2EHPA or EHEHPA as extractant and RDC as the flux measurement technique. As there no report on the extraction kinetics of Ni^{2+} by Cyanex 272, this study has been carried out. In this study, the single drop technique for F (of Ni^{2+}-transfer)-measurement has been used.

2. Materials and Methods

2.1. Reagents

Cyanex 272 (Cytec Canada Inc.) was purified by the micro-emulsion formation method [31] to 99% BTMPPA (potentiometric titration), and characterized by its density (0.9152 g/mL at 298 K) and viscosity (120 mN/m at 298 K) [32]. Aliphatic colorless kerosene distilling over 200°C - 260°C was used as diluent. $NiSO_4 \cdot 6H_2O$ (Fluka, >99%) was used as a source of Ni^{2+}. Other chemicals were of reagent grade and used as received.

2.2. Analytical

The $[Ni^{2+}]$ in the aqueous phase was determined by the bromine-dimethylglyoxime method [33] at 445 nm using a WPA S104 Spectrophotometer and occasionally by the AAS method using a Shimadzu AA-6800 Spectrophotometer, especially when its concentration was low. The stock solution of Ni^{2+} was prepared by dissolving 22.39 g $NiSO_4 \cdot 6H_2O$ in water to make 1 L solution and standardized by EDTA-titration. The solution was found to contain 4.99 g/L Ni^{2+}. The acidity of the aqueous solutions was measured by a Mettler Toledo MP 220 pH meter on calibration by double buffers of pH 4 and 7.

2.3. Procedure with the Single Drop Apparatus

The construction of single drop apparatus is described elsewhere [34]. Its schematic diagram is in **Figure 1**. A falling drop apparatus was used. In the experiment, the continuum was the organic phase and drops of aqueous solution were allowed to fall through the continuum and collected continuously from the bottom of the column, leaving a pool of *ca* 2 - 3 drops of aqueous phase to avoid entrainment. For each experiment, the volume of 100 collected drops was estimated by the density-mass method; so that the volume of a single drop could be calculated. In the actual experiments, an uncounted

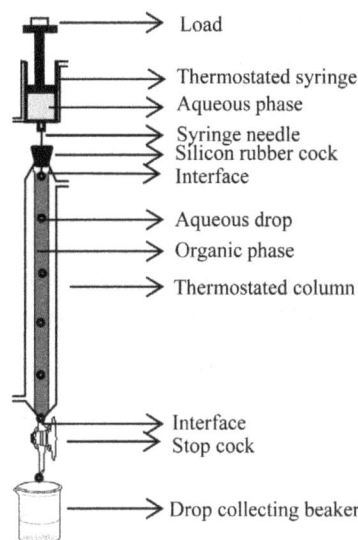

Figure 1. A schematic diagram of a single (falling) drop apparatus. Distance between two interfaces represent C.H. Thermostatic water circulation is aided by water circulating pump.

number of aqueous drops (internally circulating and slightly oscillating) of diameter (1.81 ± 0.03) mm were allowed to fall, collected in a previously weighed dry beaker and the volume of the collected aqueous phase (*ca* 2.5 mL) was determined by the density-mass method. The $[Ni^{2+}]$ in the collected mass was then estimated. On knowing the volume of a drop (determined previously), the number of drops in actual experiment could be determined. The cumulative time for 10 separate drops falling one after another was determined to get the average drop fall time, which was mostly dependent of column height and only slightly dependent on the composition of phases.

2.4. Theory of Rate Measurements by Flux (F)-Method

At a particular temperature, (F) of Ni^{2+} transfer can be represented as [35]:

$$F\left(kmol/m^2s\right) = 3.52\Delta\left[Ni^{2+}\right]\left(v/N\right)^{1/3}\left(1/t\right) \times 10^{-8} \quad (1)$$

The quantity, F, at a constant temperature is related to the concentration terms as:

$$\left(F\right) = \left(k_f\right)\left[Ni^{2+}\right]^a\left[H^+\right]^b\left[H_2A_2\right]_{(o)}^c\left[SO_4^{2-}\right]^d\left[Ac^-\right]^e \quad (2)$$

where, the unit of (k_f) depends on the values of a, b, c, d and e. Equation (2) can be rewritten as:

$$\log\left(F\right) = \log\left(k_f\right) + a\log\left[Ni^{2+}\right] - bpH$$
$$+ c\log\left[H_2A_2\right]_{(o)} + d\log\left[SO_4^{2-}\right] + e\log\left[Ac^-\right] \quad (3)$$

Equation (3) states that if pH, $[H_2A_2]$, $\left[SO_4^{2-}\right]$ and

[Ac⁻] are kept constant at pH, [H₂A₂], $\left[SO_4^{2-}\right]$ and [Ac⁻], respectively; and (F)-values are determined for various concentrations of [Ni²⁺], then the plot of log(F) vs log [Ni²⁺] will be a straight line with $s = 1$ and

$$I = \log\left(k_f\right) - b\text{pH} + c\log\left[H_2A_2\right]_{(o)}$$
$$+ d\log\left[SO_4^{2-}\right] + e\log\left[Ac^-\right]$$

From I-value, (k_f) can be calculated after determining the values of b, c, d and e. Similarly, the values of b, c, d and e together with four sets of (k_f)-values can be determined from the log(F) vs pH, log(F) vs log[H₂A₂]$_{(o)}$, log(F) vs log$\left[SO_4^{2-}\right]$ and log(F) vs log[Ac⁻] plots, respectively. The temperature dependence data can be treated by Arrhenius equation and Activated complex theory [36].

3. Results and Discussion

3.1. Characterization of Rate Measurement by Single Drop Experimentation

The plot of $a_{Ni^{2+}}$ from a drop vs t (obtained by using different C.H) [37] is a straight line which cuts the time axis at −0.5 s ($\Delta t = 0.5$ s). This time is designated as end correction term (attributed to time for drop formation and coalescence). In F-calculation, Δt term must be added to t; otherwise, error appears as demonstrated below:

When F_f' and F_f are calculated by neglecting and considering Δt value, respectively, then it is seen that log F_f' is decreased, whilst log F_f remains unchanged with increasing C.H and at any C.H, $\log F_f < \log F_f'$ [37]. It is concluded that F will be independent of C.H if Δt is added to t; and any C.H. can be used if F_f (not F_f') is calculated.

3.2. Rate Measurements

The log(F_f, kmol/m²·s) vs log([Ni²⁺], kmol/m³) plots are displayed in **Figure 2**. In all cases, straight lines are obtained with $s = (1.01 \pm 0.03)$ and I as typed on the body of figure. The unity s indicates that the rate of forward extraction of Ni²⁺ by Cyanex 272 is directly proportional to initial [Ni²⁺]. In other words, the reaction order *wrt* [Ni²⁺]$_{(ini)}$ is unity (*i.e.*, $a = 1$).

The logF_f vs pH$_{(ini)}$ plots are shown in **Figure 3** The experimental points for a particular system fall on a curve having higher slope in *lpHr* and lower slope in *hpHr*. The experimental points for a particular set of parameters fall on curve represented by:

$$\log F_f = \text{constant} - \log\left(1 + 10^{6.32}\left[H^+\right]\right) \quad (4)$$

where, constant = −6.382 (for 0.025 mol/L [H₂A₂]$_{(o,ini)}$ system), −6.062 (for 0.10 mol/L [H₂A₂]$_{(o,ini)}$ system) or,

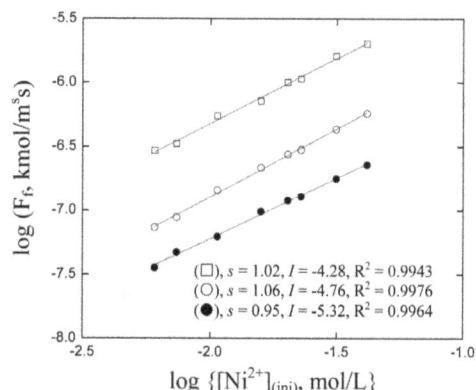

Figure 2. Effect of [Ni(II)]$_{(ini)}$ on flux [Ac⁻] = 0.25 mol/L, *Temp.* = 303 K, $\left[SO_4^{2-}\right]$ = 0.042 mol/L, C.H = 0.66 m. (O), pH$_{(ini)}$ = 6.70, [H₂A₂]$_{(o,ini)}$ = 0.025 mol/L; (●), pH$_{(ini)}$ = 6.00, [H₂A₂]$_{(o,ini)}$ = 0.025 mol/L; (□), pH$_{(ini)}$ = 6.70, [H₂A₂]$_{(o,ini)}$ = 0.30 mol/L.

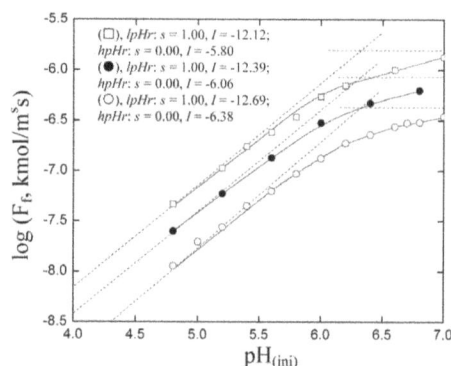

Figure 3. Effect of pH$_{(ini)}$ on flux. [Ni²⁺]$_{(ini)}$ = 1.3405 g/L, $\left[SO_4^{2-}\right]$ = 0.05 mol/L, C.H = 1.2/0.9/0.66 m. (O), [H₂A₂]$_{(o,ini)}$ = 0.025 mol/L; (●), [H₂A₂]$_{(o,ini)}$ = 0.10 mol/L; (□), [H₂A₂]$_{(o,ini)}$ = 0.30 mol/L. Other parameters are as in Figure 2.

−5.80 (for 0.30 mol/L [H₂A₂]$_{(o,ini)}$ system) and $10^{6.32}$ is a proportionality constant resulting from non-linear curve fitting. Its unit is L/mol. I-values of the asymptotic lines are embodied in figure. It is concluded that the rate of Ni²⁺ extraction is independent of [H⁺] in *lpHr*; whereas, inversely proportional to [H⁺] in *hpHr*. In other words, the reaction order *wrt* [H⁺] is −1 ($b = 1$) and 0 ($b = 0$) in *lpHr* and *hpHr*, respectively.

Figure 4 displays logF_f vs log[H₂A₂]$_{(o,ini)}$ plots. For each pH system, the plot is a straight line whose s and I are given. The s-values indicate that the rate of forward extraction is directly proportional to the square root of the extractant concentration (*i.e.*, $c = 0.5$).

The nature and extent of variations of F_f with $\left[SO_4^{2-}\right]$ are displayed in **Figure 5**. The experimental points for a particular set of parameters fall on a curve represented by:

$$\log F_f = \text{constant} - \log\left(1 + 6.30\left[SO_4^{2-}\right]\right) \quad (5)$$

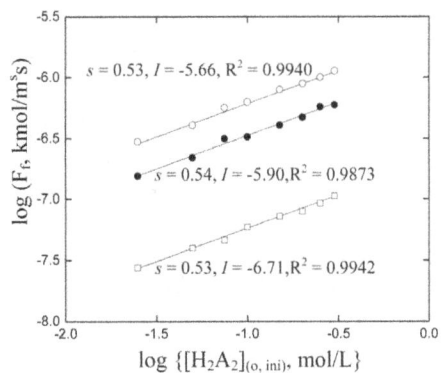

Figure 4. Effect of extractant concentration on flux $[Ni^{2+}]_{(ini)}$ = 1.3405 g/L, $\left[SO_4^{2-}\right]$ = 0.05 mol/L. (O), $pH_{(ini)}$ = 6.70, C.H = 0.66 m; (●), $pH_{(ini)}$ = 6.10, C.H = 0.90 m; (□), $pH_{(ini)}$ = 5.20, C.H = 1.20 m. Other parameters are as in Figure 2.

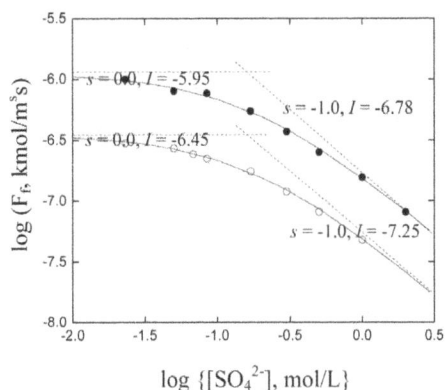

Figure 5. Effect of $\left[SO_4^{2-}\right]$ on flux $[Ni^{2+}]$ = 1.3405 g/L. (O), $pH_{(ini)}$ = 6.70, $[H_2A_2]_{(o,ini)}$ = 0.025 mol/L; (●), $pH_{(ini)}$ = 6.40, $[H_2A_2]_{(o,ini)}$ = 0.30 mol/L. The points are experimental and the solid curves are theoretical representing: $logF_f = -6.45$ (O) or -5.95 (●) $-log\left\{1+K_{SO_4^{2-}}\left[SO_4^{2-}\right]\right\}$, where $K_{SO_4^{2-}}$ is a proportionality constant; whose value in both cases is 6.30 L/mol by the Curve-Fitting method. Other parameters are as in Figure 2.

where, constant = -6.4 (for $pH_{(ini)}$ = 6.70, $[H_2A_2]_{(o,ini)}$ = 0.025 mol/L system) or, -5.95 (for $pH_{(ini)}$ = 6.40, $[H_2A_2]_{(o,ini)}$ = 0.30 mol/L system); and 6.30 is a proportionality constant resulted fromnon-linear curve-fitting and its unit is considered as L/mol. The intercepts of the asymptotic lines are given in figure. The rate of Ni^{2+} transfer is therefore inversely proportional to the term $\left(1+6.4\left[SO_4^{2-}\right]\right)$. This means that d is 0 at lcr of SO_4^{2-} and -1 at hcr of SO_4^{2-}.

The $log(F_f$, kmol/m^2 s) vs log[Ac$^-$], mol/L) plot for $pH_{(ini)}$ = 6.60 and $[H_2A_2]_{(o,ini)}$ = 0.025 mol/L is represented in **Figure 6**. Experimental points fall on a curve represented by:

$$\log F_f = -6.50 - \log\left(1+0.55\left[Ac^-\right]\right) \qquad (6)$$

where, 0.55 L/mol is proportionality constant whose value is originated from non-linear regression analysis. I-values of the asymptotic lines are quoted. The rate of Ni^{2+} transfer is therefore inversely proportional to the term $(1 + 0.55 [Ac^-])$. In other words, $e = 0$ at lcr of $[Ac^-]$ and $e = -1$ at hcr of $[Ac^-]$.

The $logF_f$ vs $1/T$ (Arrhenius) plots for 5-sets of experimental parameters are depicted in **Figure 7**. From top

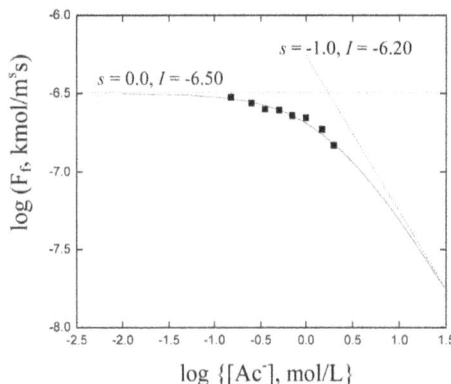

Figure 6. Effect of [Ac$^-$] on the flux $[Ni^{2+}]$ = 1.3405 mol/L, $\left[SO_4^{2-}\right]$ = 0.023 mol/L, $pH_{(ini)}$ = 6.6. Other parameters are as in Figure 2. The points are experimental and the solid curve is theoretical representing: $logF_f = -6.5 - log\left\{1+K_{AC^-}\left[Ac^-\right]\right\}$, **where K_{AC^-} is proportionality constant; and its value has been estimated as 0.55 L/mol by the Curve-Fitting method.**

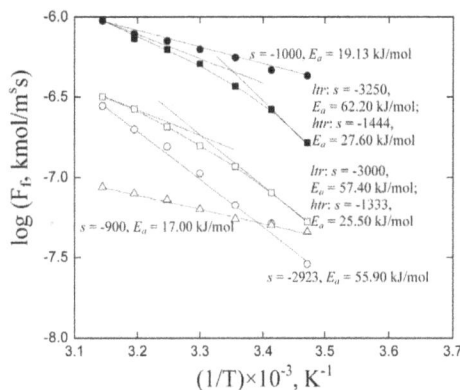

Figure 7. Effect of temperature on flux (Arrhenius plots) $[Ni^{2+}]_{(ini)}$ = 1.3405 g/L. (●), $pH_{(ini)}$ = 6.70, $[H_2A_2]_{(o,ini)}$ = 0.10 mol/L, [Ac$^-$] = 0.25 mol/L, $\left[SO_4^{2-}\right]$ = 0.05 mol/L, C.H = 0.66 m; (O), $pH_{(ini)}$ = 5.20, $[H_2A_2]_{(o,ini)}$ = 0.10 mol/L, [Ac$^-$] = 0.25 mol/L, $\left[SO_4^{2-}\right]$ = 0.05 mol/L, C.H = 1.2 m; (□), $pH_{(ini)}$ = 6.70, $[H_2A_2]_{(o,ini)}$ = 0.10 mol/L, [Ac$^-$] = 0.25 mol/L, $\left[SO_4^{2-}\right]$ = 1.00 mol/L, C.H = 1.2 m; (■), $pH_{(ini)}$ = 6.60, $[H_2A_2]_{(o,ini)}$ = 0.30 mol/L, [Ac$^-$] = 2.00 mol/L, $\left[SO_4^{2-}\right]$ = 0.023 mol/L, C.H = 1.2 m; (△), $pH_{(ini)}$ = 6.70, $[H_2A_2]_{(o,ini)}$ = 0.10 mol/L, [Ac$^-$] = 2.00 mol/L, $\left[SO_4^{2-}\right]$ = 1.0 mol/L, C.H = 1.2 m.

to bottom, 1st, 4th and 5th systems yield straight lines and s of these lines give E_a values of 19, 56 and 17 kJ/mol, respectively. On the other hand, for the 2nd and 3rd systems, curves are obtained. From limiting s of the curves, E_a values of 25.5 kJ/mol and 57.5 kJ/mol are obtained at htr and ltr, respectively for the 3rd system; whereas, 27.5 kJ/mol and 62.0 kJ/mol are obtained at htr and ltr respectively, for the 2nd system.

The temperature dependence rate data have also been treated by the Activated Complex Theory to estimate the $\Delta H^{\ddagger}$ and $\Delta S^{\ddagger}$. The plots of log ($F_f h/kT$) vs ($1/T$) are given in **Figure 8**. Natures of plots are similar to those of Arrhenius plots. The "s", "I", $\Delta H^{\ddagger}$ and $\Delta S^{\ddagger}$ values are embodied in the figure. In calculating $\Delta S^{\ddagger}$ values, log$f(R)$-values are needed which are calculated using the relation:

$$\log f(R) = \log\left[Ni^{2+}\right]_{(ini)} - \log\left(1 + 10^{6.32} \times 10^{-pH}\right)$$
$$+ 0.5\log\left[H_2A_2\right]_{(o,ini)} - \log\left(1 + 6.3\left[SO_4^{2-}\right]\right) \quad (7)$$
$$- \log\left(1 + 0.55\left[Ac^-\right]\right)$$

The calculated $\Delta H^{\ddagger}$ value varies within 17 - 65 kJ/mol; whereas, $\Delta S^{\ddagger}$ values are always negative.

3.3. Elucidation of the Value of k_f

From "I" of the straight lines or the asymptotic lines in **Figures 2-6**, the average value of logk_f at 303 K in presence of 3% (v/v) octan-1-ol in the organic phase has been evaluated to be −3.742, with *stand. dev.* of 0.04. The

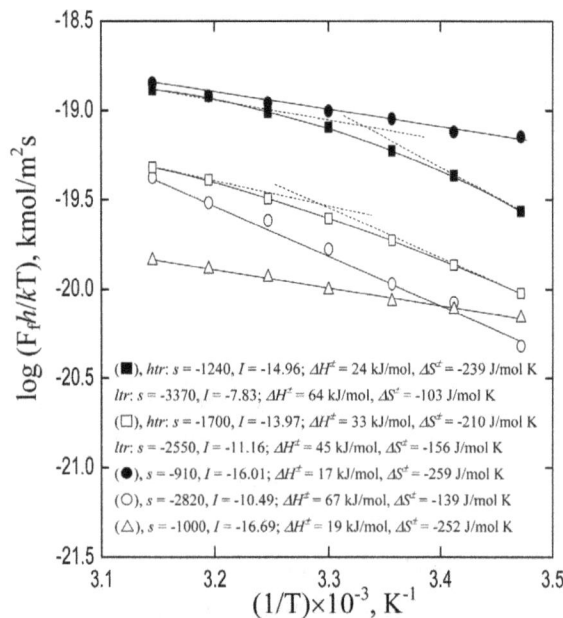

value of logk_f has also been obtained graphically. As the flux equation can be represented as: logF_f = logk_f + log$f(R)$, the plot of logF_f vs log$f(R)$ should be a straight line with s = 1 and I equaling to the value of logk_f. The plot is given in **Figure 9**. A good fit Least Squares straight line is obtained with s = 1.0288 (should be 1) and I = −3.6781. The latter value corresponding to log k_f is comparable to that obtained above. Hereafter, $k_f = 10^{-3.7}$ m$^{5/2}$/kmol$^{1/2}$·s will be considered in discussion.

3.4. Mechanism of Forward Extraction

Based on the results obtained, F in this system at 303 K can be expressed as:

$$F_f = 10^{-3.7}\left[Ni^{2+}\right]_{(ini)}\left[H_2A_2\right]_{(o,ini)}^{0.5}\left(1 + 10^{6.32}\left[H^+\right]\right)^{-1}$$
$$\times\left(1 + 6.3\left[SO_4^{2-}\right]\right)^{-1}\left(1 + 0.55\left[Ac^-\right]\right)^{-1} \quad (8)$$

Equation (8) is a too much complicated equation. It can be changed to a number of simplified flux equations depending on the concentration regions of H$^+$, SO$_4^{2-}$ and Ac$^-$. Here, following two extreme cases will be considered for discussion:

1) At hcr of H$^+$, but lcr of SO$_4^{2-}$ and Ac$^-$

$$F_f = 10^{-10.02}\left[Ni^{2+}\right]_{(ini)}\left[H_2A_2\right]_{(o,ini)}^{0.5}\left[H^+\right]^{-1} \quad (9)$$

where, $10^{-10.02} = 10^{-3.7} \times 10^{-6.32}$; and

2) At lcr of H$^+$ but hcr of SO$_4^{2-}$ and Ac$^-$

$$F_f = 10^{-4.24}\left[Ni^{2+}\right]\left[H_2A_2\right]_{(o,ini)}^{0.5}\left[SO_4^{2-}\right]^{-1}\left[Ac^-\right]^{-1} \quad (10)$$

where, $10^{-4.24} = 10^{-3.7}/6.3 \times 0.55$.

In the present case, as the reaction order *wrt* extractant concentration is a one-half, the monomeric model of extractant will be applicable [35]. The monomeric model of H$_2$A$_2$ is:

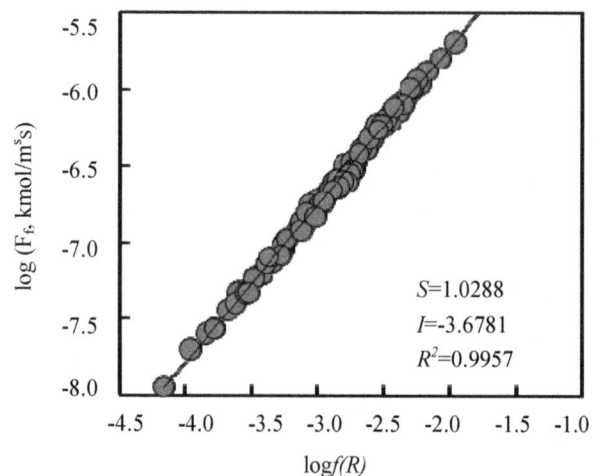

Figure 8. The log{($F_f h/kT$), kmol/m²s} vs ($1/T$) plots. Legends are as in Figure 8. (●), log$f(R)$ = −2.4679; (○), log$f(R)$ = −3.2299; (□), log$f(R)$ = −2.9834; (■), log$f(R)$ = −2.4682; (△), log$f(R)$ = −3.4670.

Figure 9. The logF_f vs log$f(R)$ plot at 303 K.

$$[H_2A_2]_{(o)} = \left(K_2^{0.5}[H_2A_2]_{(o)}\right)^2 = \left(K_2^{0.5}P_{HA}[HA]\right)^2$$
$$= K_2 P_{HA}^2 K_{a_{HA}}^{-2}\left[A^-\right]^2\left[H^+\right]^2 \quad (11)$$

Combination of Equation (9) with Equation (11) yields the flux equation as:

$$F_f = 10^{-10.02}K_2^{0.5}P_{HA}K_{a_{HA}}^{-2}\left[Ni^{2+}\right]\left[A^-\right] \quad (12)$$

Equation (12) gives the slow reaction step occurring in the bulk aqueous phase as:

$$Ni^{2+} + A^- \rightarrow \left[NiA\right]^+ \quad (13)$$

In this experimental parametric condition, Ni^{2+} extraction by Cyanex 272 is therefore chemically controlled and this statement is supported by high E_a (56 kJ/mol) obtained at the investigated hcr of H^+ (pH = 5) and lcr of SO_4^{2-} (0.05 mol/L) and Ac^- (0.25 mol/L).

The chemically controlled rate-determining step: $\left(Ni^{2+} + A^- \rightarrow NiA^+\right)$ may occur either by an S_N1 or S_N2 mechanism [38]. For S_N2 mechanism, the bimolecular reaction step may be shown as:

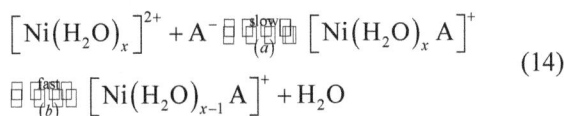

$$\left[Ni(H_2O)_x\right]^{2+} + A^- \underset{(a)}{\overset{slow}{\rightleftharpoons}} \left[Ni(H_2O)_x A\right]^+$$
$$\underset{(b)}{\overset{fast}{\rightleftharpoons}} \left[Ni(H_2O)_{x-1}A\right]^+ + H_2O \quad (14)$$

with the rate expression :

$$F_f = k_f\left[\left[Ni(H_2O)_x\right]^{2+}\right]\left[A^-\right] \quad (15)$$

Equation (15) is identical to Equation (12). Consequently in an S_N2 mechanism, the attachment of an additional ligand (A^-) to the restricted co-ordination sphere of Ni^{2+} acts as the rate determining step. The other is the S_N1 mechanism which a unimolecular process as follows:

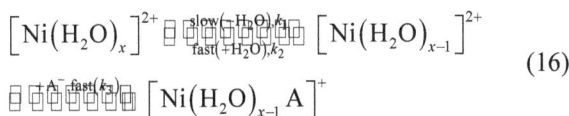

$$\left[Ni(H_2O)_x\right]^{2+} \underset{fast(+H_2O),k_2}{\overset{slow(-H_2O),k_1}{\rightleftharpoons}} \left[Ni(H_2O)_{x-1}\right]^{2+}$$
$$\underset{}{\overset{A^-,fast(k_3)}{\rightleftharpoons}} \left[Ni(H_2O)_{x-1}A\right]^+ \quad (16)$$

The steady state approximation results the rate expression for the S_N1 mechanism as:

$$F_f = k_1k_3\left[\left[Ni(H_2O)_x\right]^{2+}\right]\left[A^-\right]\Big/\left(k_2 + k_3\left[A^-\right]\right) \quad (17)$$

and if $k_2 \gg k_3\left[A^-\right]$, then the Equation (17) takes form of Equation (15); whereby (k_1k_3/k_2) will represent k_f.

Thus, it is possible to explain the same rate data by both S_N1 and S_N2 mechanisms; and as a result, it is difficult to decide whether the reaction proceeds via Equation (14) or (16). But this difficulty may effectively be overcome by the use of the thermodynamic data of the activated state, especially the $(\Delta S^{\pm})$ data for the system.

The solution effect dominates the entropy of activation

where charged ions are involved. If the solvent molecules are tightly attached around Ni^{2+} ions, their entropy is lost *i.e.* $\Delta S^{\pm}$ becomes negative. On the other hand, if the solvent molecules dissociate from the metal ions, their entropy is increased; and so, $\Delta S^{\pm}$ becomes positive. Thus for an S_N2 mechanism, where the ligand (A^-) co-ordinates to the metal ion, $[Ni(H_2O)_x]^{2+}$ to form the higher co-ordinated activated complex, $[Ni(H_2O)_x \cdot A]^+$, the value of $\Delta S^{\pm}$ would be expected to be more negative than the ground state. But for the S_N1 mechanism, where the formation of lower co-ordinated activated complex, $[Ni(H_2O)_{x-1}]^{2+}$ takes place, $\Delta S^{\pm}$ should be positive. In the present case, $\Delta S^{\pm}$ at all experimental parameters are highly negative; and so the rate controlling chemical reaction step represented by Equation (13) occurs via an S_N2 mechanism.

On the other hand, at lcr of H^+ but hcr of SO_4^{2-} and Ac^-, the existing Ni^{2+} species may be considered as $[Ni(OH)(SO_4)(Ac^-)]^{2-}$. So Equation (10) takes the form:

$$F_f = 10^{-4.24}\left[Ni(OH)(SO_4)(Ac^-)\right]^{2-}$$
$$\cdot[H_2A_2]_{(o,ini)}^{0.5}\left[SO_4^{2-}\right]^{-1}\left[Ac^-\right]^{-1} \quad (18)$$

And with the help of β_1 and β_2, Equation (18) takes the form:

$$F_f = 10^{-4.24}\beta_1\beta_2\left[Ni^{2+}\right]\left[H^+\right]^{-1}[H_2A_2]_{(o,ini)}^{0.5} \quad (19)$$

Monomeric model of $H_2A_{2(o)}$ *i.e.* Equation (11) transforms Equation (19) to

$$F_f = 10^{-4.24}\beta_1\beta_2 K_2^{0.5}P_{HA}K_{a_{HA}}^{-1}\left[Ni^{2+}\right]\left[A^-\right] \quad (20)$$

This equation suggests the rate controlling extraction reaction step given in Equation (13) is also the rate determining chemical reaction step in the latter set of condition. But E_a of 17 kJ/mol obtained at lcr of $[H^+]$ (*i.e.* high pH: 6.7) and hcr of SO_4^{2-} (1 mol/L) and Ac^- (2 mol/L) suggests that the diffusion of a reactant to the reaction site or the product from the reaction site to the bulk organic phase is slower than the reaction step given in Equation (13).

Thus depending on the extraction condition, the Ni^{2+} extraction in the present system by Cyanex 272 may be either 1) pure chemical controlled (at low pH, $\left[SO_4^{2-}\right]$ and $[Ac^-]$) or 2) pure diffusion controlled (at high pH, $\left[SO_4^{2-}\right]$ and $[Ac^-]$) or 3) mixed (intermediate) controlled. In most of the cases (moderate pH and/or, $\left[SO_4^{2-}\right]$ and/or $[Ac^-]$) at 303 K, the process is mixed controlled which may be chemically controlled at ltr and diffusion controlled at htr.

4. Conclusions

The end effect in the single drop experimentation is 0.50 s and this time is needed to be summed up with drop fall time to calculate F of independent C.H. At 303 K, the empirical flux equation is:

$$\log(F) = 10^{-3.7} \left[Ni^{2+} \right]_{(ini)} \left[H_2 A_2 \right]_{(o,ini)}^{0.5} \left(1 + 10^{6.32} \left[H^+ \right] \right)^{-1}$$

$$\times \left(1 + 6.3 \left[SO_4^{2-} \right] \right)^{-1} \left(1 + 0.55 \left[Ac^- \right] \right)^{-1}.$$

E_a and $\Delta H^{\ddagger}$ values depend on experimental condition and are found to vary within 17 - 58 kJ/mol and 17 - 67 kJ/mol. $\Delta S^{\ddagger}$ value is always negative. At low pH, $\left[SO_4^{2-} \right]$ and $[Ac^-]$, the process is under chemical control; whereas, at high pH, $\left[SO_4^{2-} \right]$ and $[Ac^-]$, the process is under diffusion control. But in most cases, the process is under intermediate control; which may be chemically controlled at *ltr* and diffusion controlled at *htr*. The rate determining chemical reaction step is identified as the formation of 1:1 complex between Ni^{2+} and anion (A^-) of the dimeric extractant. Moreover, negative $\Delta S^{\ddagger}$ value indicates that the chemical rate determining step occurs through an S_N2 mechanism.

REFERENCES

[1]   N. B. Devi, K. C. Nathsarma and V. Chakravorthy, "Separation and Recovery of Cobalt(II) and Nickel(II) from Sulphate Solution Using Sodium Salts of D2EHPA, PC 88A and Cyanex 272," *Hydrometallurgy*, Vol. 49, No. 1-2, 1998, pp. 47-61.
doi:10.1016/S0304-386X(97)00073-X

[2]   K. Sarangi, B. R. Reddy and R. P. Das, "Extraction Studies of Cobalt(II) and Nickel(II) from Chloride Solutions Using Na-Cyanex 272: Separation of Co(II)/Ni(II) by the Sodium Salts of D2EHPA, PC 88A and Cyanex 272 and Their Mixtures," *Hydrometallurgy*, Vol. 52, No. 3, 1999, pp. 253-265. doi:10.1016/S0304-386X(99)00025-0

[3]   P. V. R. Bhaskara Sarma and B. R. Reddy, "Liquid-Liquid Extraction of Nickel at Macrolevel Concentration from Sulphate/Chloride Solutions Using Phosphoric Acid Based Extractants," *Minerals Engineering*, Vol. 15, No. 6, 2002, pp. 461-464.
doi:10.1016/S0892-6875(02)00063-8

[4]   J. S. Preston, "Solvent Extraction of Cobalt and Nickel by Organophosphorous Acids: Comparison of Phosphoric, Phosphonic and Phosphinic Acid Systems," *Hydrometallurgy*, Vol. 9, No. 2, 1982, pp. 115-133.
doi:10.1016/0304-386X(82)90012-3

[5]   B. R. Reddy, D. N. Priya and K. H. Park, "Separation and Recovery of Cadmium(II), Cobalt(II) and Nickel(II) from Sulphate Leach Liquors of Spent Ni-Cd Batteries Using Phosphorous Based Extractants," *Separation and Purification Technology*, Vol. 50, No. 2, 2006, pp. 161-166.
doi:10.1016/j.seppur.2005.11.020

[6]   C. A. Nogueira and F. Delmas, "New Flowsheet for the Recovery of Cadmium, Cobalt and Nickel from Spent Ni-Cd Batteries by Solvent Extraction," *Hydrometallurgy*, Vol. 52, No. 3, 1999, pp. 267-287.
doi:10.1016/S0304-386X(99)00026-2

[7]   I. Van de Voorde, L. Pinoy, E. Courtijn and F. Verpoort,

"Equilibrium Studies of Nickel(II), Copper(II) and Cobalt(II) Extraction with Aloxime 800, D2EHPA, and Cyanex Reagents," *Solvent Extraction and Ion Exchange*, Vol. 24, No. 6, 2006, pp. 893-914.
doi:10.1080/07366290600952717

[8]   B. Gajda and M. B. Bogacki, "Effect of Tributylphosphate on the Extraction of Nickel(II) and Cobalt(II) Ions with Di(2-ethylhexyl) Phosphoric Acid," *Physicochemical Problems of Mineral Processing*, Vol. 41, 2007, pp. 145-152.

[9]   D. H. Fatmehsari, D. Darvishi, S. Etemadi, A. R. E. Hollagh, E. K. Alamdari and A. A. Salardini, "Interaction between TBP and D2EHPA during Zn, Cd, Mn, Cu, Co and Ni Solvent Extraction: A Thermodynamic and Empirical Approach," *Hydrometallurgy*, Vol. 98, No. 1-2, 2009, pp. 143-147. doi:10.1016/j.hydromet.2009.04.010

[10] C. Y. Cheng, K. R. Barnard, W. Zhang and D. J. Robinson, "Synergistic Solvent Extraction of Nickel and Cobalt: A Review of Recent Developments," *Solvent Extraction and Ion Exchange*, Vol. 29, No. 5-6, 2011, pp. 719-754. doi:10.1080/07366299.2011.595636

[11] B. K. Trait, "Cobalt-Nickel Separation: The Extraction of Cobalt(II) and Nickel(II) by Cyanex 301, Cyanex 302 and Cyanex 272," *Hydrometallurgy*, Vol. 32, No. 3, 1993, pp. 365-372. doi:10.1016/0304-386X(93)90047-H

[12] B. R. Reddy, D. N. Priya, S. V. Rao and P. Radhika, "Solvent Extraction and Separation of Cd(II), Ni(II) and Co(II) from Chloride Leach Liquors of Spent Ni-Cd Batteries Using Commercial Organo-Phosphorous Extractants," *Hydrometallurgy*, Vol. 77, No. 3-4, 2005, pp. 253-261. doi:10.1016/j.hydromet.2005.02.001

[13] W. A. Rickelton, D. S. Flett and D. W. West, "Cobalt-Nickel Separation by Solvent Extraction with Bis(2,4,4-trimethylpentyl) Phosphinic Acid," *Solvent Extraction and Ion Exchange*, Vol. 2, No. 6, 1984, pp. 815-838.
doi:10.1080/07366298408918476

[14] Z. Lenhard, "Extraction and Separation of Cobalt and Nickel with Extractants Cyanex 302, Cyanex 272 and Their Mixtures," *Chemistry in Industry* (*Kemija u Industriji*), Vol. 57, No. 9, 2008, pp. 417-423.

[15] P. K. Parhi, S. Panigrahi, K. Sarangi and K. C. Nathsarma, "Separation of Cobalt And Nickel from Ammoniacal Sulphate Solution Using Cyanex 272," *Separation and Purification Technology*, Vol. 59, No. 3, 2008, pp. 310-317. doi:10.1016/j.seppur.2007.07.026

[16] K. C. Sole and J. B. Hiskey, "Solvent Extraction Characteristics of Thiosubstituted Organophosphinic Acid Extractants," *Hydrometallurgy*, Vol. 30, No. 1-3, 1992, pp. 345-365. doi:10.1016/0304-386X(92)90093-F

[17] L. Luo, J-H. Wei, G-Y. Wu, F. Joyohisa and S. Atsushi, "Extraction Studies of Co(II) and Ni(II) from Chloride Solution Using PC 88A," *Transactions of Nonferrous Metals Society of China*, Vol. 16, No. 3, 2006, pp. 687-692. doi:10.1016/S1003-6326(06)60122-2

[18] R. A. Kumbasar, "Selective Extraction and Concentration of Cobalt from Acidic Leach Solution Containing Cobalt and Nickel through Emulsion Liquid Membrane Using PC 88A as Extractant," *Separation and Purification Technology*, Vol. 64, No. 3, 2009, pp. 273-279.

doi:10.1016/j.seppur.2008.10.011

[19] C. Bourget, B. Jakovljeivic and D. Nucciarone, "Cyanex® 301 Binary Extractant System in Cobalt/Nickel Recovery from Acidic Sulphate Solutions," *Hydrometallurgy*, Vol. 77, No. 3-4, 2005, pp. 203-218. doi:10.1016/j.hydromet.2004.12.005

[20] I. Yu. Fleitlikh, G. L. Pashkov, N. A. Grigorieva, L. K. Nikiforova, M. A. Pleshkov and Y. M. Shneerson, "Cobalt and Nickel Recovery from Sulphate Media Containing Calcium, Manganese and Magnesium with Mixture of Cyanex 301 and a Trialkylamine," *Solvent Extraction and Ion Exchange*, Vol. 29, No. 5-6, 2011, pp. 782-799. doi:10.1080/07366299.2011.595627

[21] D. S. Flett, "Solvent Extraction in Hydrometallurgy: The Role of Organophosphorous Extractants," *Journal of Organometallic Compounds*, Vol. 690, No. 10, 2005, pp. 2426-2438.

[22] B. R. Reddy, S. V. Rao and K. H. Park, "Solvent Extraction Separation and Recovery of Cobalt and Nickel from Sulphate Medium Using Mixtures of TOPS 99 and TIBPS Extractants," *Minerals Engineering*, Vol. 22, No. 5, 2009, pp. 500-505. doi:10.1016/j.mineng.2009.01.002

[23] R. K. Biswas, A. K. Karmakar and M. S. Rahman, "Extraction Equilibrium of Ni(II) in the Ni^{2+}-SO_4^{2-}-Ac^-(Na^+, H^+)-Cyanex 272(H_2A_2)-Kerosene-3% (v/v) Octan-1-ol System," *Journal of Scientific Research*, Vol. 4, No. 1, 2012, pp. 83-97.

[24] K. Akiba and H. Freiser, "Equilibrium and Kinetics of Nickel Extraction with 2-Hydroxy-5-nonylbenzophenone-oxime," *Separation Science and Technology*, Vol. 17, No. 5, 1982, pp. 745-750. doi:10.1080/01496398208068565

[25] A. Hokura, J. M. Perera, F. Grieser and G. W. Stevens, "A Kinetic Study of Nickel Ion Extraction by Kelex 100 at the Liquid-Liquid Interface," *Solvent Extraction and Ion Exchange*, Vol. 16, No. 2, 1998, pp. 619-636. doi:10.1080/07366299808934543

[26] H. Watarai, M. Takahashi and K. Shibata, "Interfacial Phenomena in the Extraction Kinetics of Nickel(II) with 2'-Hydroxy-5'-nonylacetophenone Oxime," *Bulletin of the Chemical Society of Japan*, Vol. 59, No. 11, 1986, pp. 3469-3473. doi:10.1246/bcsj.59.3469

[27] A. Buch, M. Stambouli and D. Pareaus, "Kinetics of Nickel(II) Extraction by 2-Ethylhexanal Oxime in Ammonium Nitrate Solutions," *Separation and Purification Technology*, Vol. 60, No. 2, 2008, pp. 120-127. doi:10.1016/j.seppur.2008.01.023

[28] T. Sana, K. Shiomori and Y. Kawano, "Extraction Rate of Nickel with 5-Dodecylsalicylaldoxime in a Vibro-Mixer," *Separation and Purification Technology*, Vol. 44, No. 2, 2005, pp. 160-165. doi:10.1016/j.seppur.2005.01.005

[29] D. B. Dreisinger and W. C. Cooper, "The Kinetics of Cobalt and Nickel Extraction Using EHEHPA," *Solvent Extraction and Ion Exchange*, Vol. 4, No. 2, 1986, pp. 317-344. doi:10.1080/07366298608917869

[30] D. B. Dreisinger and W. C. Copper, "The Kinetics of Zinc, Cobalt and Nickel Extraction in the D2EHPA-Heptane-HCl System Using the Rotating Diffusion Cell Technique," *Solvent Extraction and Ion Exchange*, Vol. 7, No. 2, 1989, pp. 335-360. doi:10.1080/07360298908962312

[31] Z. S. Hu, Y. Pan, W. W. Ma and X. Fu, "Purification of Organophosphorous Acid Extractants," *Solvent Extraction and Ion Exchange*, Vol. 13, No. 5, 1995, pp. 965-976. doi:10.1080/07366299508918312

[32] R. K. Biswas, M. A. Habib and H. P. Singha, "Colorimetric Estimation and Some Physicochemical Properties of Purified Cyanex 272," *Hydrometallurgy*, Vol. 76, No. 1-2, 2004, 97-104.

[33] E. B. Sandell, "Colorimetric Determination of Trace Metals," 3rd Edition, Intersciences, New York, 1959, p. 668.

[34] R. J. Whewell, M. A. Hughes and C. Hanson, "The Kinetics of the Solvent Extraction of Copper(II) with LIX Reagents-III. The Effects of LIX 63N in LIX 64N," *Journal of Inorganic and Nuclear Chemistry*, Vol. 38, No. 11, 1976, pp. 2071-2075. doi:10.1016/0022-1902(76)80471-X

[35] R. K. Biswas, M. A. Hanif and M. F. Bari, "Kinetics of Forward Extraction of Manganese(II) from Acidic Chloride Medium by D2EHPA in Kerosene Using the Single Drop Technique," *Hydrometallurgy*, Vol. 42, No. 3, 1996, pp. 399-409. doi:10.1016/0304-386X(95)00102-M

[36] R. K. Biswas, M. A. Habib and A. K. Karmakar, "Kinetics of Solvent Extraction of Iron(III) from Sulphate Medium by Purified Cyanex 272 Using a Lewis Cell," *Solvent Extraction and Ion Exchange*, Vol. 25, No. 1, 2007, pp. 79-98. doi:10.1080/07366290601067838

[37] R. K. Biswas, M. R. Ali, A. K. Karmakar and M. Kamruzzman, "Kinetics of Solvent Extraction of Copper(II) by Bis(2,4,4-trimethylpentyl)phosphinic Acid Using the Single Drop Technique," *Chemical Engineering and Technology*, Vol. 30, No. 6, 2007, pp. 774-781. doi:10.1002/ceat.200600284

[38] T. Sato, T. Yoshino, T. Nakamura and T. Kudo, "The Kinetics of Al(III) Extraction from Acidic Solutions by Di-(2-ethylhexyl) Phosphoric Acid," *Journal of Inorganic and Nuclear Chemistry*, Vol. 40, No. 8, 1978, pp. 1571-1574. doi:10.1016/0022-1902(78)80470-9

List of Symbols and Abbreviations Used

a, b, c, d, e: Reaction orders w.r.t $[Ni^{2+}]$, $[H^+]$, $[H_2A_2]_{(o)}$, $\left[SO_4^{2-}\right]$ & $[Ac^-]$, respectively

$a_{Ni^{2+}}$: Amount of Ni^{2+} transferred, kmol

β_1: Stability constant of $NiOH^+$: $[NiOH^+]$ $[H^+]$/ $[Ni^{2+}]$

β_2: Stability constant of $NiOHSO_4Ac^{2-}$: $[[NiOHSO_4Ac]^{2-}]/[NiOH^+]\left[SO_4^{2-}\right]$ $[Ac^-]$

C.H: Column (better to say continuum) height, m

$\Delta[Ni^{2+}]$: Concentration change in aqueous drop during travel, mg/L

Δt: End correction term, s

[]: Sign of concentration

A^-: Anion of monomeric BTMPPA

Ac^-: Acetate ion

BTMPPA, H_2A_2: Dimeric bis(2,4,4-trimethylpentyl)phosphinic acid

$\Delta H^{\pm}$: Enthalpy change in activation, kJ/mol

$\Delta S^{\pm}$: Entropy change in activation, kJ/mol K

E_a: Activation energy, kJ/mol

F: Ni^{2+} Transfer flux, $kmol/m^2 \cdot s$

$f(R)$: Function of reactants

h: Planck's constant (6.625×10^{-37} kJ·s)

hcr: High concentration region

$hpHr$: High pH region

htr: High temperature region

HA: Monomer of BTMPPA

I: Intercept

k: Boltzman constant (1.38×10^{-26} kJ/K)

$K_{a_{HA}}$: Ionization constant of HA, $kmol/m^3$

K_2: Dimerization constant of BTMPPA, $m^3/kmol$

k_f: Rate constant in forward extraction, $m^{5/2}/kmol^{1/2} \cdot s$

lcr: Low concentration region

$lpHr$: Low pH region

ltr: Low temperature region

N: Number of collected drop

P_{HA}: Distribution constant or partition coefficient of HA

RDC: Rotating diffusion cell

s: Slope

S_N2: Substitution nucleophilic bimolecular mechanism

S_N1: Substitution nucleophilic unimolecular mechanism

t: Drop fall time, s

T: Temperature, K

v: Volume of collected drop, cm^3

wrt: With respect to

Subscript

f: Forward

(ini): Initial

(int): Interface

(o): Organic

28

Equilibrium of the Extraction of V(IV) in the V(IV)-SO_4^{2-} (H^+, Na^+)—Cyanex 302-Kerosene System

Ranjit Kumar Biswas, Aneek Krishna Karmakar

Department of Applied Chemistry & Chemical Engineering, Rajshahi University, Rajshahi, Bangladesh

Email: rkbiswas53@yahoo.com

ABSTRACT

The title system has been investigated from the equilibrium point of view. Significant extraction occurs above pH 2. Equilibration time is 20 min. The extraction ratio (D) remains constant with increasing [V(IV)] of at least 0.50 g/L. It is inversely proportional to $[H^+]^2$, $[H^+]$ and $[H^+]^{0.3}$ in the lower pH (<2.25), medium pH (~2.90) and higher pH (~4.0) regions, respectively. Moreover, it is proportional to $[Cyanex\ 302]^2$; and $[SO_4^{2-}]^0$ and $[SO_4^{2-}]^{-1}$ in the lower $[SO_4^{2-}]$ (<0.05 mol/L) and higher $[SO_4^{2-}]$ (>1 mol/L) regions, respectively. The apparent extraction equilibrium constant (K_{ex}) in 0.02 mol/L SO_4^{2-} medium and at 303 K is found to vary from $10^{-3.447}$ to $10^{1.508}$ with increasing equilibrium pH from 2.25 to 4.00. Various sulphated, hydrolyzed, hydrated and mixed sulphated hydrolyzed species of V(IV) have been considered at different extraction conditions to propose the extraction equilibrium reactions to form always $[VO(HA_2)_2]$ as the extractable species. The system is highly temperature dependent with ΔH value of ~90 kJ/mol and ~25 kJ/mol in lower and higher temperature regions, respectively. The calculated loading capacity is low (4.05 g V(IV)/100 g Cyanex 302). Kerosene is a better diluent over $CHCl_3$, Cyclo-C_6H_{12} and CCl_4; but much better solvents are C_6H_6, $C_6H_5CH_3$, n-C_7H_{16}, $C_6H_4(CH_3)_2$, petroleum benzin, 1,2-$C_2H_4Cl_2$, C_6H_5Cl. Mineral acids (1 mol/L) are able to strip off V(IV) from the organic phase in a single-stage. Using Cyanex 302, almost complete separations of V(IV) from Cu(II) at pH 1.0 and from Ni(II) at pH$_{(eq)}$ 4.5 are possible in a single-stage of extraction; whereas, its separation from Zn(II) at pH$_{(eq)}$ 2.5, Co(II) at pH$_{(eq)}$ 3.5, Fe(III) at pH$_{(eq)}$ 2.0 and Ti(IV) at pH$_{(eq)}$ 2.5 will require counter-current multi-stage extractions.

Keywords: Extraction Equilibrium; Vanadium(IV); Cyanex 302; Kerosene; Sulphate

1. Introduction

Vanadium is used for alloying steel and the manufacture of oxidative catalyst. The rich deposits of its ores, *viz.* patronite (V_2S_3), vanadinite ($3Pb_3(VO_4)_2 \cdot PbCl_2$), carnotite ($K_2U_2V_2O_{11} \cdot 3H_2O$) etc. are rare on the earth's crust now. Consequently, it is necessary to develop extraction processes for low grade ores and waste materials (tar sand, waste desulphurization catalyst etc.). Solvent extraction technique is convenient for such purpose. The technique can build up concentration by using low (O/A) ratio in extraction and high O/A ratio in stripping. Works on the solvent extraction of V(IV) by various extractants prior to 1976 were summarized by Sekine and Hasegawa [1]. Di-2-ethylhexyl phosphoric acid (D2EHPA) is a promising extractant for V(IV) and V(V) [2-8]. Vanadium-(IV) and (V) have also been extracted by EHEHPA (2-ethylhexyl phosphonic acid mono-2-ethylhexyl ester) [9]. A recent development in the field of solvent extraction is the use of organophosphinic acid derivatives and their sulphur analogues (Cyanex reagents) introduced by

American Cyanamid Company and Cytec Canada Inc. Cyanex 302 and Cyanex 301 are the mono- and disulphide analogues of Cyanex 272 (di-2,4,4-trimethylpentylphosphinic acid). The sulphur substitution decreases pK$_a$ values (*viz.* 6.4, 5.6 and 2.6 for Cyanex 272, Cyanex 302 [10,11] and Cyanex 301 [12], respectively) which permits to work at lower pH [13]. Cyanex reagents differ from other commercial organophosphorous reagents (e.g. D2EHPA, DDPA, TBP, EHPEHPA etc.) in that the former reagents contain P-C bonding, whereas the latter reagents contain P-O-C bonding. The presence of P-C bonding in Cyanex reagents renders them to be less susceptible to hydrolysis and less soluble in water [14].

In recent past, the extraction behaviors of V(IV) from sulphuric acid solution by Cyanex 272 [15,16] and of V(IV) and V(V) from hydrochloric acid solution by Cyanex 272 and Cyanex 301 [14] had been reported. There is no report on the extraction behavior of V(IV) from any acid solution using Cyanex 302. The present paper reports the extraction characteristics of V(IV) from sulphate medium by Cyanex 302 dissolved in kerosene. The effects of the aqueous and organic phase variables (including diluent variation) have been investigated to determine the dependence of variables and to calculate the extraction equilibrium constant; and also to propose mechanism of extraction. The loading capacity is also elucidated. Finally, the possibility of separation of V(IV) from some cations of the first transition series in binary mixtures has been predicted.

2. Materials and Methods

2.1. Materials

Cyanex 302 was collected from Cytec Canada Inc. as a gift. It contains 78% - 80% R_2PSOH, 10% - 12% R_3PO, 2% - 3% R_2PO_2H, 2% R_2PS_2H and 8% unknown compounds [17] and has been used without further purification as R_3PO, R_2PO_2H and R_2PS_2H—all have the extracting power. Kerosene is bought from the local market and distilled to collect the colorless aliphatic fraction distilling over 200°C - 260°C. Ammonium vanadate (99%, Riedel-deHaen) and vanadium(IV) sulphate oxide (99.9%, Alfa Aesar-Johnson-Mathey), hydrogen peroxide (30%, Merck-Germany) are used in this study without further purification. Diluents other than kerosene are the products of Riedel-deHaen and E Merck-India; all are more than 99% pure.

2.2. Analytical

Concentrations of V(IV) in aqueous solutions have been measured by the HNO_3 oxidative-H_2O_2 method [18] at 450 nm using a UV-visible spectrophotometer (UV-1650 PC, Shimadzu, Japan). For standard and test solutions preparations NH_4VO_3 and $VOSO_4 \cdot 5H_2O$, respectively, are used. A Mettler Toledo pH meter (model 320) is used for pH measurement and adjustment (by addition of either anhydrous Na_2CO_3 or dilute H_2SO_4 solution).

2.3. Procedure

Extraction and stripping procedures are given elsewhere [19]. In both cases, two phases are agitated at O/A=1 (O = 20 mL) and 303 K (otherwise stated) for a predetermined time (20 min in extraction and 1 h in stripping). The phase separation is quick; and the aqueous phase after equilibration is analyzed for its equilibrium pH and V(IV)-content. Then the value of extraction ratio or stripping percentage is calculated as usual [19].

2.4. Notations and Abbreviations

K_{ex}, Extraction equilibrium constant;
D, Extraction or distribution ratio;
^CD, D at a constant equilibrium pH and extractant concentration;

β, Stability constant;
K_{SO_4}, A proportionality constant in sulphate dependence study;
H_2A_2, Cyanex 302 (dimeric);
A^-, Anion of monomeric Cyanex 302;
L^-, Uni-negative anion existing in the aqueous phase viz. OH^- and HSO_4^-;
ΔH, Apparent enthalpy change;
R, CH_3-$C(CH_3)_2$-CH_2-$CH(CH_3)$- CH_2^-;
l.c.r, Lower concentration region;
h.c.r, Higher concentration region;
l.pH.r, Lower pH region;
h.pH.r, Higher pH region;
l.t.r, Lower temperature region;
h.t.r, Higher temperature region;
Suffix (o), Organic phase;
(ini), Initial;
(eq), Equilibrium.

2.5. Treatment of Extraction Equilibrium Data

The main constituent R_2PSOH of Cyanex 302 is dimeric in non-polar diluents [17,20]. In the aqueous solution, V(IV) virtually exists as VO^{2+} (owing to high charge to radius ratio) which can form complex with co-existing OH^- and HSO_4^- (L^-). Therefore, the extraction equilibrium at a constant temperature can be represented by:

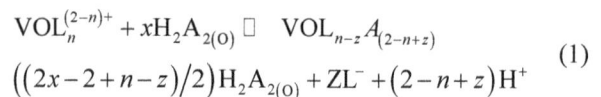

$$VOL_n^{(2-n)+} + xH_2A_{2(O)} \ \square \ VOL_{n-z}A_{(2-n+z)}$$
$$((2x-2+n-z)/2)H_2A_{2(O)} + ZL^- + (2-n+z)H^+ \quad (1)$$

The K_{ex} of Equation (1) can be expressed as:

$$\log D = \log K_{ex} + (2-n+z)pH$$
$$+ x\log[H_2A_2]_{(O)} - z\log[L^-] \quad (2)$$

Where, $D = [VO^{2+}]_{(o, eq)}/[VO^{2+}]_{(eq)}$. Equation (2) represents the basic equation for chelate forming solvent extraction system by a dimeric acidic extractant. All concentrations and pH terms in Equation (2) refer to the equilibrium values. Consequently, Equation (2) represents that the value of log D should be independent of $[VO^{2+}]$ at a set of constant equilibrium pH, [extractant] and [anion]. Corrected D-values (i.e. ^CD) at a set of constant equilibrium pH and [extractant] can be calculated by mass-balance (Equation (3)):

$$\log {}^CD = \log D + m\left(pH_{(ini)} - pH_{(eq)}\right)$$
$$+ x\left[\log[H_2A_2]_{(o,ini)} - \left\{\log[H_2A_2]_{(o,ini)}\right.\right.$$
$$\left.\left. - x[VO^{2+}]_{(o,eq)}\right\}\right] \quad (3)$$

Where, m = $(2 - n + z)$ = pH dependence = 2 (upto $pH_{(eq)}$ = 1.5) and <2 ($pH_{(eq)}$ >1.5) and x = extractant dependence

= 2.00 and all concentration terms are in mol/L. Moreover, as K_{ex} is related to temperature by the Van't Hoff equation, $\log {}^{C}D$ will also depend on temperature.

3. Results and Discussion

3.1. Extraction Equilibrium

Through running some preliminary experiments, it has been found that V(IV) is extractable by Cyanex 302 at pH around 3.0. When 0.20 g/L V(IV) existing in the aqueous phase at $pH_{(ini)}$ 4.00 containing 0.02 mol/L SO_4^{2-} was extracted with 0.15 mol/L H_2A_2 in kerosene at 303 K and O/A = 1, then it was found that the $[V(IV)]_{(o)}$ was increased up to phase contact of 18 min. Therefore the equilibration time is about 18 min; but 20 min has been used subsequently in order to ensure equilibrations.

The variation of extraction ratio with [V(IV)] was found out at two different set of experimental parameters. It is found in both cases that the $[V(IV)]_{(o)}$ is increased, but the value of D is decreased continuously with increasing $[V(IV)]_{(ini)}$. This is contrary to the general principle of solvent extraction chemistry as suggested by Equation (2) which is valid at constant $[H_2A_2]_{(o,eq)}$ and $pH_{(eq)}$. The observed decreasing behavior might be due to the non-constancy of $[H_2A_2]_{(o,eq)}$ and $pH_{(eq)}$ for various extents of V(IV) extraction. On calculating $\log {}^{C}D$ (by Equation (3)) on considering n = 0.60 (tangential slope at $pH_{(eq)}$ of 3.41 for 0.10 mol/L Cyanex 302 system) or 0.85 (tangential slope at $pH_{(eq)}$ of 3.10 for 0.20 mol/L Cyanex 302 system) and m = 2.00 (for both systems), the $\log {}^{C}D$ vs. $\log ([V(IV)]_{(ini)}$, mol/L) plots are drawn in **Figure 1**. The plots are horizontal up to at least $[V(IV)]_{(ini)}$ of 0.50 g/L. The decreasing behavior of $\log {}^{C}D$ value with increasing $[V(IV)]_{(ini)}$ over 0.50 g/L indicates the non-ideality of V(IV) bearing aqueous phase which might be due to the polymerization of V(IV) in the aqueous phase and/or in both phases.

At a constant equilibrium [extractant], the plot of $\log {}^{C}D$ vs. $pH_{(eq)}$ should be a straight line (cf. Equation (2)) with slope equaling to "2 − n + z" (the number of H^+ liberated). The $\log {}^{C}D$ values are calculated accordingly. **Figure 2** represents $\log {}^{C}D$ vs. $pH_{(eq)}$ plots at constant $[H_2A_2]_{(o,eq)}$ of 0.10 and 0.20 mol/L. In both cases, straight lines are not obtained. Curves with limiting slopes of ~2 are obtained at l.pH.r (below pH 2.25), whilst the tangential slope at pH = 4.0 is ~0.30 in both cases. At pH ~2.9, the tangential slope is unity. The unity sloped tangential lines have intercepts of −3.04 and −2.34 for 0.10 and 0.20 mol/L Cyanex 302 systems, respectively. The respective intercepts of lines having slope 2 are −5.485 and −4.845. It is concluded from this result that the pH dependence is dependent of equilibrium pH range used;

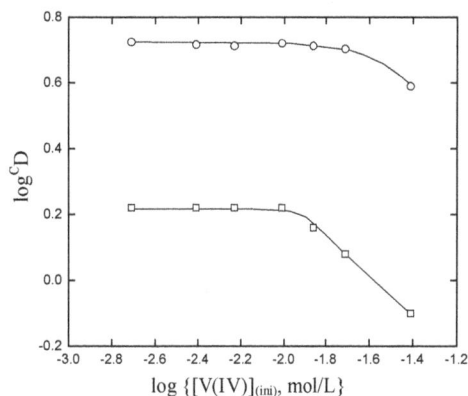

Figure 1. Effect of [V(IV)]on ${}^{C}D \cdot [SO_4^{2-}] = 0.02$ mol/L, Temp. = (303 ± 0.5) K, Equilibration time = 20 min, O/A = 1 (O = 20 mL). (□) [Cyanex 302] = 0.10 mol/L, $pH_{(ini)}$= 4.30, $pH_{(eq)}$ const. = 3.41; (○) [Cyanex 302] = 0.20 mol/L, $pH_{(ini)}$= 3.80, $pH_{(eq)}$ const. = 3.10.

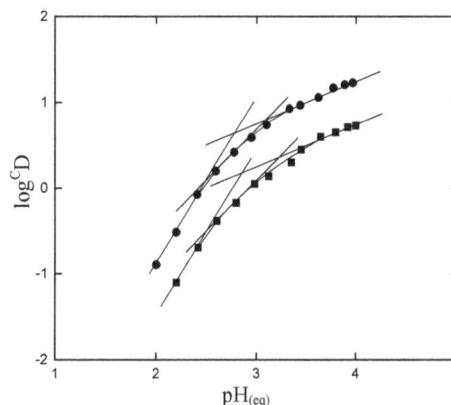

Figure 2. Effect of $pH_{(eq)}$. $[V(IV)]_{(ini)}$ = 200 mg/L, Temp. = 303 ± 0.5 K, Eq. time = 20 min, O/A = 1 (O = 20 mL), $[SO_4^{2-}]$ = 0.02 mol/L. (□) [Cyanex 302] = 0.10 mol/L; S = 2 ($pH_{(eq)}$< 2.25), 1 ($pH_{(eq)}$ = 2.9) and 0.3 ($pH_{(eq)}$ = 4.0); I = −5.485 (when S = 2) and −3.04 (when S = 1); (○) [Cyanex 302] = 0.20 mol/L; S = 2 ($pH_{(eq)}$< 2.25), 1 ($pH_{(eq)}$ = 2.9) and 0.3 ($pH_{(eq)}$ = 4.0); I = −4.845 (when S = 2) and −2.34 (when S = 1).

two H^+ ions are liberated per V(IV) being extracted whence $pH_{(eq)}$ is kept below 2.25; and with increasing $pH_{(eq)}$ value, the number of H^+ ions liberated per V(IV) being extracted is decreased from 2 to 1 at $pH_{(eq)}$ of 2.9 and 0.3 at $pH_{(eq)}$ of 4.0.

According to Equation (2), the plot of $\log {}^{C}D$ vs. $\log [H_2A_2]_{(o,eq)}$ at a particular constant $pH_{(eq)}$ should be a straight line with slope giving the mole ratio (x) of Cyanex 302/V(IV) in extractable species. The $\log {}^{C}D$ vs.$\log \{[Cyanex 302]_{(o,eq)}$, mol/L} plots at two sets of parametric conditions are shown in **Figure 3**. Straight lines are obtained with slopes equaling to ~2 in both cases. The intercepts of the lines are 2.07 and 2.30 for $pH_{(ini)}$ systems of 3.8 and 4.2, respectively. It is therefore

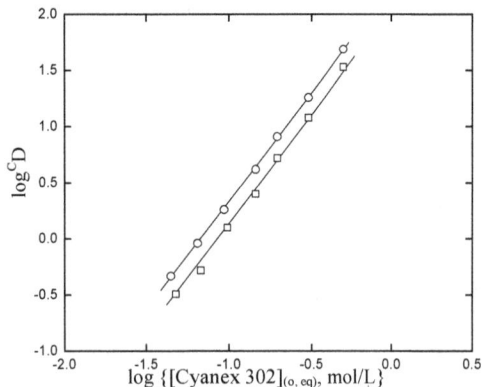

Figure 3. Effect of [Cyanex 302]. [V(IV)] = 200 mg/L, Temp. = (303 ± 0.5) K, Eq. time = 20 min, O/A = 1 (O = 20 mL), [SO_4^{2-}] = 0.02mol/L. (□) $pH_{(ini)}$ = 3.8, $pH_{(eq)}$ const. = 3.12; S = 2.00, I = 2.07; (○) $pH_{(ini)}$ = 4.2, $pH_{(eq)}$ const. = 3.35; S = 2.00, I = 2.30.

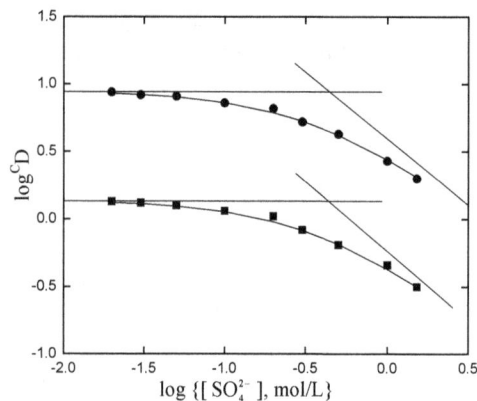

Figure 4. Effect of [SO_4^{2-}]. $pH_{(ini)}$ = 4.2, [V(IV)] = 200 mg/L, Temp. = (303 ± 0.5) K, Eq. time = 20 min, O/A = 1 (O = 20 mL). (■), [Cyanex 302] = 0.10 mol/L, $pH_{(eq)}$ const. = 3.85; (●), [Cyanex 302] = 0.20 mol/L, $pH_{(eq)}$ const. = 3.79

concluded that for the V(IV)-chelate extraction, 2 moles of Cyanex 302 is needed to extract 1 g ion of V(IV) *i.e.* the value of "*x*" in Equations (2) and (3) is 2.

Co-existing anion in the aqueous phase often affects the extraction characteristics of a metal ion by an extractant, particularly when the extraction occurs via the ion-pair formation and solvation mechanisms. In chelate forming extraction systems, the co-existing anion may take part in chelate formation and also the chelate formation may be hindered by the prior formation of metal-co-existing anion complex. Since the extraction has been carried out from sulphate medium, the effect of [SO_4^{2-}] on extraction has been studied. The related plot is displayed in **Figure 4**. Experimental points fall on a curve rather than on a straight line. In *l.c.r* of SO_4^{2-}, ^CD is seldom changed, whilst in the *h.c.r* of SO_4^{2-}, it is considerably decreased with increasing[SO_4^{2-}]. Curves in figure are theoretical and represented by:

$$\log {}^CD = 0.14(■), 0.95(●) - \log\left(1 + 2.24\left[SO_4^{2-}\right]\right) \quad (4)$$

Where, 2.24 is the value of K_{SO_4} as derived by the curve fitting method. Intercepts of the asymptotes at *h.c.r* of sulphate ion are −0.24 and 0.59 for 0.10 and 0.20 mol/L Cyanex 302 systems, respectively, whilst the respective intercepts at *l.c.r* of SO_4^{2-} are 0.14 and 0.95.

The Van't Hoff plots for the investigated system at two sets of experimental parameters are shown in **Figure 5**. In both cases, it is found that the extraction ratio is increased with increasing temperature but the straight line relationship does not hold. Slopes of the lines at *h.t.r* are −1100 and −1400 and at *l.t.r* are −4600 and −4400 for 0.10 and 0.20 mol/L Cyanex 302 systems, respectively. From the slopes of the plots, the apparent ΔH values have been calculated as 21.80 and 27.7 kJ/mol at *h.t.r* and 91.1 and 87.1 kJ/mol at *l.t.r* for 0.10 and 0.20 mol/L Cyanex

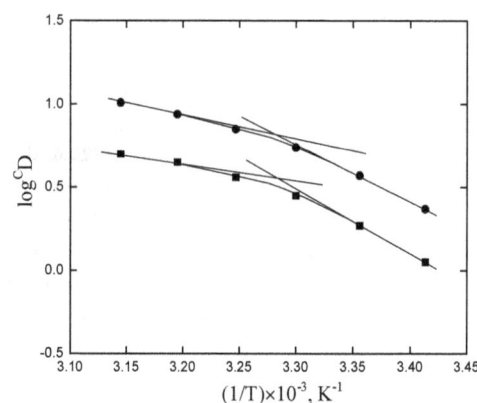

Figure 5. Effect of temperature. [V(IV)] = 200 mg/L, Eq. time = 20 min, [SO_4^{2-}] = 0.02 mol/L, O/A = 1 (O = 20 mL). (■), $pH_{(ini)}$ = 4.40, $pH_{(eq)}$ const. = 3.45, [Cyanex 302] = 0.10 mol/L; S = −1100, ΔH = 21.80 kJ/mol (*h.t.r*); S = −4600, ΔH = 91.1 kJ/mol (*l.t.r*); (●), $pH_{(ini)}$ = 3.80, $pH_{(eq)}$ const. = 3.10, [Cyanex 302] = 0.20 mol/L; S = −1400, ΔH = 27.70 kJ/mol (*h.t.r*); S = −4400, ΔH = 87.1 kJ/mol (*l.t.r*).

302 systems, respectively. The extraction of V(IV) by Cyanex 302 is therefore, extensively increased with increasing temperature (endothermic) with ΔH value of ~25 kJ/mol at *h.t.r* and of ~90 kJ/mol at *l.t.r*. It is evident from these studies that the value of "*x*" is 2 irrespective of the experimental parameter but the value of "*z*" is 0 at low [SO_4^{2-}] and 1 at high [SO_4^{2-}]. The value of "2 − *n* + *z*" is 2 in low $pH_{(eq)}$, 1 in intermediate $pH_{(eq)}$, and 0.3 in high $pH_{(eq)}$. At *l.c.r* of SO_4^{2-} and at low $pH_{(eq)}$, 2 − *n* + *z* = 2 implies that *n* = 0; but at intermediate $pH_{(eq)}$, 2 − *n* + *z* = 1 implies that *n* = 1, and finally at high $pH_{(eq)}$, 2 − *n* + *z* = 0.30 implies that *n* = 1.70. On the other hand, at *h.c.r* of SO_4^{2-} and at low $pH_{(eq)}$, 2 − *n* + *z* = 2 implies *n* = 1; but at intermediate $pH_{(eq)}$, 2 − *n* + *z* = 1 implies *n* = 2 (possibly representing one bisulphate and one hydroxide

being complexed with VO^{2+}) and at high $pH_{(eq)}$, $2 - n + z = 0.30$ implies n = 2.70 (possibly representing almost one/two bisulphate and two/one hydroxide being complexed with VO^{2+}).

3.2. Evaluation of Extraction Equilibrium Constant

The foregoing experimental results give the equation for $^C D$ at 303 K as:

$$\log{}^C D = \log K_{ex} + x pH_{(eq)} + 2\log\left[Cyanex\ 302\right]_{(o,eq)} \\ - \log\left(1 + 2.24\left[SO_4^{2-}\right]\right) \quad (5)$$

Based on the Equation (5), the value of $\log K_{ex}$ has been evaluated from intercepts of the straight lines or asymptotic lines in **Figures 2-4**. The evaluated values of $\log K_{ex}$ from different parametric studies are shown in **Table 1**. It is observed that the value of $\log K_{ex}$ is extensively increased with increasing the value of $pH_{(eq)}$ or decreasing the value of "x". The $\log K_{ex}$ value of -3.447 for $x = 2$ is increased to -0.972 for $x = 1$ and to 1.508 for $x = 0.3$. The variation of the value of $\log K_{ex}$ with $pH_{(eq)}$ is given in **Figure 6**. The figure also shows the variation of "x" with $pH_{(eq)}$. It is seen that as the value of "x" decreases, the value of $\log K_{ex}$ increases tremendously.

3.3. Extraction Mechanism

The foregoing results lead to the following expression relating the equilibrium constant with extraction ratio in the extraction of 0.20 g/L V(IV) (which will also be valid up to 700 mg/L V(IV)) at 303 K:

$$K_{ex} = {}^C D\left[H^+\right]^x \left(1 + 2.24\left[SO_4^{2-}\right]\right)\Big/\left[H_2A_2\right]_{(O)}^2 \quad (6)$$

At $l.c.r$ of $[SO_4^{2-}]$, $1 >> 2.24$ $[SO_4^{2-}]$, so that Equation (6) becomes:

$$K_{ex} = {}^C D\left[H^+\right]^x \Big/ \left[H_2A_2\right]_{(O)}^2 \quad (7)$$

Equation (7) suggests the following general chemical reaction as the extraction equilibrium reaction:

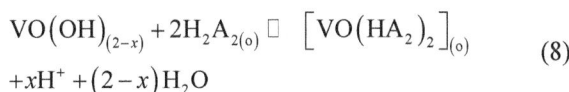

$$VO(OH)_{(2-x)} + 2H_2A_{2(o)} \ \square\ \left[VO(HA_2)_2\right]_{(o)} \\ + xH^+ + (2 - x)H_2O \quad (8)$$

When $x = 2$ i.e. at $pH_{(eq)} \leq 2.25$, Equation (8) becomes:

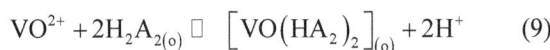

$$VO^{2+} + 2H_2A_{2(o)} \ \square\ \left[VO(HA_2)_2\right]_{(o)} + 2H^+ \quad (9)$$

and when $x = 1$ i.e. at $pH_{(eq)} \approx 2.90$, Equation (8) becomes:

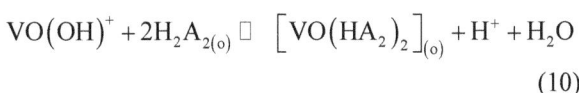

$$VO(OH)^+ + 2H_2A_{2(o)} \ \square\ \left[VO(HA_2)_2\right]_{(o)} + H^+ + H_2O \quad (10)$$

and when $x = 0$ i.e. at $pH_{(eq)}$ somewhere greater than 4, Equation (8) becomes:

$$VO(OH)_2 + 2H_2A_{2\,(o)} \ \square\ \left[VO(HA_2)_2\right]_{(o)} + 2H_2O \quad (11)$$

It is therefore seen that the hydrolysis of VO^{2+} starts around pH 2.25 and consequently the pH dependence starts to decrease. It is reported that the values of $\beta_{VO(OH)^+}$ and $\beta_{VO(OH)_2}$ are $10^{7.9}$ and $10^{18.31}$, respectively [21]. As evaluated be- fore, the equilibrium constant for the reaction given by Equation (9) is $10^{-3.447}$ and that by Equation (10) is $10^{-0.972}$. For Equation (11), the equilibrium constant will be somewhat greater than $10^{1.508}$. At $h.c.r$ of SO_4^{2-}, $1 << 2.24$ $[SO_4^{2-}]$; so that Equation (6) becomes:

Table 1. Evaluation of the approximate apparent K_{ex} values at various $pH_{(eq)}$ and at 303 K ([V(IV)] = 0.20 g/L).

Fig. No.	$pH_{(eq)}$	$[H_2A_2]_{(o)}$, mol/L	$[SO_4^{2-}]$, mol/L	Intercept (I)	Apparent $\log K_{ex}$	Avg. $\log K_{ex}$
\multicolumn{7}{l}{1) $pH_{(eq)} \leq 2.25$; pH dependence = 2.0}						
2	variable	0.10	0.02	−5.485	−3.466	−3.447
		0.20	0.02	−4.845	−3.428	
\multicolumn{7}{l}{2) $pH_{(eq)} = 2.50$; pH dependence = 1.5}						
2	variable	0.10	0.02	−4.295	−2.276	−2.250
		0.20	0.02	−3.640	−2.223	
\multicolumn{7}{l}{3) $pH_{(eq)} = 2.90$; pH dependence = 1.0}						
2	variable	0.10	0.02	−3.040	−1.021	−0.972
		0.20	0.02	−2.340	−0.923	
\multicolumn{7}{l}{4) $pH_{(eq)} = 3.10$; pH dependence = 0.8}						
2	variable	0.10	0.02	−2.355	−0.336	
		0.20	0.02	−1.740	−0.323	
3	3.12	variable	0.02	2.070	−0.400	−0.360
				0.140 (l.c.r)	−0.356	
4	3.12	0.10	variable	−0.240 (h.c.r)	−0.386	
\multicolumn{7}{l}{5) $pH_{(eq)} = 3.35$; pH dependence = 0.6}						
2	variable	0.10	0.02	−1.61	0.409	
		0.20	0.02	−1.07	0.347	
3	3.35	variable	0.02	2.30	0.309	0.346
				0.95 (l.c.r)	0.338	
4	3.35	0.20	variable	0.59 (h.c.r)	0.328	
\multicolumn{7}{l}{6) $pH_{(eq)} = 4.00$; pH dependence = 0.30}						
2	variable	0.10	0.02	−0.465	1.554	1.508
		0.20	0.02	0.045	1.462	

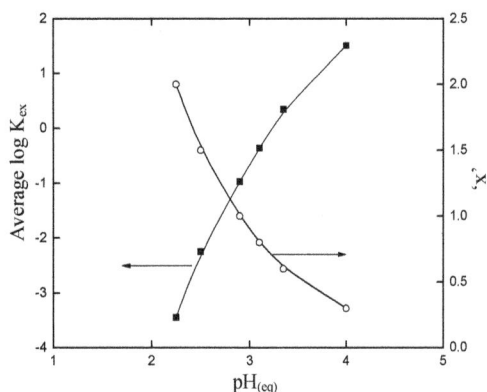

Figure 6. The approximate apparent extraction equilibrium constant (K_{ex}) at various equilibrium pH (pH$_{(eq)}$) at 303 K and for [V(IV)] of 0.20 g/L.

$$K_{ex} = 2.24\,{}^{C}D\left[H^{+}\right]^{x}\left[SO_4^{2-}\right]/\left[H_2A_2\right]^2_{(O)} \qquad (12)$$

Equation (12) suggests the liberation of sulphate ion during extraction reaction. But at a certain pH, the values of K_{ex} obtained at *l.c.r* and *h.c.r* of sulphate are identical. It is suggested that at *h.c.r* of SO_4^{2-}, the general Equation (8) also represents the extraction equilibrium reaction. But in this case, as the sulphate concentration increases, the free non-sulphated/bisulphated V(IV)-species concentration decreases during the progress of extraction and this gradual depletion is probably compensated through dissociation of sulphated/bisulphated V(IV)-species. It appears therefore that the equilibrium shift occurs between suphated/bisulphated and non-sulphated/bisulphated species as suggested below

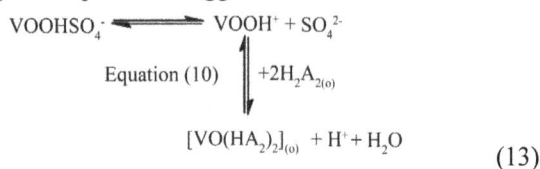

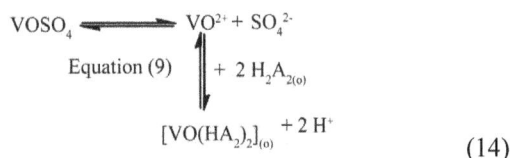

$$VOOHSO_4^- \rightleftharpoons VOOH^+ + SO_4^{2-}$$
$$\text{Equation (10)} \quad \big\Updownarrow +2H_2A_{2(o)}$$
$$[VO(HA_2)_2]_{(o)} + H^+ + H_2O \qquad (13)$$

$$VOSO_4 \rightleftharpoons VO^{2+} + SO_4^{2-}$$
$$\text{Equation (9)} \quad \big\Updownarrow + 2\,H_2A_{2(o)}$$
$$[VO(HA_2)_2]_{(o)} + 2\,H^+ \qquad (14)$$

at pH ~2.90 and ≤ 2.25, respectively. It is reported that $\beta_{VOHSO_4^-}$ is $10^{2.44}$ [21].

3.4. Effect of Diluent

As the diluent may tremendously affect the metal-ion distribution in a solvent extraction process, the extraction ratios have been measured when V(IV) in the same aqueous phase has been extracted by 0.15 mol/L H_2A_2 solutions dissolved in different diluents keeping all other parametric conditions identical. The results are represented in **Table 2**. It is observed that the extraction ratio

increases in the following order with the variation of diluent: $CHCl_3$ (D = 0.85) < CCl_4 (D = 3.01) = cyclo-C_6H_{12} (D = 3.01) < kerosene (D = 3.45) < C_6H_5Cl (D = 4.01) = n-C_7H_{16} (D = 4.01) = 1,2-$C_2H_4Cl_2$ (D = 4.01) < petroleum benzin (D = 4.72) = C_6H_4-$(CH_3)_2$ (D = 4.72) < C_6H_6 (D = 5.68) = C_6H_5-CH_3 (D = 5.68). The study helps draw the conclusion that C_6H_6 and C_6H_5-CH_3 are very good diluents followed by petroleum benzin and C_6H_4-$(CH_3)_2$ for the extraction of V(IV) by Cyanex 302. Kerosene is a better diluent over $CHCl_3$, CCl_4 and cyclo-C_6H_{12}. 77.54% V(IV) extraction in kerosene phase can be increased to about 85.02% V(IV) extraction in C_6H_6 or C_6H_5-CH_3 phase whilst reduced to ~46% V(IV) extraction in the $CHCl_3$ phase.

3.5. Loading of Cyanex 302 with V(IV)

The loading of V(IV) into the kerosene solution of Cyanex 302 is presented in **Figure 7**. It is observed that the loading of the organic phase with V(IV) is ended up at the 11th contact. An aliquot of 1 L Cyanex 302 solution of concentration 0.20 mol/L is saturated with 4.96 g V(IV) and so the loading capacity is calculated to be about 4.05 g V(IV) per 100 g of Cyanex 302. The loading capacity is considerably low for the system, and so it cannot be recommended for a large scale separation of V(IV) from an aqueous solution. The extraction of 4.96 g V(IV)/L by 1 L 0.20 molar Cyanex 302 at saturated loading implies the Cyanex 302/V(IV) mole ratio of 2.05 which is slightly higher than that (2.00) obtained from the extractant dependence studies. This slight variation may be due to extractant loss for partitioning (aqueous solubility) on repeated contact with fresh amounts of the

Table 2. Effect of diluent on extraction. [V(IV)]$_{(ini)}$ = 200 mg/L, pH$_{(ini)}$ = 4.10, [Cyanex 302] = 0.15 mol/L, [SO_4^{2-}] = 0.02 mol/L, Temp. = (303 ± 0.5) K, Equilibration time = 1 h, O/A = 1 (O = 20 mL).

Diluent	[V(IV)]$_{(eq)}$, mg/L	[V(IV)]$_{(o\ eq)}$, mg/L	D	% of extraction
Carbon tetrachloride	49.9	150.1	3.01	75.04
Chlorobenzene	39.9	160.1	4.01	80.03
Petroleum benzin	34.9	165.1	4.72	82.53
n-Heptane	39.9	160.1	4.01	80.03
Benzene	30.0	170.0	5.68	85.02
Toluene	30.0	170.0	5.68	85.02
Cyclohexane	49.9	150.1	3.01	75.04
1,2-dichloroethane	39.9	160.1	4.01	80.03
Xylene	34.9	165.1	4.72	82.53
Chloroform	108.2	91.8	0.85	45.92
Kerosene	44.9	155.1	3.45	77.54

aqueous phase by the same organic phase. The loading results indicate that the mechanism of extraction at high loading is not changed from that suggested at low loading i.e. in equilibrium studies.

3.6. Stripping of Ti(IV)-Loaded Organic Phase by Mineral Acids

The maximum V(IV) loaded organic phase containing 4.96 g/L V(IV) with theoretically no free-extractant, after proper dilution and adjustment of free extractant concentration, has been subjected for stripping study with 0.1, 0.3 and 1.0 mol/L H_2SO_4, HNO_3 and HCl solutions at 303 K and at O/A of 1. The stripping results are given in **Table 3**. It is found that stripping percentage is more or less acceptable in all three acids used alone. In all cases, stripping percentage is increased with increasing concentration of acid. It is seen that 90% stripping by 0.10

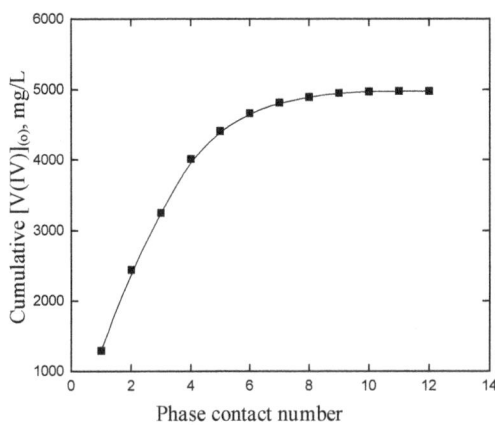

Figure 7. Loading of V(IV) in the organic phase. $[V(IV)]_{(ini)}$ = 2000 mg/L, [Cyanex 302] = 0.20 mol/L, pH = 6.0, $[SO_4^{2-}]$ = 0.04 mol/L, Equilibration time = 20 min, Temp. = (303 ± 0.5) K, O/A = 1 (O = 100 mL).

Table 3. Stripping of V(IV) loaded organic phase using different acid solutions. $[V(IV)]_{(o)}$ = 200 mg/L, $[Cyanex 302]_{(o)}$ = 0.10 mol/L, Equilibration time = 1 h, Temp. = (303 ± 0.5) K, O/A = 1 (O = 20 mL).

Stripping agent	Acid concentration, mol/L	$[V(IV)]_{(aq)}$, mg/L	% of V(IV) stripped
H_2SO_4	0.10	180.0	90.00
	0.30	187.0	93.58
	1.00	200.0	100.00
HCl	0.10	160.0	80.00
	0.30	170.0	85.00
	1.00	190.0	95.00
HNO_3	0.10	172.0	86.42
	0.30	180.0	90.00
	1.00	200.0	100.00

mol/L H_2SO_4 is increased to 100% stripping by 1.00 mol/L H_2SO_4. Similarly, 80% stripping by 0.10 mol/L HCl is increased to 95% stripping with 1.00 mol/L HCl; whereas, 86.42% stripping by 0.10 mol/L HNO_3 is increased to 100% stripping by 1.00 mol/L HNO_3. Sulphuric acid or nitric acid (1 mol/L) is sufficient to strip off V(IV) quantitatively. Hydrochloric acid can also be used in stripping if more than one-stage stripping is practiced or more concentrated solution being used.

3.7. Separation Ability of V(IV) from Some Other Metal ions

In order to examine the effectiveness of Cyanex 302 towards the mutual separations of V(IV) from some 3d-block metal ions viz. Ti(IV), Fe(III), Co(II), Ni(II), Cu(II) and Zn(II), the extraction percentages of these metal ions have been estimated while 0.20 g/L metal ion being extracted from 0.10 mol/L SO_4^{2-} (or, $[SO_4^{2-}]$ = H_2SO_4 when $[H_2SO_4] > 0.10$ mol/L) medium at different $pH_{(eq)}$ values by 0.10 mol/L extractant (in kerosene) at 303 K and O/A = 1 (O = 20 mL) after phase mixing of 1 h. The extraction results given in **Table 4**, which predict the following:

Table 4. Extraction data of some 3d-block elements by Cyanex 302 dissolved in kerosene. [Cyanex 302] = 0.10 mol/L (in kerosene); [Metal ion] = 0.2 g/L; $[SO_4^{2-}]$ = $[H_2SO_4]$ or 0.10 mol/L, Temp = 303 K, O/A = 1 (O = 20 mL), Equilibration time = 1 h.

$pH_{(eq)}$	V(IV)	Ti(IV)	Fe(III)	Co(II)	Ni(II)	Cu(II)	Zn(II)
0.0	NE	NE	4.0	NE	NE	CE	NE
0.5	NE	NE	17.0	NE	NE	CE	NE
1.0	NE	2.6	50.0	NE	NE	0.1	2.0
1.5	0.1	18.6	88.0	NE	NE	0.2	18.0
2.0	3.0	52.3	93.0	NE	NE	0.3	67.0
2.5	20.0	78.5	97.0	1.0	NE	0.4	92.0
3.0	50.0	92.0*	99.0*	8.0	NE	0.5	98.0
3.5	72.5	98.0	CE	26.0	NE	0.6	99.5
4.0	94.0*	99.5	0.7	56.0	NE	0.8	CE
4.5	98.6	CE	0.9	83.0	NE	0.10	0.11
5.0	99.0	0.12	0.13	94.0	2.0	0.14	0.15
5.5	CE	0.16	0.17	99.0	8.0	0.18	0.19
6.0	0.20	0.21	0.22	CE	21.0	0.23	0.24
6.5	0.25	0.26	0.27	0.28	70.0	0.29	0.30
7.0	0.31	0.32	0.33	0.34	92.0	0.35	0.36

NE: non-extractable, CE: complete extraction, *aqueous solution becomes cloudy before extraction but becomes clear after extraction.

1) V(IV) can be completely separated from Cu(II) at pH 1.0 in a single-stage extraction.

2) V(IV) can be separated almost completely from Ni(II) at $pH_{(eq)}$ 4.50 in a single-stage extraction.

3) It is possible to separate V(IV) from Zn(II) at $pH_{(eq)}$ 2.50, Co(II) at $pH_{(eq)}$ 3.50, Fe(III) at $pH_{(eq)}$ 2.00 and Ti(IV) at $pH_{(eq)}$ 2.50 on using counter current multi-stage extractions.

4. Conclusions

The following conclusions are drawn:

1) Vanadium(IV) can be extracted by Cyanex 302 at pH above 3.0. The equilibration time is 20 min. Up to at least 0.7 g/L V(IV), the extraction ratio (D) is independent of [V(IV)] in the aqueous phase.

2) D is found to be proportional to $[H^+]^{(-2)-(-0.3)}$, $[Cyanex\ 302]^2$ and $(1+2.24\ [SO_4^{2-}])^{-1}$. Apparent K_{ex} value is found to be dependent of pH; it varies from $10^{-3.447}$ at $pH_{(eq)} \leq 2.25$ to $10^{1.508}$ at $pH_{(eq)} = 4.0$.

3) The extraction is highly sensitive to temperature, particularly at $l.t.r$ with ΔH value of ~90 kJ/mol; but at $h.t.r$ it is ~25 kJ/mol.

4) At various concentration levels of experimental parameters, the extraction equilibrium reactions have been proposed; and it is seen that at all conditions $[VO(HA_2)_2]$ is the extractable species though reacting V(IV) species in the aqueous phase may vary with its concentration and pH levels.

5) The loading capacity has been determined to be 4.05 g V(IV) per 100 g Cyanex 302; and it indicates that the mechanism of extraction at high loading does not change from that suggested at low loading (extracted species being $[VO(HA_2)_2]$).

6) Aromatic diluents appear as better diluent over other categories; kerosene is a better diluent over $CHCl_3$, $1,2-C_2H_4Cl_2$ and CCl_4.

7) The V(IV)-loaded organic phase can be quantitatively stripped by 1 mol/L H_2SO_4 and HNO_3 in a single stage.

8) Almost complete separations of V(IV) from Cu(II) at pH 1.0 and from Ni(II) at $pH_{(eq)}$ 4.5 are possible in a single-stage of extraction; whereas, its separation from Zn(II) at $pH_{(eq)}$ 2.5, Co(II) at $pH_{(eq)}$ 3.5, Fe(III) at $pH_{(eq)}$ 2.0 and Ti(IV) at $pH_{(eq)}$ 2.5 will require counter-current multi-stage extractions.

5. Acknowledgements

One of the authors (A. K. Karmakar) acknowledges the National Science Information and Communication Technology (NSICT) Division of the Ministry of Science and Technology, Government of the People's Republic of Bangladesh (Secretariat, Dhaka) for awarding a fellowship to complete this work.

REFERENCES

[1] T. Sekine and Y. Hasegawa, "Solvent Extraction Chemistry: Fundamentals and Applications," Marcel Dekker, Inc., New York, 1977, pp. 564-567.

[2] F. Islam and R. K. Biswas, "The Solvent Extraction of Vanadium(IV) with HDEHP in Benzene and Kerosene: The Solvent Extraction of Vanadium(IV) from Sulphuric Acid Solution with Bis-(2-ehylhexyl) Phosphoric Acid in Benzene and Kerosene," *Journal of Inorganic and Nuclear Chemistry*, Vol. 42, No. 3, 1980, pp. 415-420. doi:10.1016/0022-1902(80)80018-2

[3] F. Islam and R. K. Biswas, "Kinetics of Solvent Extraction of Metal Ions with HDEHP-II: Kinetics and Mechanism of Solvent Extraction of V(IV) from Acidic Aqueous Solutions with Bis-(2-ethylhexyl)phosphoric Acid in Benzene," *Journal of Inorganic and Nuclear Chemistry*, Vol. 42, No. 3, 1980, pp. 421-429. doi:10.1016/0022-1902(80)80019-4

[4] T. Sato, T. Nakamura and M. Kawamura, "The Extraction of Vanadium(IV) from Hydrochloric Acid Solutions by Di-(2-ethylhexyl)-phosphoric Acid," *Journal of Inorganic and Nuclear Chemistry*, Vol. 40, No. 5, 1978, pp. 853-856. doi:10.1016/0022-1902(78)80164-X

[5] J. P. Brunette, F. Rastegar and M. J. F. Leroy, "Solvent Extraction of Vanadium(V) by Di-(2-ethylhexyl)-phosphoric Acid from Nitric Acid Solutions," *Journal of Inorganic and Nuclear Chemistry*, Vol. 41, No. 5, 1979, pp. 735-737. doi:10.1016/0022-1902(79)80364-4

[6] M. A. Hughes and R. K. Biswas, "The Kinetics of Vanadium(IV) Extraction in the Acidic Sulphate-D2EHPA-n-heptane System Using the Rotating Diffusion Cell Technique," *Hydrometallurgy*, Vol. 26, No. 3, 1991, pp. 281-297. doi:10.1016/0304-386X(91)90005-7

[7] R. S. Juang and R. H. Lo, "Stoichiometry of Vanadium(IV) Extraction from Sulfate Solutions with Di(2-Ethylhexyl) Phosphoric Acid Dissolved in Kerosene," *Journal of Chemical Engineering of Japan*, Vol. 26, 1993, pp. 219-222. doi:10.1252/jcej.26.219

[8] R. K. Biswas and M. G. K. Mondal, "Kinetics of VO^{2+} Extraction by D2EHPA," *Hydrometallurgy*, Vol. 69, No. 1-3, 2003, pp. 117-133. doi:10.1016/S0304-386X(02)00208-6

[9] J. Saji and M. L. P. Reddy, "Solvent Extraction Separation of Vanadium(V) from Multivalent Metal Chloride Solution Using 2-Ethylhexyl Phosphonic Acid Mono-2-Ethylhexyl Ester," *Journal of Chemical Technology and Biotechnology*, Vol. 77, No. 10, 2002, pp. 1149-1156. doi:10.1002/jctb.690

[10] J. Saji and M. L. P. Reddy, "Selective Extraction and Separation of Titanium(IV) from Multivalent Metal Chloride Solutions Using 2-Ethylhexyl Phosphonic Acid Mono 2-Ethylhexyl Ester," *Separation Science and Technology*, Vol. 38, No. 2, 2003, pp. 427-441. doi:10.1081/SS-120016583

[11] J. Saji, J. K. Saji and M. L. P. Reddy, "Liquid-Liquid Extraction of Tetravalent Titanium from Acidic Chloride Solutions by Bis(2,4,4-trimethylpentyl)phosphinic Acid," *Solvent Extraction and Ion Exchange*, Vol. 18, No. 5, 2000, pp. 877-894. doi:10.1080/07366290008934712

[12] M. Ulewicz and W. Walkowiak, "Selective Removal of Transition Metal Ions in Transport through Polymer Inclusion Membranes with Organophosphorus Acid," *Environment Protection Engineering*, Vol. 31, No. 3-4, 2005, pp. 74-81.

[13] W. A. Rickelton, "Novel Uses for Thiophosphinic Acids in Solvent Extraction," *Journal of Metals*, Vol. 44, No. 5, 1992, pp. 52-54.

[14] A. Saily and S. N. Tandon, "Liquid-Liquid Extraction Behavior of V(IV) Using Phosphinic Acids as Extractants," *Fresenius' Journal of Analytical Chemistry*, Vol. 360, No. 2, 1998, pp. 266-270.

[15] P. Zhang, K. Inoue and H. Tsuyama, "Recovery of Molybdenum and Vanadium from Spent Hydrodesulfurization Catalysts by Means of Liquid-Liquid Extraction," *Kagaku Kogaku Ronbunshu*, Vol. 21, 1995, pp. 451-456. doi:10.1252/kakoronbunshu.21.451

[16] P. Zhang, K. Inoue, K. Yoshizuka and H. Tsuyama, "Solvent Extraction of Vanadium(IV) from Sulfuric Acid Solution by Bis(2,4,4-trimethylpentyl) Phosphinic Acid in Exxsol D80," *Journal of Chemical Engineering of Japan*, Vol. 29, No. 1, 1996, pp. 82-87. doi:10.1252/jcej.29.82

[17] K. C. Sole and J. B. Hiskey, "Solvent Extraction Characteristics of Thio Substituted Organophosphinic Acid Extractants," *Hydrometallurgy*, Vol. 30, No. 1-3, 1992, pp. 345-365. doi:10.1016/0304-386X(92)90093-F

[18] J. Bassett, R. C. Denney, G. H. Jeffery and J. Mendham, "Vogel's Textbook of Quantitative Inorganic Analysis Including Elementary Instrumental Analysis," 4th Edition, ELBS and Longman, London, 1979, pp. 752-753.

[19] M. R. Ali, R. K. Biswas, S. M. A. Salam, A. Akhter, A. K. Karmakar and M. H. Ullah, "Cyanex 302: An Extractant for Fe^{3+} from Chloride Medium," *Bangladesh Journal of Scientific and Industrial Research*, Vol. 46, No. 4, 2011, pp. 407-414.

[20] E. Paatero, T. Lantto and P. Ernola, "The Effect of Trioctylphosphine Oxide on Phase and Extraction Equilibria in Systems Containing Bis(2,4,4-tri-methylpentyl) Phosphinic Acid," *Solvent Extraction and Ion Exchange*, Vol. 8, No. 3, 1990, pp. 371-388. doi:10.1080/07366299008918006

[21] R. M. Smith and A. E. Martell, "Critical Stability Constant," In: *Inorganic Complexes*, Plenum Press, New York, 1976.

29

Aminododecyldiphosphonic Acid for Solvent Extraction of Bismuth Ions

Baghdad Medjahed[1], M'Hamed Kaid[1], Mohamed Amine Didi[1*], Didier Villemin[2]

[1]Laboratory of Separation and Purification Technology, Department of Chemistry,
Faculty of Sciences, Tlemcen University, Tlemcen, Algeria
[2]Laboratoire de Chimie Moléculaire et Thioorganique, UMR CNRS 6507,
INC3M, FR 3038, Labex EMC3, ENSICAEN, Caen, France
Email: *madidi13@yahoo.fr

ABSTRACT

The extraction of Bi(III) in nitrate media has been investigated using aminododecyldimethylenediphosphonic acid, ADDMDPA, which was previously synthesized and characterized. The extraction of the cation was carried out in different media with the addition of CH_3COONa, KNO_3 and HNO_3. The maximum extraction yield for Bismuth is 70% after addition of 0.01 M of potassium nitrate at pHi = 2.9, in one step.

Keywords: Bismuth(III); Aminododecyldimethylenediphosphonic Acid; Solvent Extraction

1. Introduction

Bismuth is used in the cosmetics industry for the preparation of creams and hair dyes, while some of its colloidal salts (subcitrate and subgallate), due to their antiseptic, astringent and diuretic properties, have important applications in pharmaceutical preparations and are employed as anti-ulcer, antibacterial, anti-HIV and radiotherapeutic agents [1].

World reserves of bismuth are usually obtained as a sub-product in lead, copper, tin and gold ores [2-4]. During the industrial metallurgical process of these ores, leaching stages with H_2SO_4, HCl and HNO_3 are involved, and highly acidic solutions with base metals and bismuth are obtained [5,6].

Bismuth is a curious metal and could be toxic in an unsuitable form [2,7]. Its metal extraction is a major challenge in the metal addressing the environmental pollution, its mode of physiopathological action was little studied and it is not yet understood [6,7].

The synthesis of new organophosphorus extractants which form stable complexes with metallic species is of great importance for improving existing hydrometallurgycal processes for their recovery [8,9].

In fact, we have synthesized aminododecyldimethylenediphosphonic acid for such a purpose. The characterization of this product was achieved using various spectroscopic methods including (^{1}H NMR, ^{31}P NMR, ^{13}C NMR, FTIR and the like).

We have studied this acid in the recovery of metal species Bi^{+3} under the optimal conditions in different media, in one step.

2. Experimental

2.1. Reagents and Solutions

The reagents used in this work were dodecan, phosphorus acid (35%) from (Aldrich). HCHO and Bismuth nitrate was purchased from Merck. The aqueous solutions concentrations of Bi(III) in nitrate medium taken as 0.5 $mmol \cdot L^{-1}$ and the organic (ADDMDPA) solution concentration was taken from 0.5 to 10 $mmol \cdot L^{-1}$.

2.2. Instrumentation

^{13}C {-[1]H}, ^{31}P {-^{1}H} and ^{1}H NMR spectra were measured on a Bruker AC 250 working at 250 MHz in a $CDCl_3$ solution. Infrared spectra were measured on a Perkin Elmer 16 PC-FTIR equipped with a thermostat to maintain the temperature of the sample at 25.0°C ± 0.1°C. A Phywe WTM 320 with combined glass electrode was used to measure the pH of the aqueous solution before and after extraction. In a water-acetone mixture (15:5), a known mass of the ADDMDPA was titrated with a solution of NaOH (5 $mmol \cdot L^{-1}$). Metal ions were determined using the electrothermal atomic absorption GFAAS, system GBC 932, and system 3000 automated graphite furnace (GBC Scientific Equipment, Dandenong, Australia). Background correction was made with a deuterium lamp

*Corresponding author.

and pyrolytic graphite tubes were used. Settings such as lamp current, wavelength, temperature programs, and slit width were those recommended in the operating manual.

2.2.1. Synthesis of the Extractant and Characterization

ADDMDPA was synthesized following a method first described by Largman et al. [6] with an original modification developed in our laboratory [7]. The product presented the following properties, and its formula is shown in **Figure 1**. Formula: CH_3-$(CH_2)_{11}$ $N(CH_2 P(O)(OH)_2)_2$ IR υ (cm^{-1}): 2750(vas PO), 1120 (vs PO), 1015 (vas P-OH), 966 s P-OH); 1H NMR (D_2O, Na_2CO_3) δ/TMS (ppm): 1.35 (s, 4H, CH_2), 1.75 (m, 12H, CH_2), 3.125 (d, $^2J_{HP}$ = 8.92 NCH_2-P), 3.35 (m, 2H, N-CH_2); ^{31}P NMR(D_2O, Na_2CO_3): δ/H_3PO_4 (ppm) 8.78; ^{13}C NMR (D_2O, Na_2CO_3) (ppm):13.5 (s, C_1), 17.14(s, C_2), 30 (s, C_3), 50 (d, $^2J_{CP}$ = 138.7, NCH_2-P).

The obtained pKai: 3.25, 8.4 and 9.3 indicated that in water-acetone medium, the fourth acidity is not attained corresponding to very weak acidity.

The presence of P = O wide band indicates hydrogen intermolecular bonds P = O·H-OP and N|·> H-O-P.

The ADDMDPA is soluble in most organic solvents, and shows intermolecular hydrogen bonding, forming polymers depending on the solvent polarity [10].

Thus, in chloroform that we used in our study, the ADDMDPA is generally present in a dimeric form, as shown in **Figure 1** [11,12].

2.2.2. Extraction Experiments

The extraction experiments were carried out with ADD-MDPA dissolved in chloroform. After various preliminary tests with different solvents, chloroform has been chosen since it dissolves the extractant without trouble or emulsion.

The appropriate volume of the aqueous solution (10 mL) containing the metal ion and the ADDMDPA (aqueous/organic volume ratio of 1:1) were mixed in glass flasks.

The mixtures were shaken in a moderate way, for at least 6 min for Bi(III). All the experiments were carried out at 20°C.

3. Results and Discussion

The results of the extraction experiments will be dis-cussed in terms of the distribution coefficient, D, and the extraction yield, Y [13].

$$D = \left(\frac{m_i - m_f}{m_f} \right) \times \frac{V_{aq}}{V_{org}} \quad (1)$$

$$Y(\%) = \frac{m_i - m_f}{m_i} \times 100 \quad (2)$$

with the variables being as follows: m_i: initial mass of metal ion in the aqueous phase; m_f: mass of metal ion in the aqueous phase after extraction; V_{org}: volume of the organic phase; V_{aq}: volume of the aqueous phase.

The molar ratio Q is defined as the ratio of the number of moles of the ligand in the organic phase versus the number of moles of metal in the aqueous phase before extraction.

$$Q = \frac{n_{extractant}}{n_{metal}} \quad (3)$$

3.1. Extraction Kinetics

The equilibrium times were 6 min for Bi(III). The results are shown in **Figure 2**.

Figure 2 shows that the maximum extraction yield was obtained after 6 min of moderate shaking

3.2. Effect of the Molar Ratio

Figure 3 shows that the optimal extractant concentration that gives the maximum yield (52%) was 2 that corre-spond to extactant concentration equal to 1 mmol.

The stoichiometric coefficients obtained from the plots ln D vs ln [AADDMDP] and ln D vs pHeq, shown in **Figures 4** and **5** respectively, may suggest the reaction mechanism in neutral media. This result led us to pro-pose the following extraction equilibrium.

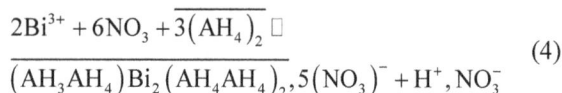

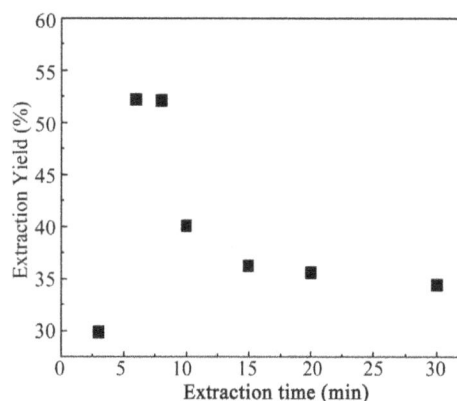

$$\frac{2Bi^{3+} + 6NO_3 + \overline{3(AH_4)_2}}{\overline{(AH_3AH_4)Bi_2(AH_4AH_4)_2}, 5(NO_3)^- + H^+, NO_3^-} \quad (4)$$

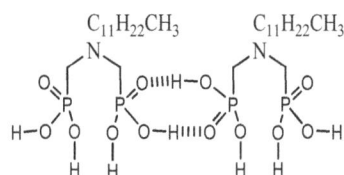

Figure 1. Dimeric form of ADDMDPA.

Figure 2. Extraction kinetics of Bi(III); $[Bi^{+3}]_0$ = 5 × 10^{-4} M, $[ADDMDPA]_0$ = 10^{-3} M, Vaq/Vorg = 1, T = 20°C, pHi = 2.9.

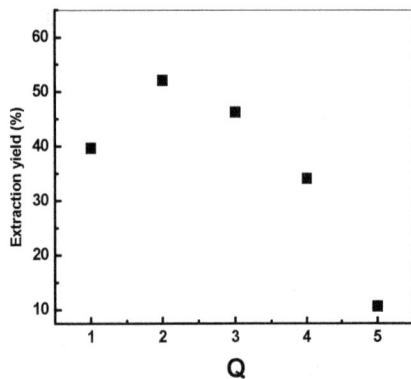

Figure 3. Effect of the molar ratio on the extraction yield of Bi (III) in nitrate medium; $[Bi^{3+}]_0 = 5 \times 10^{-4}$ M, $Vaq/Vorg = 1$, $t = 6$ min, $T = 20°C$.

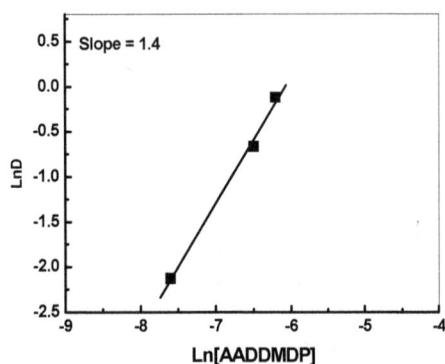

Figure 5. Effect of equilibrium pH on the distribution ratio of Bi(III).

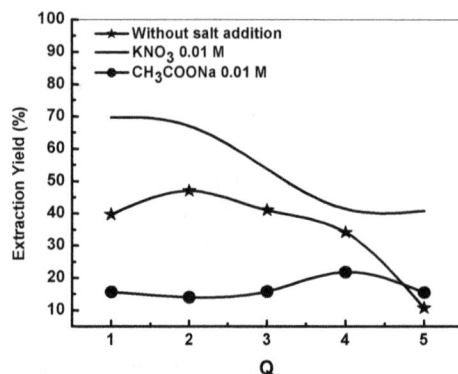

Figure 4. Effect of extractant concentration on the distribution ratio of Bi(III). $T = 20°C$, $Vaq/Vorg = 1$.

Figure 6. Effects of molar ratio on the extraction yield, with and without salt addition.

The AADDMDP extracts in cationic exchange mode with substantial yields because the extractions are done to only one cycle.

3.3. Influence of the Ionic Strength on the Extraction of Bi^{3+}

The influence of the ionic strength on the extraction yields of Bi(III) with AADDMDP diluted in chloroform, was studied by adding potassium nitrate and sodium acetate to the aqueous phase at the same concentration 0.01 M.

The yield of extraction of Bi(III) decreases with the increase of concentration of the sodium acetate in aqueous phase.

According to these results, it is noticed that the addition of CH_3COONa decreases the extraction yield, on the other hand the addition of KNO_3 increased it to 70%. This is due probably to the basic medium imposed by the sodium acetate; consequently we were interested to the study of the evolution of the yield according to the variation of the concentration of KNO_3 in the aqueous phase.

3.4. Effect of the Addition of KNO_3

In order to study the influence of the ionic force on the

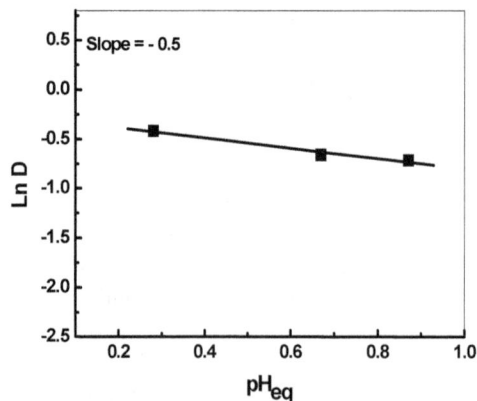

extraction of Bi^{3+}, three quantities of KNO_3 were added to the aqueous phase before extraction.

According to these results, after the addition of KNO_3 0.01 M, the yield is 70% with Q = 1. In the interval of 0.1 - 1 M, the increase of the concentration of KNO_3 will have a negative effect on the extraction yield, because of the competition between the two cations (K^+, Bi^{3+}).

The stoichiometric coefficients obtained from the plots Log E vs. Log [AADDMDP] and Log E vs. pHeq, shown in **Figures 8** and **9** respectively may suggest the reaction mechanism of bismuth extraction.

The slopes are respectively equal to (0.75) near to 1 for extractant and 1 for the pH.

The equilibrium equation is written as follows:

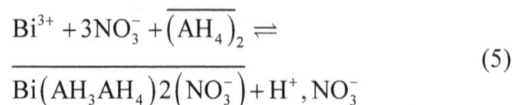

$$Bi^{3+} + 3NO_3^- + \overline{(AH_4)_2} \rightleftharpoons$$
$$\overline{Bi(AH_3AH_4)2(NO_3^-)} + H^+, NO_3^- \quad (5)$$

3.5. Effect of Nitric Acid on the Extraction of Bi (III)

We note that in acidic media, the extraction yield decreases drastically reaching a minimum of 5%.

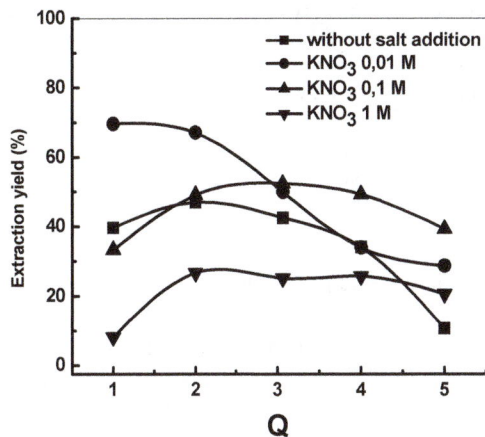

Figure 7. Effects of molar ratio on the extraction yield after the addition of potassium nitrate $[Bi^{3+}] = 5 \times 10^{-4}$ M, Vaq/Vorg = 1, t = 6 min, T = 20°C.

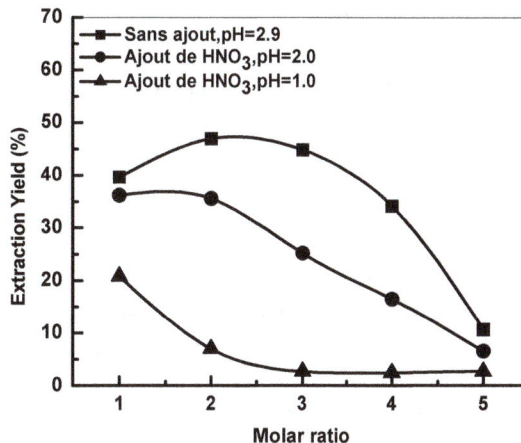

Figure 10. Effect of pH on the extraction yield $[Bi^{3+}] = 5 \times 10^{-4}$ M, Vaq/Vorg = 1, t = 6 min, T = 25°C.

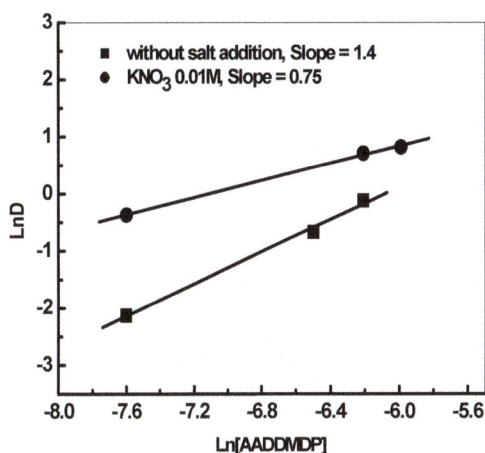

Figure 8. Effect of extractant concentration on the distribution ratio of Bi(III) before and after addition of potassium nitrate. $[Bi^{3+}] = 5.10^{-4}$M, Vaq/Vorg = 1, t = 6 min, T = 20°C.

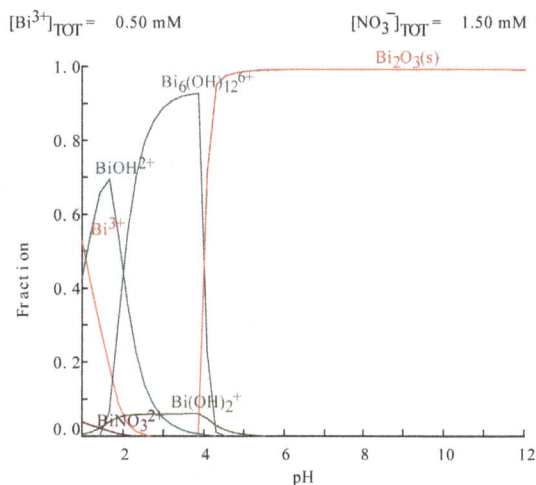

Figure 11. Distribution diagrams of bismuth ion (0.5 mmol·L^{-1}) in nitrate media using Medusa and Hydra programs [14].

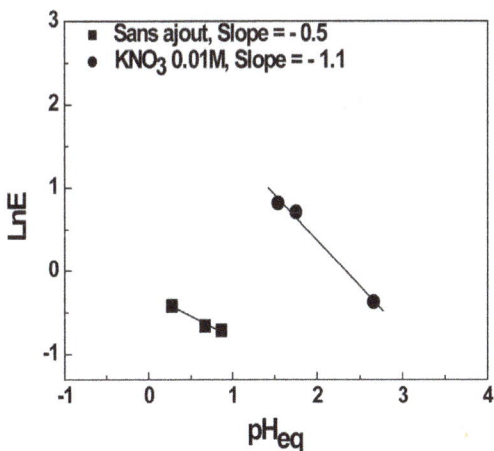

Figure 9. Effect of equilibrium pH on the distribution ratio of Bi(III) before and after addition of potassium nitrate. $[Bi^{3+}]_0 = 5 \times 10^{-4}$M, Vaq/Vorg = 1, t = 6 min, T = 20°C.

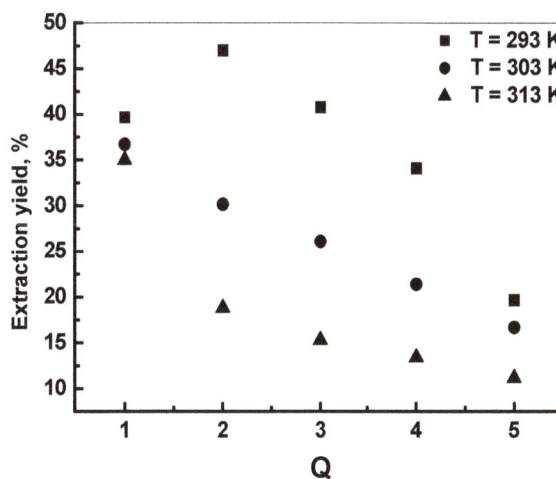

Figure 12. Effet of temperature on the extraction yield $[Bi^{3+}]_0 = 5 \times 10^{-4}$ M, Vaq/Vorg = 1, t = 6 min, T = 20°C.

3.6. Effect of Temperature

The effect of temperature on the extraction of bismuth (III) ions was studied under optimum conditions.

Different thermodynamic parameters were computed using Van't Hoff equation in the form.

$$InKc = -\frac{\Delta H}{RT} + \frac{\Delta S}{R} \qquad (6)$$

$$\Delta G = RTLn\ Kc \qquad (7)$$

Where ΔH, ΔS, ΔG, and T are the enthalpy, entropy, Gibbs free energy, and temperature in Kelvin, respectively. The values of equilibrium ratio (Kc), were calculated at each temperature using the relationship.

$$Kc = Fe/(1 - Fe) \qquad (8)$$

Where F_e is the fraction of Bi(III) ions extracted at equilibrium.

$$\Delta H = -7.47 KJ \cdot mol^{-1}$$

$$\Delta S = -29.33 J \cdot mol^{-1} \cdot K^{-1}$$

The plot of log Kc vs 1/T is a straight line as shown in **Figure 13** with correlation coefficient r = 0.9947. The numerical values of ΔH, ΔS are computed from the slope. The negative value of Gibbs free energy as shown in **Table 1** indicates the spontaneous nature of extraction, while negative value of ΔH reflects the exothermic extraction behavior. The negative value of ΔS indicates the complex stability

4. Conclusions

The solvent extraction of the species Bi(III) with ADD-MDPA dissolved in chloroform was explained taking into account the formation of different complexes.

In nitrate medium, the extraction kinetic is very fast. The optimal extraction parameters for metal ions concentration of 0.5 mmol·L^{-1} in the aqueous phase and 1

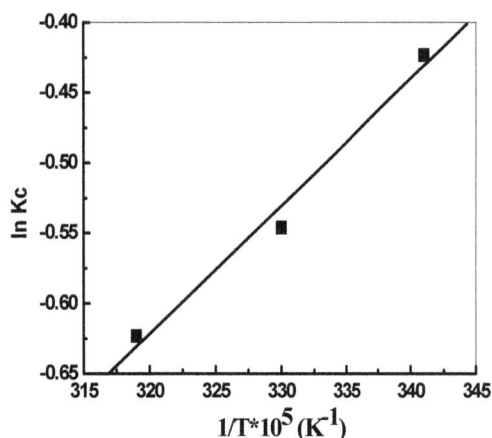

Figure 13. Variation of log Kc with 1/T for the extraction of bismuth(III).

Table 1. Thermodynamic constants of the extraction of bismuth (III)) ions

Thermodynamic parameters	Values
ΔH (KJ/mol)	− 7.47
ΔS (J/mol K)	− 29.33
ΔG (KJ/mol)	− 1.029 (T = 293˚C)

mmol·L^{-1} in the organic phase, are Q = 2, Vaq/Vorg = 1, T = 25˚C.

The addition of CH_3COONa decreased the yield of extraction much while the addition of KNO_3 to the same concentration of 0.01M with increased the extrability of our cation up to 70% with Q = 1.

The increase in the concentration of H$^+$ of the aqueous phase had an antagonistic effect on the yield of extraction

5. Acknowledgements

We gratefully acknowledge the "Ministère de la Recherche et des Nouvelles Technologies", CNRS (Centre National de la Recherche Scientifique) and the Program TASSILI 10MDU799 for their financial support.

REFERENCES

[1] M. Burguera, J. L. Burguera, C. Rondon and M. I. Garcia, "Determination of Bismuth in Biological Tissues by Electrothermal Atomic Absorption Spectrometry Using Platinum and Tartaric Acid as Chemical Modifier," *Journal of Analytical Atomic Spectrometry*, Vol. 16, No. 10, 2001, pp. 1190-1195.

[2] F. Habashi, "Arsenic, Antimony, and Bismuth Production," *Encyclopedia of Materials*: *Science and Technology*, 2008, pp. 332-336.

[3] J. D. Jorgenson, "US Geological Survey," Mineral Commodity Summaries, Eastern Region, Reston, 2003, p. 37.

[4] J. A. Reyes-Aguilera, M. P. Gonzalez, R. Navarro, T. I. Saucedo and M. Avila-Rodriguez, "Supported Liquid Membranes (SLM) for Recovery of Bismuth from Aqueous Solutions," *Journal of Membrane Science*, Vol. 310, No. 1-2, 2008, pp. 13-19.
doi:10.1016/j.memsci.2007.10.020

[5] E. M. Donaldson and M. Wang, "Determination of Silver, Antimony, Bismuth, Copper, Cadmium and Indium in Ores, Concentrates and Related Materials by Atomicabsorption Spectrophotometry after Methyl Isobutyl Ketone Extraction as Iodides," *Talanta*, Vol. 33. No. 3, 1986, pp. 233-242. doi:10.1016/0039-9140(86)80057-1

[6] J. G. Yang, J. Y. Yang, M. T. Tang, C. B. Tang and W. Liu, "The Solvent Extraction Separation of Bismuth and Molybdenum from a Low Grade Bismuth Glance Flotation Concentrate," *Hydrometallurgy*, Vol. 96, No. 4, 2009, pp. 342-348. doi:10.1016/j.hydromet.2008.12.006

[7] K. Campos, R. Domingo, T. Vincent, M. Ruiz and A. M. S. Guibal, "Bismuth Recovery from Acidic Solutions Us-

ing Cyphos IL-101 Immobilized in a Composite Biopolymer Matrix," *Water Research*, Vol. 42, No. 14, 2008, pp. 4019-4031. doi:10.1016/j.watres.2008.07.024

[8] E. P. Horwitz, H. Diamond, K. A. Martin and R. Chiarizia, "Extraction of Americium (III) from Chloride Media by Octyl(phenyl)-N,N-diisobutylcarbamoylmethylphosphine Oxide," *Solvent Extraction and Ion Exchange*, Vol. 5, No. 3, 1987, pp. 419-446. doi:10.1080/07366298708918575

[9] D. Villemin, B. Moreau, A. Elbilali, M. A. Didi, M. Kaid and P. A. Jaffres, "Green Synthesis of Poly(aminomethylenephosphonic) Acids," *Phosphorus, Sulfur and Silicon*, Vol. 185, No. 12, 2010, pp. 2511-2519. doi:10.1080/10426501003724897

[10] A. Buch, M. Stambouli, D. Pareau and G. Durang, "Solvent Extraction of Nickel (II) by Mixture of 2-Ethylhexanal Oxime and Bis(2-ethylhexyl) Phosphoric Acid," *Solvent Extraction and Ion Exchange*, Vol. 20, No. 1, 2002, pp. 49-66. doi:10.1081/SEI-100108824

[11] M. A. Didi, A. Elias and D. Villemin, "Effect of Chain Length of Alkane-1-hydroxy-1,1'-methyl Diphosphonics Acids on the Iron (III) Liquid-Liquid Extraction," *Solvent Extraction and Ion Exchange*, Vol. 20, No. 4-5, 2002, pp. 407-415. doi:10.1081/SEI-120004813

[12] J. R. Ferraro, A. W. Herlinger and R. Chiarizia, "Correlation of the Asymmetric and Symmetric POO Frequencies with the Ionic Potential of the Metal Ion in Compounds of Organophosphorus Acid Extractants: A Short Review," *Solvent Extraction and Ion Exchange*, Vol. 16, No. 3, 1998, pp. 775-794. doi:10.1080/07366299808934552

[13] P. E. Body, P. R. Dolan and D. E. Mulcahy, "Environmental Lead: A Review," *Critical Reviews in Environmental Control*, Vol. 20, No. 5-6, 1991, pp. 299-310. doi:10.1080/10643389109388403

[14] I. Puigdomenech, "HYDRA (Hydrochemical Equilibrium-Constant Database) and MEDUSA (Make Equilibrium Diagrams Using Sophisticated Algorithms) Programs," Royal Institute of Technology, Sweden. http://www.ke- mi.kth.se/medusa

Permissions

List of Contributors

Chuan-Wei Oo, Hasnah Osman, Sharon Fatinathan and Maizatul Akmar Md. Zin
School of Chemical Sciences, Universiti Sains Malaysia, Penang, Malaysia

Fathi Habashi
Department of Mining, Metallurgical, and Materials Engineering, Laval University, Quebec City, Canada

Weifeng Tong, Xiaofeng Wu, Xingxiang Fan and Yuedong Wu
Kunming Institute of Precious Metals, Kunming, China

Weidong Xing, Jiachun Zhao and Bojie Li
Sino-Platinum Metals Co. Ltd., Kunming, China

Haigang Dong
State Key Laboratory of Advanced Technology of Comprehensive Utilization of Platinum Metals, Kunming, China

Ranjit Kumar Biswas and Aneek Krishna Karmakar
Department of Applied Chemistry and Chemical Engineering, Rajshahi University, Rajshahi, Bangladesh

Martín A. Encinas-Romero, Guillermo Tiburcio-Munive and Jesús L. Valenzuela-García
Departamento de Ingeniería Química y Metalurgia, Universidad de Sonora, Hermosillo, México

Tatjana V. Brovman
Tver State Technical University, Tver, Russia

Shaohua Yin, Wenyuan Wu, Xue Bian, Yao Luo and Fengyun Zhang
School of Materials and Metallurgy, Northeastern University, Shenyang, China

Eskandar Keshavarz Alamdari
Department of Mining and Metallurgical Engineering, Amirkabir University of Technology, Tehran, Iran
Research Center for Materials and Mining Industries Technology, Amirkabir University of Technology, Tehran, Iran

Sayed Khatiboleslam Sadrnezhaad
Department of Materials Science and Engineering, Sharif University of Technology, Tehran, Iran

Vanesa Bazan
CONICET—Instituto de Investigaciones Mineras, Universidad Nacional de San Juan, San Juan, Argentina

Elena Brandaleze and Leandro Santini
Metallurgical Department and Technology and Materials Develop Center, DEYTEMA-Universidad Tecnológica Nacional, Facultad Regional de San Nicolás, Colón, Argentina

Pedro Sarquis
Instituto de Investigaciones Mineras, Universidad Nacional de San Juan, San Juan, Argentina

Bo Zhang and Zheng-Liang Xue
Key Laboratory for Ferrous Metallurgy and Resources Utilization of Ministry of Education, Wuhan University of Science and Technology, Wuhan, China

Mona Mahmoud Abd El-Latif, Marwa S. Showman and Rania R. Abdel Hamide
Fabrication Technology Department, Advanced Technology and New Materials Research Institute (ATNMRI), City of Scientific Research and Technological Applications (SRTA-City), Previously "Mubarak City for Scientific Research and Technology Applications (MuCSAT)", Alexandria, Egypt.

Amal M. Ibrahim
Dharyya College for Arts and Sciences, Qassim University, Al Qassim, Saudi Arabia
Surface Chemistry and Catalysis Laboratory, Physical Chemistry Department, National Research Center, Cairo, Egypt

Irina Baksheyeva
Siberian Federal University, Krasnoyarsk, Russia

Margaret Sviridova
Institute of Chemistry and Chemical Technology of Siberian Branch of Russian Academy of Sciences, Krasnoyarsk, Russia

Victor Bragin
Siberian Federal University, Krasnoyarsk, Russia
Institute of Chemistry and Chemical Technology of Siberian Branch of Russian Academy of Sciences, Krasnoyarsk, Russia

Ii Ahmadi
Department of Mining Engineering, Isfahan University of Technology, Isfahan, Iran

Pratima Meshram, Sushanta Kumar Sahu, Banshi Dhar Pandey and Vinay Kumar
Metal Extraction & Forming Division, CSIR-National Metallurgical Laboratory, Jamshedpur, India

Tilak Raj Mankhand
Department of Metallurgical Engineering, IIT BHU, Varanasi, India

Nomampondo P. Magwa, Gareth M. Watkins and Zenixole R. Tshentu
Department of Chemistry, Rhodes University, Grahamstown, South Africa

Eric Hosten
Department of Chemistry, Nelson Mandela Metropolitan University, Port Elizabeth, South Africa

Ali Reza Eivazi Hollagh
Mining & Metallurgical Engineering, Amirkabir University of Technology, Tehran, Iran

Eskandar Keshavarz Alamdari
Mining & Metallurgical Engineering, Amirkabir University of Technology, Tehran, Iran
Research Center for Materials and Mining Industries Technology, Amirkabir University of Technology, Tehran, Iran

Davooud Moradkhani
Faculty of Engineering, Zanjan University, Zanjan, Iran

Ali Akbar Salardini
Materials and Energy Research Center, Tehran, Iran

Natalia Shkatulyak
South Ukrainian National Pedagogical University after K.D. Ushinskii, Odessa, Ukraine

S. Chitra and K. Paramasivan
Centralised Waste Management Facility, Bhabha Atomic Research Centre Facilities, Kalpakkam, India

P. K. Sinha
Waste Management Division, Bhabha Atomic Research Centre (BARC), Mumbai, India

Jinming Liu and Yinghui Zhang
School of Material Science and Engineering, Jiangxi University of Science and Technology, Ganzhou, China

Cuiping Guo
Department of Materials Science and Engineering, University of Science and Technology Beijing, Beijing, China

A. S. Bolokang
Department of Engineering Metallurgy, University of Johannesburg, Johannesburg, South Africa
Transnet Rail Engineering, Pretoria, South Africa

M. J. Phasha
Transnet Rail Engineering, Pretoria, South Africa

D. E. Motaung
DST/CSIR Nanotechnology Innovation Centre, National Centre of Nano-Structured Materials, Council for Scientific and Industrial Research, Pretoria, South Africa

Alok Prasad Das
Centre of Biotechnology, Siksha 'O' Anusandhan University, Bhubaneswar, India

Lala Behari Sukla and Nilotpala Pradhan
Institute of Minerals and Materials Technology, Bhubaneswar, India

Hossein Kamran Haghighi and Mohammad Mehdi Salarirad
Department of Mining and Metallurgical Engineering, Amirkabir University of Technology, Tehran, Iran

Davood Moradkhani
Faculty of Engineering, University of Zanjan, Zanjan, Iran

Sajad Ahmad Ganai, Javid Ahmad Banday and Tabassum Ara
Department of Chemistry, National Institute of Technology, Srinagar, Kashmir

Abid Hussain Shalla
Islamic University of Science and Technology, Awantipora, India

Lianpeng Song and Zhimin Yin
School of Materials Science and Engineering, Central South University, Changsha, China

Yuan Yuan
School of Materials Science and Engineering, Central South University, Changsha, China
China Railway Construction Electrification Bureau Group Kang Yuan New Materials Co., LTD., Jiangyin, China

Satit Praipruke
Program of Petrochemistry, Faculty of Science, Chulalongkorn University, Bangkok, Thailand

Korbratna Kriausakul
Green Chemistry Research Laboratory, Department of Chemistry, Faculty of Science, Chulalongkorn University, Bangkok, Thailand

Supawan Tantayanon
Green Chemistry Research Laboratory, Department of Chemistry, Faculty of Science, Chulalongkorn University, Bangkok, Thailand
National Center of Excellence for Petroleum, Petrochemicals and Advanced Materials (NCE-PPAM), Bangkok, Thailand

Abhilash, Kapil Deo Mehta and Bansi Dhar Pandey
CSIR-National Metallurgical Laboratory, Jamshedpur, India

Ranjit Kumar Biswas, Aneek Krishna Karmakar and Muhammad Saidur Rahman
Department of Applied Chemistry and Chemical Engineering, Rajshahi University, Rajshahi, Bangladesh

Baghdad Medjahed, M'Hamed Kaid and Mohamed Amine Didi
Laboratory of Separation and Purification Technology, Department of Chemistry, Faculty of Sciences, Tlemcen University, Tlemcen, Algeria

Didier Villemin
Laboratoire de Chimie Moléculaire et Thioorganique, UMR CNRS 6507, INC3M, FR 3038, Labex EMC3, ENSICAEN, Caen, France

www.ingramcontent.com/pod-product-compliance
Lightning Source LLC
Chambersburg PA
CBHW080637200326
41458CB00013B/4664